BEST PRACTICES FOR VALIDATING CO_2 GEOLOGICAL STORAGE

OBSERVATIONS AND GUIDANCE FROM THE IEAGHG WEYBURN-MIDALE CO_2 MONITORING AND STORAGE PROJECT

EDITOR: BRIAN HITCHON
Hitchon Geochemical Services Ltd.

Geoscience Publishing
Geoscience Publishing, Box 79088, Sherwood Park, Alberta T8A 5S3, Canada

©Geoscience Publishing, 2012. All rights reserved.
(Geoscience Publishing is a Division of Hitchon Geochemical Services Ltd.)

National Library of Canada Cataloguing in Publication Data
Best Practices for Validating CO_2 Geological Storage
    Observations and Guidance from the IEAGHG
    Weyburn-Midale CO_2 Monitoring and Storage Project
Includes references and index
ISBN 978-0-9680844-7-2
    1. Carbon dioxide–validating geological storage.
    2. Geological storage of carbon dioxide.
    3. Saskatchewan, Canada–Weyburn Field.

All rights reserved. No part of this publication may be reproduced, stored in a retrieval system, or transmitted in any form or by any means, electronic, mechanical, photocopying, recording, or otherwise, without the prior permission of the copyright owner.

Book and cover design: Kevin Zak
Set in: Minion Pro 9.5/13
Printed in Canada by: McCallum Printing Group Inc.
Printed on Enviro Paper; 100% post-consumer recycled fibres, acid- and chlorine-free.

Cover photograph: A canola field near Weyburn, Saskatchewan, with a covered CO_2 injection well (left – back cover) and production well (right – front cover). Photograph by **Bob Buzash** of *RMB Photography*; courtesy of PTRC.

CONTENTS

FOREWORD	vi
EXECUTIVE SUMMARY	vii
ACKNOWLEDGEMENTS	xii

1. INTRODUCTION — 1
- Purpose, Scope, and Context — 1
- History of the Project — 3
- Weyburn and Midale CO_2-EOR Operations — 4
- Industry Experience with Large-scale CCS and Similar Operations — 4
- Organization of the Chapters — 6

2. CHARACTERIZATION — 9
- Summary — 9
- Introduction — 10
- Geology — 16
- Hydrogeology — 33
- Containment — 52
- Natural Analogues — 67
- Geomechanics — 70
- Geochemistry — 74
- Recommendations — 75

3. STORAGE PERFORMANCE PREDICTIONS — 79
- Summary — 79
- Introduction — 80
- CO_2 Migration — 83
- Potential CO_2 Migration Above the Reservoir — 90
- Invasion Percolation Modelling — 91
- Storage Capacity and Mass Partitioning — 105
- Containment — 111
- Recommendations — 115

4. GEOCHEMICAL MONITORING — 119
- Summary — 119
- Introduction — 120
- Groundwater Sampling — 121

Soil Gas	124
Reservoir Fluids	142
Recommendations	152

5. GEOPHYSICAL MONITORING — 155

Summary	155
Introduction	156
Geophysical Characterization of the Rock–Fluid System	160
Geophysical Characterization of the Weyburn Reservoir	162
Feasibility Studies	172
Downhole Monitoring Methods	180
3D Seismic Methods	192
3D Seismic Monitoring Without a Baseline	207
Recommendations	210

6. HISTORY MATCHING AND PERFORMANCE VALIDATION — 215

Summary	215
Introduction	216
Coreflood Experiments	218
Simulation Model	223
Fracture Flow Experiments	228
Field-scale Studies	235
Seismic Simulation and History Matching	242
Markov Chain Monte Carlo Stochastic Inversion	246
Summary of Results	250
Recommendations	253

7. WELL INTEGRITY — 255

Summary	255
Introduction	256
Well Integrity Assessment	257
New Well Design Considerations	263
Well Remediation and Conversion	268
Well Abandonment Considerations	269
Well Integrity Monitoring	272
Well Integrity Field Testing	274
Conclusions	277
Recommendations	284

8. RISK ASSESSMENT — 289
Summary — 289
Introduction — 290
Risk Evaluation Criteria — 296
Assessing the Risks for CO_2 Storage — 300
Analysis of Risks — 304
Evaluation of Risks — 306
Conclusions — 316
Recommendations — 319

9. COMMUNITY OUTREACH — 321
Summary — 321
Introduction — 322
Communications and the Project — 323
Risk Assessment Meeting — 325
Crisis Communications — 326

REFERENCES — 331
INDEX — 343

Recommended format for citation of chapters in this book.

Rostron, B.J. et al. (2012). Characterization. *In* B. Hitchon (Editor), *Best Practices for Validating CO_2 Geological Storage*, 9–77. Geoscience Publishing.

FOREWORD

The IEAGHG Weyburn-Midale CO_2 Monitoring and Storage Project has now been active since 2000 and has therefore established an extensive database of knowledge over the 11 year period of its operation. The knowledge gained is now being shared through this book to the benefit of carbon capture and storage (CCS) project proponents and practitioners around the world.

The IEAGHG Weyburn-Midale CO_2 Monitoring and Storage Project was the second commercial scale CO_2 injection project to come on board after that in the Sleipner oilfield, offshore Norway in 1996. The project though is very different from Sleipner and therefore provides an extended research and monitoring data set that is currently unique in its own right. The target reservoir is an onshore carbonate formation, about 1500 metres deep, that has been operated as an oilfield since the 1950s. The CO_2 flood that commenced in 2000 will extend the life of the field by some 25 years and thus has provided a field laboratory to test monitoring and modelling tools.

Enhanced oil recovery using CO_2 is becoming an attractive option to stimulate carbon storage projects in many parts of the world. This book, coming from a project with a strong international reputation for scientific credibility and excellence, is both timely and internationally important. The fact that the IEAGHG Weyburn-Midale CO_2 Monitoring and Storage Project was able to counteract claims of leakage was a result of the thorough monitoring baselines established, the continuity of researchers working on the project, and the scientific quality of their research institutions. This book will stand as a prime example of their work for many years to come.

John Gale
General Manager
IEA Greenhouse Gas R&D Programme
Cheltenham, United Kingdom

EXECUTIVE SUMMARY

Introduction

During the past 12 years, teams of international scientists have been hard at work in southern Saskatchewan in an effort to determine the best methods for securely storing anthropogenic CO_2 in the subsurface. Their world-leading research has been part of a project known as the IEAGHG Weyburn-Midale Monitoring and Storage Project, a project which has used enhanced oil recovery operations (CO_2-EOR) as a means of studying the geological storage of CO_2 on an industrial scale.

The neighbouring Weyburn and Midale oilfields, operated by Cenovus Energy Inc. and Apache Canada Ltd., respectively, collectively contain in excess of 20 million tonnes of anthropogenic CO_2. This makes these operations the largest existing CCS project in the world.

This 'Best Practices Manual' provides a summary of key knowledge gained from research in these fields. The project has been managed by the Petroleum Technology Research Centre (PTRC).

Research was carried out in two distinct phases. The first phase, which ran from 2000 to 2004, demonstrated that the Weyburn reservoir provided a suitable site for CO_2 geological storage. Findings from this phase were reported in the proceedings of the 7th International Conference on Greenhouse Gas Control Technologies in 2004. The second phase, conducted from 2005 to 2012, incorporated the Midale oilfield. Among many technical accomplishments, this phase achieved advances in monitoring technologies, methods for field testing for wellbore integrity, and detailed profiling of storage and associated environmental risks.

The aim of this book is to provide technical guidance to future project operators, regulators, and other stakeholders with an interest in the validation of safe and permanent geological storage of CO_2. Although the book has been developed from applied research undertaken in the context of CO_2-EOR operations, many of the lessons learned and recommendations given are equally applicable to other CO_2 geological storage scenarios, including in deep saline formations (aquifers).

Similarly, while some of the research and associated findings are specific to carbonate reservoirs, such as those at Weyburn and Midale, much of the work reported here will prove equally relevant to storage in other reservoir types. More detailed research about storage in other types of reservoirs can be found elsewhere: specifically in the 2009 publication *Pembina Cardium CO_2 Monitoring Pilot: A CO_2-EOR project, Alberta, Canada*, which provides an account of research into a sandstone reservoir, also in the context of CO_2-EOR.

This book is divided into technical sections which deal with (a) characterization and storage performance predictions, (b) geochemical and geophysical monitoring, (c) history matching and performance validation, (d) wellbore integrity, (e) risk assessment, and (f) public outreach.

Technologies and methodologies used in the Weyburn-Midale Project are described in each section, with summaries of recommended best practices, based on the project experiences and results.

Operational CO_2-EOR aspects – including CO_2 capture, transport, and surface facilities – are not covered in this book. Post-closure issues, such as long-term liability and monitoring, are also not addressed in detail.

Characterization and Storage Performance Prediction

Geological characterization is a critical component of storage assessment. It needs to be considered at a variety of scales, being governed by the required storage capacity, and constrained by the availability of data. Definition of the storage complex – that is, the intended reservoir(s) – is required, as is also detailed characterization of any surrounding strata that will contain the injected CO_2, including seals. Characterization of strata above, below, and adjacent to the storage complex is required to enable long-term performance predictions and for risk assessments. The importance of regional characterization data is a key differentiator from technical assessments for 'traditional' CO_2-EOR operations.

Techniques used for geological characterization include standard assessments of geological maps and cross-sections developed from well logs and seismic information. Fundamental structural elements such as faults and other lineaments require careful consideration, because they represent potential conduits for natural or induced fluid flow. Geological models must be constructed so that they can adequately inform hydrogeological characterization and predictive modelling of storage performance, together with assessment of potential leakage scenarios.

Hydrogeological characterization is of critical importance to geological storage projects, and needs to be considered at the same range of regional and site scales as those provided by the base geological models. Hydrostratigraphy – considering the hydraulic properties of all strata, including reservoirs and seals – and hydrochemistry – assessing potential properties of formation waters – should both be determined. Identification and assessment of seals are of paramount importance. Information utilized may include the following: core analysis for measurement of porosity and permeability (with care over scaling effects), pressure gradient plots, permeability and salinity distribution maps, fracture mapping (for example, using seismic data), and comparison with natural analogue sites (for example, nearby reservoirs with high CO_2 contents in natural gas accumulations). Physical properties of all strata, especially seals, should also be characterized in the context of regional stress fields.

The geological and hydrogeological characterization data which are collected form the framework for predictive modelling of storage performance. Long-term storage security can be assessed using a variety of approaches and modelling codes. The Weyburn-Midale Project used both reactive transport and invasion percolation approaches, and the successful application of these methodologies is described in detail within this book. Assessments of storage capacity, linked to mass partitioning of CO_2 between free phase, aqueous, and mineral species, are also described.

Geochemical and Geophysical Monitoring

Monitoring of CO_2 geological storage can be broadly divided into two types. The first, the deep focused type, is used for monitoring CO_2 distribution in the storage complex, as well as for monitoring seal integrity and pressure evolution in response to injection. The second, the shallow focused type, is useful for leakage detection in order to protect sensitive environmental receptors, such as shallow groundwater and ecosystems.

A wide variety of deep focused techniques, both geochemical and geophysical, has been assessed by project research, and is detailed in this book. Time-lapse surface seismic surveys have provided compelling results for the identification of CO_2 in the reservoir. They have also yielded semi-quantitative information on CO_2 concentrations, pressure evolution, and potential leakage detection. Characterization of rock physics assists greatly with these interpretations, and model-based analysis helps to reduce the uncertainties associated with non-unique interpretations of geophysical data. Several 'downhole' techniques have also been employed. They include continuous passive seismic monitoring. This technology requires adequate geomechanical characterization, and has been of limited use for tracking CO_2 distribution in the Weyburn reservoir. However, it has yielded important information for risk assessment and also for stakeholder assurance about potential risks associated with induced seismicity.

Geochemical monitoring of reservoir fluids, while restricted to available wells, can provide useful monitoring data for reservoir fluid evolution in response to injection. Monitoring of reservoir formation water evolution at Weyburn, for example, allowed simulation studies to infer the significant role of fracture flow for reservoir fluid migration.

Geochemical monitoring of the shallow subsurface environment, principally groundwater and soil gas, proved to be an important element in the overall monitoring program. Acquisition of baseline data is very important for determining natural conditions and seasonal variations. As with all monitoring programs, design will be governed partly by site-specific factors. However, general principles based on project experience of sampling point distribution, sampling methodologies, monitoring frequency, and choice of determinants are discussed in this book.

Geochemical monitoring of shallow groundwater is also described, in the context of project research findings. As with soil gas, sampling campaigns must seek to establish baseline conditions in the context of available monitoring points (wells), and with an understanding of shallow groundwater distribution and vulnerability. Where continued access to existing wells is problematic, a dedicated monitoring network may be required.

History Matching and Performance Validation

History matching is an essential aspect of storage assessment. Monitoring data enable calibration and refinement of predictive modelling, not just as an academic exercise, but as a likely regulatory requirement in many jurisdictions. Project research incorporated both laboratory (coreflood experiments) and field-scale history matching. It also included the development of an algorithm to identify reservoir permeability models which could best fit geochemical and seismic monitoring data.

Wellbore Integrity

Because new, purpose-built wells for CO_2 geological storage sites will be constructed to suitably high standards, research focused on the integrity of existing wells. The most important consideration for many such wells is the condition of cement plugs and, in cased wells, cement sheaths. This information is a key component in surveys of existing wells. It is required prior to injection, and may draw together information from both desk- and field-based investigations. This manual includes details of all wellbore-related research in the Weyburn-Midale Project, including the development of a field testing tool to determine the integrity of cement sheaths in existing wells.

Risk Assessment

Risk assessment is part of the wider risk management process. In developing storage standards and regulatory regimes, it is being recognized around the world as an over-arching technical driver for storage assessment, in all phases of the project life cycle. Project research in its first phase broke new ground for CO_2 geological storage by linking performance assessment with risks, using both deterministic and stochastic assessment. In its second phase, application of a methodology known as RISQUE, based on expert judgement, enabled comprehensive assessment of project and environmental risks. These included ground-breaking work on the assessment of potential environmental impacts that could result from leakage scenarios. The principles in this work are applicable to a wide range of potential storage projects, and are fully described in this book.

Public Outreach

Public outreach activities within the project included local community liaison and involvement in the risk assessment process, development of a communications website, and development of a crisis communications strategy. These activities are all summarized here. The importance of this work in relation to new CCS projects should not be underestimated.

ACKNOWLEDGEMENTS

The Petroleum Technology Research Centre has received support from both industry and government, and now thanks the following for their contribution to the Final Phase of the IEAGHG Weyburn-Midale CO_2 Monitoring and Storage Project.

Apache Canada Ltd., Aramco Services Company, Cenovus Energy Inc., Chevron Energy Technology Corp., Dakota Gasification Company, Natural Resources Canada, Government of Alberta (Alberta Innovates – Energy and Environment Solutions), Government of Saskatchewan (Enterprise Saskatchewan, and Ministry of Energy and Resources), Nexen Inc., OMV Austria Exploration & Production Gmbh, Research Institute of Innovative Technology for the Earth (RITE, Japan), Saskatchewan Power Corporation, Schlumberger Canada Ltd., Shell International Exploration and Production B.V., and U.S. Department of Energy.

Final Phase research was led by the following organizations and individuals, all of whom contributed directly to the compilation of the research reported in this book.

Alberta Innovates – Technology Futures (E.H. Perkins, S. Talman, M. Uddin), British Geological Survey (Dave Jones), Cameco (Harm Maathuis), Empresa Communications (Richard Fink), Fugro, Chevron (Mark Meadows), Institute for Corrosion and Multiphase Technology, Ohio University (Yoon-Soek Choi), Lawrence Livermore National Laboratory (Susan Carroll, Walt McNab, Abe Ramirez, Rick Ryerson), Permedia Research Group, Haliburton (Andrew Cavanagh), Saskatchewan Ministry of Energy and Resources (Gavin Jensen, Erik Nickel), Saskatchewan Research Council (Peng Luo), T.L. Watson and Associates, Inc. (Theresa Watson), University of Alberta (Doug Schmidt), Opsens Solutions Inc. (Nathan Deisman), University of Bristol (James Verdon), University of Calgary (Ian Hutcheon, Bernhard Mayer, Maurice Shevalier), University of California Irvine (Russ Detwiler), URS (Donna Pershke), St. Francis Xavier University (Dave Risk) and University of Saskatchewan (Tom Kotzer, Igor Mozorov,)

PTRC also thanks Barbara Dietiker (Geological Survey of Canada) for compiling and formatting the illustrations.

INTRODUCTION

1.1 Purpose, Scope, and Context

1.1.1 Purpose

This book records lessons learned during 11 years of research (from 2000 to 2012) undertaken by the International Energy Agency Greenhouse Gas R&D Programme, and titled (IEAGHG) Weyburn-Midale CO_2 Monitoring and Storage Project (Weyburn-Midale Project). The project was associated with the CO_2 enhanced oil recovery (CO_2-EOR) operations at the Weyburn and Midale fields in southeastern Saskatchewan, Canada.

These oilfields were discovered in 1953–1954, with commercial CO_2-EOR operations initiated first at Weyburn in 2000 and then at Midale in 2005. The CO_2 stream for the CO_2-EOR operations is purchased from the Great Plains Synfuels Plant of Dakota Gasification Company in Beulah, North Dakota. The gas is about 95% CO_2 and is transported through a 320-km pipeline to the Weyburn and Midale fields.

Technical guidance is given for future CO_2 storage projects and for CO_2-EOR operations in which the operator wishes to validate CO_2 geological storage, for example to achieve CO_2 emission credits. Although the project was carried out in the context of an operating CO_2-EOR site, the technical aspects are relevant to a wide range of CO_2 storage scenarios, including carbon capture and storage (CCS) projects in which the intended storage is in deep aquifers with saline formation waters or in depleted oil or gas fields.

1.1.2 Scope

The major objective of the project was to assess the integrity of the geosphere at the Weyburn and Midale fields. Particular emphasis was placed on technical work that provides guidance on (1) site characterization, (2) modelling,

Authors: Jørg E. Aarnes (Det Norske Veritas, Oslo, Norway) and **Neil Wildgust** (Petroleum Technology Research Centre, Regina, Saskatchewan).

(3) monitoring, and (4) risk management for this and other projects seeking to validate CO_2 storage. Validation means providing evidence to meet the following storage parameters.
1. The capacity and injectivity are sufficient to accept the required annual and total quantities of CO_2.
2. The geological conditions ensure long-term containment of injected CO_2 and displaced formation fluids.
3. The measurement, monitoring and verification (MMV) plan ensures safety and environmental protection.
4. There is conformance between the predicted and observed reservoir behaviour.

The research also directed attention to the necessity for public consultation and outreach. Because the project was conducted in producing oilfields, neither closure nor post-closure issues such as long-term liability and monitoring – nor the capture and transport of CO_2 and surface facilities associated with CO_2 injection – were addressed.

The intended audience for the book includes operators, regulators, and stakeholders with an interest in the technical validation of safe and permanent CO_2 storage, and the process by which this validation can be achieved.

1.1.3 Context

The project was carried out in the context of both the Weyburn and Midale CO_2-EOR operations, and of existing and emerging regulations applicable to CO_2 storage and CO_2-EOR. Emerging regulations for CO_2 storage projects include more robust requirements for geological characterization, monitoring, performance validation, and reporting, than those required by the current regulations for CO_2-EOR and similar operations. These differences are being evaluated by regulatory authorities and it is envisioned that CO_2-EOR operators may, in some circumstances, be required to meet some or all of the additional requirements for CO_2 storage projects. For example, the U.S. Environmental Protection Agency requires that projects currently operating under regulations applicable to CO_2-EOR operations (Class II) must meet the regulatory requirements applicable to CO_2 storage (Class VI) if there is an increased risk to what is defined as 'underground sources of drinking water'. This may be the case if the oilfield is being pressurized, or if CO_2 injection continues beyond the oil-producing life of the field.

1.2 History of the Project

The project was initiated in 2000 by the Petroleum Technology Research Centre (PTRC), Regina, Saskatchewan, and was endorsed by IEAGHG in recognition of the potential global significance of CCS technology. PTRC has been the technical manager from the beginning.

Research was undertaken in two phases, referred to as Phase 1 (2000–2004) and Final Phase (2005–2011). Phase 1 focused on the Weyburn Field and dealt with efforts to predict and to verify the ability of a deep carbonate oil reservoir to securely store CO_2 in conjunction with an active CO_2-EOR operation. This phase was organized into four main categories or 'themes'.

1. Geological characterization of the geosphere and biosphere.
2. Prediction, monitoring, and verification of CO_2 movement.
3. Prediction of CO_2 storage capacity and distribution, and the determination of economic limits.
4. Assessment of long-term risks at the storage site.

Phase 1 results demonstrated that the site was suitable for CO_2 storage and also provided a comprehensive data set for future reference. Results obtained from the first phase of research were reported in 2004 under the proceedings of the 7th International Conference on Greenhouse Gas Control Technologies in Vancouver, British Columbia, Canada. They have been used extensively in the compilation of this book; readers may therefore wish to refer to the Phase 1 summary report (Wilson and Monea, 2004), which is also freely available online at http://ptrc.ca/weyburn_first.php.

The Final Phase included research at both the Weyburn and Midale fields, with reorganized themes, as follows.

1. Geological integrity.
2. Wellbore integrity.
3. Geophysical and geochemical storage monitoring methods.
4. Risk assessment.

Additional non-technical efforts focused on (1) reviewing and comparing domestic and international legal and regulatory frameworks, (2) developing and carrying out strategies for public communication and outreach, and (3) assessing the business environment for CO_2-EOR.

1.3 **Weyburn and Midale CO_2-EOR Operations**

The Weyburn Field contains about 220 Mm^3 (1.4 billion barrels) of oil, and is currently producing about 4450 m^3/d (28 000 barrels per day), of which 2860 m^3/d (18 000 barrels per day) is incremental production due to the CO_2 flood (see Figure 1-1). From the start of injection until the end of 2011, about 18 Mt of anthropogenic CO_2 have been stored at Weyburn. Annual injection rates are approaching 5 Mt/a, although half of that total is currently recycled from the reservoir, so the net storage rate is in the range of 1.7–2.4 Mt/a (production information courtesy Cenovus Energy Inc., formerly Encana). Cenovus Energy Inc. is the field operator.

In total, approximately 30 Mt of CO_2 are projected to be injected into the reservoir during the course of the CO_2-EOR operations, although the net rate of storage will decrease with time as the proportion of recycled CO_2 increases. An additional 25 Mt of CO_2 could be stored by maintaining injection after cessation of CO_2-EOR operations, that is, for storage purposes only. The storage capacity of the field may also be increased if CO_2-EOR operations are extended into the wider oilfield.

The Midale Field is located adjacent to the Weyburn Field. It was first field-tested for CO_2 injection in 1985, but commercial CO_2-EOR operations only commenced in the summer of 2005. The CO_2-EOR operation is expected to enhance incremental production by 9.5 Mm^3 (60 million barrels) over the next 20–25 years. The annual CO_2 injection rate is currently 0.7 Mt, of which 0.2 Mt is recycled from the reservoir. Since the start of injection, close to 3 Mt of CO_2 have been stored at the field. Expected storage is about 10 Mt of CO_2. Figure 1-2 shows both the actual and anticipated production from the Midale Field by primary recovery, waterflood, and three phases of EOR operations. The field operator is Apache Canada Ltd.

1.4 **Industry Experience with Large-scale CCS and Similar Operations**

The Global CCS Institute classifies a CCS project as large-scale if the annual injected CO_2 is not less than 800 000 t if sourced from a coal-fired power plant or not less than 400 000 t if sourced from intensive industrial facilities. For a CO_2-EOR project to qualify as a CCS project the CO_2 must be sourced from an anthropogenic source, that is, not produced from a subsurface reservoir and piped to the oilfield for CO_2 injection. Anthropogenic sources include (1) natural gas processing (stripping the gas stream to recover its CO_2 content to meet a commercial specification), (2) the capture of CO_2 from the flue stacks at coal-fired power plants, and (3) the capture of CO_2 from industrial sources such as oil refineries, ethanol production, iron and steel mills, cement factories, and fertilizer plants. The Great Plains Synfuels Plant is considered an anthropogenic industrial source.

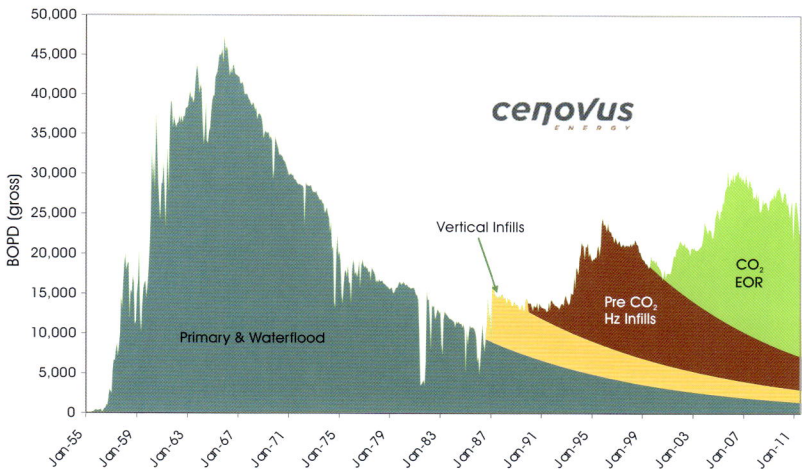

Figure 1-1. *Oil production (BOPD: barrels of oil per day) in the Weyburn Field, by primary recovery and waterflood, vertical infill water injectors, horizontal infill water injectors, and CO_2-EOR.*

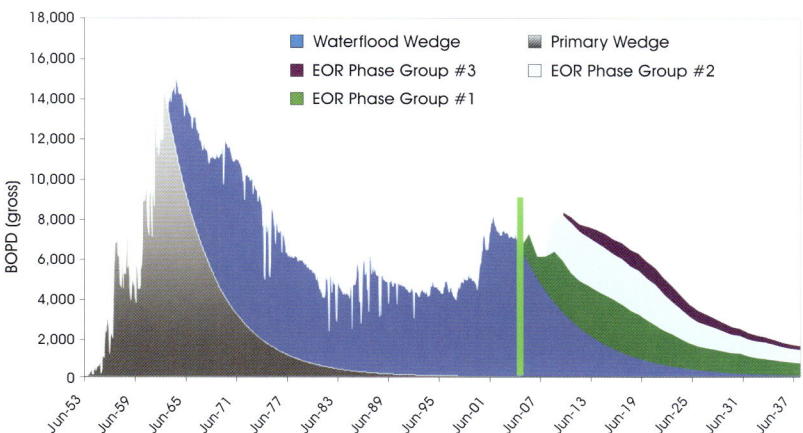

Figure 1-2. *Oil production (BOPD: barrels of oil per day) in the Midale Field, and predicted daily production until 2040, by primary recovery, waterflood, and three EOR phases involving CO_2 injection.*

As of 2011, the Institute classified eight projects as large-scale (Global CCS Institute, 2011), five of which are CO_2-EOR operations (the Weyburn and Midale CO_2-EOR operations are defined as a single project). Although further industrial-scale CCS demonstration projects are at various stages of development in North America, Australia, and Europe, the small number of operational projects emphasizes the importance of harvesting lessons learned from the experiences gained in the current suite of projects.

Although a majority of the CO_2 delivered to CO_2-EOR operations is sourced from natural reservoirs (see, for example, Advanced Resources International and Melzer Consulting, 2010), the collective experience from four decades of CO_2-EOR operations with naturally sourced CO_2 across North America has provided a large body of practical information relevant to assessment and validation of CO_2 geological storage performance. The portfolio of acid gas injection projects in Alberta and British Columbia also provides an industrial analogue.

1.5 Organization of the Chapters

The research results reported here are presented in the following seven chapters.
- Chapter 2. Characterization
- Chapter 3. Storage Performance Predictions
- Chapter 4. Geochemical Monitoring
- Chapter 5. Geophysical Monitoring
- Chapter 6. History Matching and Performance Validation
- Chapter 7. Well Integrity
- Chapter 8. Risk Assessment

Each chapter includes (1) a review of key research activities considered to be significant for the assessment and validation of storage site suitability and performance, (2) a discussion of these activities with emphasis on the value that they provided in terms of reaching the research objectives of the project, and (3) the lessons learned, formulated as recommendations for those concerned with validating other CO_2 geological storage sites.

The reader will find a greater level of technical detail in this book than in similar reports on this subject. While other related guidelines [for example, Hoek et al. (2002), IPCC (2006), World Resources Institute (2008), Onishi et al. (2009), DNV (2010, 2011), and European Commission (2011a, b)] and best practices manuals [for example, Chadwick et al. (2008), Kaldi and Gibson-Poole (2008), U.S. Department of Energy, National Energy Technology Laboratory (2009a, 2010, 2011)] have generally focused on processes, the aim here is to guide the technical work that should be performed. Thus the seven technical chapters may be used as a handbook for information on the design and execution of studies on a CO_2 storage project, whether part of a CO_2-EOR operation, or specifically designed for the sole purpose of long-term isolation of CO_2 from the atmosphere.

Following the technical chapters is a chapter that discusses experiences related to community consultation and outreach, and offers some direction and recommendations arising from the specifics of this project to communicators organizing consortia-driven research programs.

2

CHARACTERIZATION

Summary

Research results from the Weyburn-Midale Project are site specific and relevant to the CO_2-EOR operations at the Weyburn and Midale fields. However, the design, scope, and processes of the research are applicable to other CO_2-EOR operations and much of the work may also be applied to CO_2 geological storage.

Geological characterization is critical for CO_2 storage, not least in defining the storage complex. In the project, several vertical and areal scales were required to cover the geology and hydrogeology to meet the needs of other disciplines, and all strata from the Precambrian basement to the surface were characterized, though not all in the same detail. The storage complex required the maximum detail. Geological models were constructed mainly with a view to their application to numerical models rather than to illustrate complex geological situations.

Hydrogeological characterization started with the geological framework, and all vertical and areal scales mapped as part of the geological characterization went through equivalent hydrogeological characterization. The hydrostratigraphy and hydrochemistry were determined for each unit based on hydraulic head maps and cross-sections, pressure-depth plots, and 'water driving force' maps, necessary because of the variable density formation waters.

Identification of seals is of paramount importance, and methods used to characterize them included core analysis of shales, vertical pressure gradient plots, and fracture mapping using seismic amplitude versus offset and azimuth (AVOA) analysis. Aquitards identified as seals received special attention including determination of their permeability and porosity by a variety of methods, their bulk chemical composition and mineralogy, and physical properties (Young's modulus, Poisson's ratio and shear modulus).

Recommendations arising out this research begin on page 75.

Authors: B.J. Rostron (University of Alberta, Edmonton, Alberta), **S. Whittaker** (Global CCS Institute, Canberra, Australia), **C. Hawkes** (University of Saskatchewan, Saskatoon, Saskatchewan), and **D. White** (Geological Survey of Canada, Ottawa, Ontario).

2.1 Introduction

Over the course of the project, the geosphere and biosphere surrounding the project site were investigated using a variety of techniques to characterize the geological system as suitable, or not, for the long-term storage of CO_2. This chapter is divided into six sections, each describing the activities undertaken and how this research led to the 'best practices' recommendations that may be of interest to an operator of a CO_2 storage project. The work broadly covers the following topics: geology, hydrogeology, containment, natural analogues, geomechanics, and geochemistry.

Determining the integrity of the geosphere around the Weyburn Field is fundamental to assessing and validating the suitability of the field for CO_2 storage. Specifically, it is necessary to identify flow units and geological seals that are inherent characteristics of the storage system. This allows for the development of a framework for modelling, monitoring, verification, and risk assessment.

This chapter is a summary of activities within the project area that are necessary to characterize and map the regional geological and hydrogeological systems. Although some of these activities are necessarily specific to the local geology, the broad procedures are applicable to any CO_2 storage project.

2.1.1 Study Area

Characterization of the Weyburn geology and hydrogeology was undertaken both vertically and areally at various scales. In terms of vertical scales, as much detail as possible regarding the containment features was desirable. In addition, to assess the potential for CO_2 migration outside the injection zone and to understand possible migration paths, the initial geological, hydrogeological, and seismic work covered all strata from the Precambrian basement to the surface. The degree of detail was driven by data availability and the requirements for monitoring and risk assessment.

Several study 'areas' were defined during both phases of the project, from the largest (regional scale) down to smaller scales for specific purposes. With respect to CO_2 storage, this requirement arises because, although fluid migration pathways may extend for hundreds of kilometres in a sedimentary basin, the storage complex occupies only a relatively small area. At the largest scale, a 40 000 km² regional geological study area extending 100 km in all directions from the Weyburn Field was defined in Phase I (Figure 2-1 and Figure 2-2). That area was chosen so as to capture sufficient data from about 30 000 oil, gas, and water wells to develop a regional structural framework for more detailed geological mapping, including the deepest strata. The regional study area was extended farther in some cases (for example, for the hydrogeological study of Mississippian aquifers) in order to capture sufficient data in the key injection zone.

Hydrogeological investigations were carried out at several different depths and spatial scales. Regional hydrogeology and hydrochemistry were divided into 'deep' (below the base of potable water, or the biosphere) and 'shallow' (potable) investigations. Potable aquifers in a 3 × 3 township area around the Phase 1A injection site were examined in greater detail due to their importance as a local water supply.

In the Final Phase of the project, an even finer resolution of the structure within the storage complex was required for CO_2 flow-path modelling. Thus, an 'intermediate' geological study area was defined by Twp 5–8, Rge 11–15, W2Mer (Figure 2-2), utilizing about 3600 oil and gas wells.

At the smallest scale, a risk assessment region extending 10 km beyond the extent of the CO_2-EOR flood was defined within the regional geological study area. Figure 2-1 and Figure 2-2 provide an overview of the nested geological models in the context of the larger Williston Basin.

The CO_2-EOR flood was designed to be implemented over a 15-year period using 75 inverted 9-spot injector-producer well patterns (Figure 2-3). At the beginning of the project, research activities were focused on the so-called initial 19 injection well patterns, termed the Phase 1A area (Figure 2-3, yellow). Later, in 2005, the so-called Phase 1B area was introduced; it consisted of four

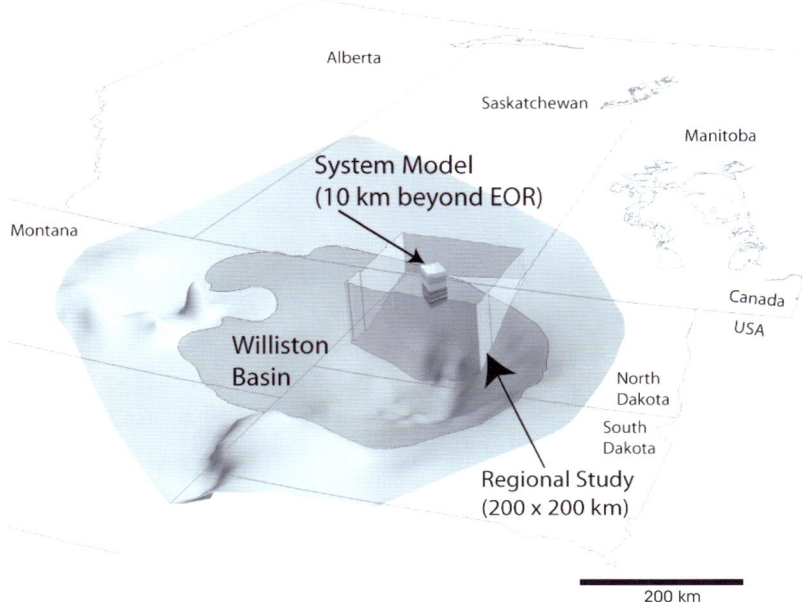

Figure 2-1. *Overview of the nested geological models in the context of the larger Williston Basin.*

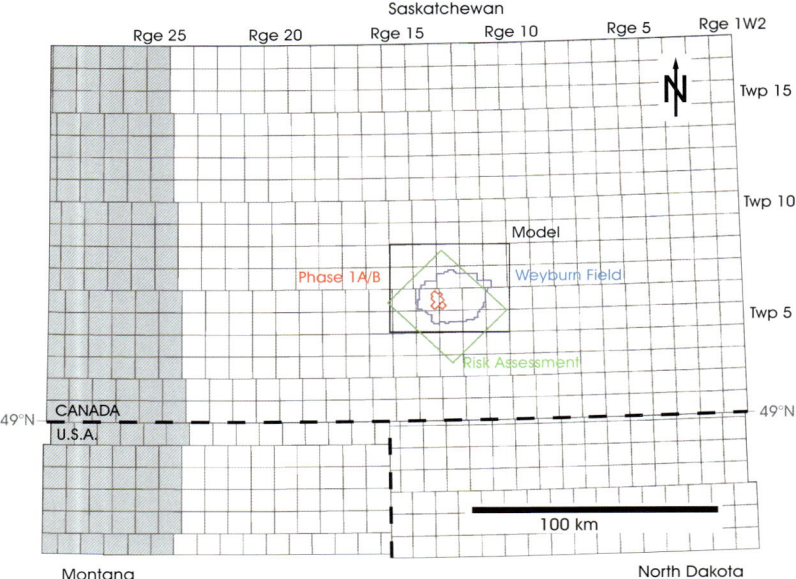

Figure 2-2. *Map of the various study areas used for the project. The Weyburn Field is outlined in blue. The 'regional area' extends approximately 100 km from the Phase 1A/B CO_2 injection area (red). The screened area to the west was used for hydrogeological studies of the Mississippian aquifers. Phase I 'risk assessment region' is shown in green. The Final Phase 'geological model study area' is outlined in black.*

additional injection well patterns (Figure 2-3, pale blue). From 2007–2011, research activities were focused mainly on the Phase 1A and 1B areas of the CO_2-EOR flood.

The regional stratigraphy from the Precambrian basement to the ground surface was mapped primarily through the interpretation of geophysical well logs and core samples, augmented by regional seismic investigations using >2000 km of 2D seismic data. This information highlighted basinal tectonic elements and basement features, with more detailed geological, mineralogical, and isotope studies on selected geological units.

Also studied was one site of several naturally occurring CO_2 accumulations about 400 km from the project area in southwestern Saskatchewan. There, accumulations containing up to 80% CO_2 are hosted in Devonian strata, with an estimated age of accumulation of 50 Ma.

2.1.2 Physiography

The study area is in the Williston Basin (Figure 2-1), a large intra-cratonic

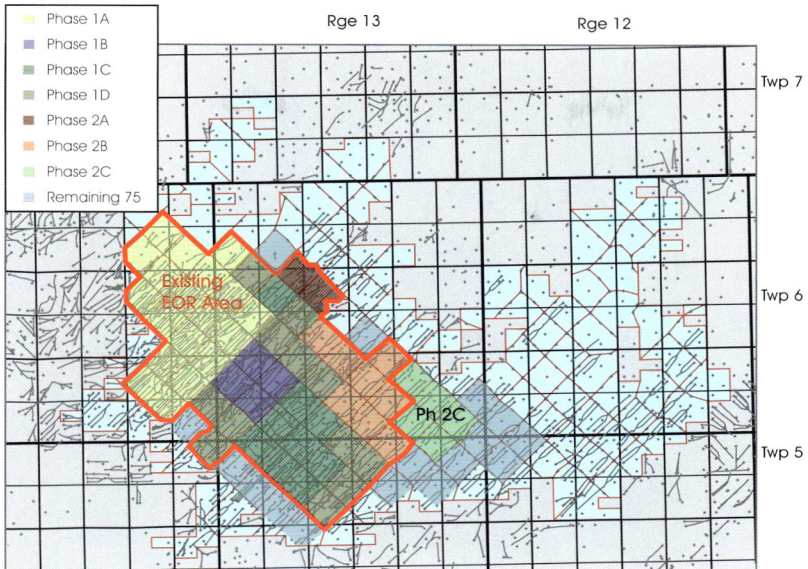

Figure 2-3. *Actual and planned CO_2-EOR patterns. Map courtesy Cenovus Energy Inc., 2010.*

sedimentary basin with minimal topographic relief forming part of the Interior Plains of North America. Included are parts of southern Saskatchewan, northwestern North Dakota and northeastern Montana, with expansive grasslands and low rolling hills suitable for crop farming and ranching.

There are two major river basins – the Assiniboine–Red–Nelson River Basin (Qu'Appelle and Souris rivers), and the Missouri–Mississippi River Basin (Wood, Poplar, and Little Missouri rivers). The Souris River and its tributaries form the main surface drainage in the study area, flowing south towards North Dakota then back across the border to Manitoba and ultimately into Hudson Bay.

Population in and near the study area is sparse. At the very northern edge is Regina (2009 population, 210 000), the provincial capital of Saskatchewan. The other two significant centres are Weyburn and Estevan, with populations of about 10 000. Smaller communities are scattered across the area.

2.1.3 Definition of the Storage Complex

Modelling of oil reservoirs is typically used only for operational purposes. These models provide a basis for evaluating alternative injection strategies to maximize oil production, and to predict the sweep efficiency of injected fluids within the reservoir.

For EOR operations, modelling is primarily focused on monitoring the CO_2–waterflood to calibrate the models and to form a basis for more robust decisions for fine tuning the injection strategy.

In contrast, the primary focus of CO_2 geological storage is to ensure that injected volumes of CO_2 and displaced formation fluids are contained within a designated subsurface volume in order to avoid potential negative impacts on the environment and economic resources. An essential component in the characterization of sites suitable for long-term geological storage of CO_2 is the assessment of the integrity of sealing units, particularly the cap rock. Consequently, emerging regulations for CO_2 geological storage include requirements to demonstrate containment of CO_2 and displaced fluids within a larger containment system, which may consist of a series of barriers to upward or downward movement of CO_2. Operators of CO_2 storage projects are further required to use modelling and baseline monitoring to demonstrate long-term containment.

We adopt here the language of the European CCS Directive (European Commission, 2009), that is, we use 'storage complex' when referring to the limits of acceptable migration of CO_2 and displaced formation fluids. The storage complex is defined in the Directive as the storage site (essentially the target storage formation) and the surrounding geological domain, which can have an effect on overall storage integrity and security. This term is analogous to the notion of a confinement zone in the *Federal Requirements Under the Underground Injection Control (UIC) Program for Carbon Dioxide (CO_2) Geologic Sequestration (GS) Wells* by the U.S. Environmental Protection Agency (2010). The confinement zone is defined as a geological formation, group of formations, or part of a formation stratigraphically overlying (and underlying) the injection zone(s).

2.1.4 Storage Complex at the Weyburn Field

If an operator of a CO_2-EOR project seeks credits for CO_2 storage, or wishes to formally transition the project to a CO_2 storage project, then the applicable regulations may require that a storage complex is formally defined. While requirements may vary slightly as different jurisdictions develop regulations, it is likely that an acceptable storage complex will need to meet the following criteria.

1. It is sufficiently extensive to provide confidence that –
    A. Planned volumes and rates of CO_2 can be effectively injected.
    B. Injected CO_2 will be contained within its areal and vertical boundaries.

2. Migration of CO_2 within the storage complex does not pose any significant risks to the environment or to subsurface economic resources.
3. It provides sufficient opportunities for (A) cost-effective monitoring to demonstrate containment, (B) early indications of loss of containment or circumstances that may lead to loss of containment, and (C) support of the appropriate risk management.

In order to meet these criteria, the project proposed that the storage complex for the Weyburn Field consist of the following geological units, in ascending stratigraphic order (Figure 2-4 and Figure 2-5).

FROBISHER BEDS, MISSION CANYON FORMATION: consisting of three distinct 'units' – the Frobisher 'Vuggy' at the base, overlain by the Frobisher 'Marly', and capped, in places, by the Frobisher Evaporite.

MIDALE BEDS, CHARLES FORMATION: the CO_2 injection zone in the Weyburn Field, and comprising a lower 'Vuggy' zone, overlain by the 'Marly' zone, and capped by the 'Three Fingers' zone.

MIDALE EVAPORITE, CHARLES FORMATION: representing the first seal for CO_2 migration out of the injection zone.

RATCLIFFE BEDS, CHARLES FORMATION: thin layers of porous carbonates interbedded between two significant evaporite layers – Oungre and Upper Ratcliffe.

POPLAR BEDS, CHARLES FORMATION: the final carbonate cycle in the Mississippian succession, consisting of thin porous intervals interbedded with two evaporite layers – Greenpoint and P3.

WATROUS FORMATION, TRIASSIC: dominantly non-marine 'red beds', providing a regional sealing layer for the many hydrocarbon pools in the Williston Basin, and a regionally extensive diagenetically altered zone cross-cutting the Charles Formation beneath the sub-Mesozoic unconformity.

This definition was developed on the basis of extensive geological, geophysical, hydrogeological, and geotechnical investigations combined with expert knowledge of the behaviour of the reservoir provided by the operator. Further details on the characterization of the storage complex are provided below.

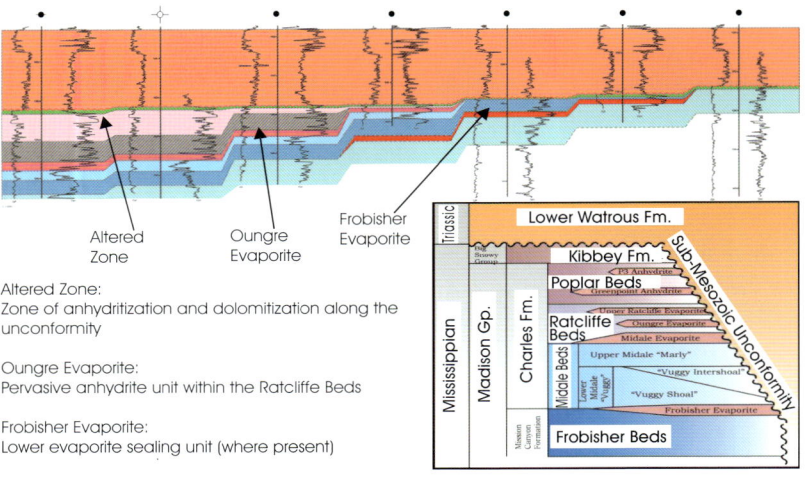

Altered Zone:
Zone of anhydritization and dolomitization along the unconformity

Oungre Evaporite:
Pervasive anhydrite unit within the Ratcliffe Beds

Frobisher Evaporite:
Lower evaporite sealing unit (where present)

Figure 2-4. *Details of the Mississippian stratigraphy in the Weyburn Field, including new layers that were included in the Final Phase geological model.*

2.2 Geology

2.2.1 Introduction

The Weyburn Field occurs in Mississippian carbonates at a depth of about 1500 m in the northeastern part of the Williston Basin. The basin centre is in North Dakota with a maximum thickness of about 4900 m, although at the Weyburn Field the succession ranges from 2800–3000 m. Initial phases of basin development began during the Late Cambrian to Early Ordovician and continued episodically until the Late Cretaceous, when a foreland basin developed along the western margin of North America (Kent and Christopher, 1994).

In the study area strata range in age from Middle Cambrian and Early Ordovician sandstones that directly overlie the Precambrian basement to Quaternary glacial deposits at the surface (Figure 2-5). Rocks of the Williston Basin can be subdivided into two broad packages.

1. Paleozoic strata dominated by carbonates, evaporites, and minor shales.
2. Mesozoic strata dominated by shales, siltstones, and sandstones.

The Mississippian Midale Beds of the Weyburn reservoir occur near the uppermost part of the Paleozoic succession (Figure 2-4) immediately below a basin-wide angular unconformity that separates the Paleozoic and Mesozoic packages and represents a depositional hiatus of about 60 Ma. The sub-Mesozoic unconformity is a very important and effective regional flow barrier throughout

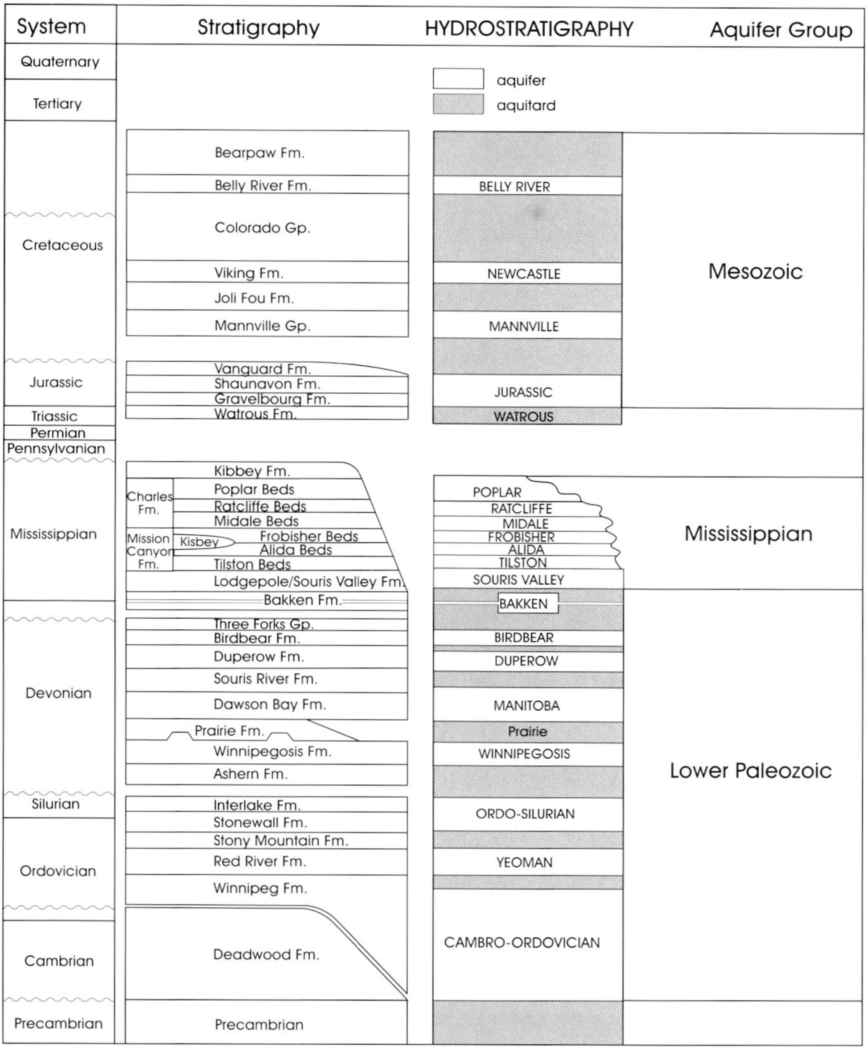

System	Stratigraphy	HYDROSTRATIGRAPHY	Aquifer Group
Quaternary		aquifer	
Tertiary		aquitard	
Cretaceous	Bearpaw Fm.		Mesozoic
	Belly River Fm.	BELLY RIVER	
	Colorado Gp.		
	Viking Fm.	NEWCASTLE	
	Joli Fou Fm.		
	Mannville Gp.	MANNVILLE	
Jurassic	Vanguard Fm. / Shaunavon Fm. / Gravelbourg Fm.	JURASSIC	
Triassic / Permian / Pennsylvanian	Watrous Fm.	WATROUS	
Mississippian	Kibbey Fm.		Mississippian
	Charles Fm. — Poplar Beds	POPLAR	
	Ratcliffe Beds	RATCLIFFE	
	Midale Beds	MIDALE	
	Mission Canyon Fm. / Kisbey — Frobisher Beds	FROBISHER	
	Alida Beds	ALIDA	
	Tilston Beds	TILSTON	
	Lodgepole/Souris Valley Fm.	SOURIS VALLEY	
	Bakken Fm.	BAKKEN	
Devonian	Three Forks Gp. / Birdbear Fm.	BIRDBEAR	Lower Paleozoic
	Duperow Fm.	DUPEROW	
	Souris River Fm.		
	Dawson Bay Fm.	MANITOBA	
	Prairie Fm.	Prairie	
	Winnipegosis Fm.	WINNIPEGOSIS	
	Ashern Fm.		
Silurian	Interlake Fm. / Stonewall Fm.	ORDO-SILURIAN	
Ordovician	Stony Mountain Fm.		
	Red River Fm.	YEOMAN	
	Winnipeg Fm.		
Cambrian	Deadwood Fm.	CAMBRO-ORDOVICIAN	
Precambrian	Precambrian		

Figure 2-5. *Composite stratigraphic and hydrostratigraphic column. Figure based on the Saskatchewan Provincial Stratigraphic Chart modified to include regional aquifers and aquitards.*

the northern part of the basin, with tilted porous Paleozoic carbonates truncated against relatively low permeability siliciclastics of the Triassic–Jurassic Watrous Formation. In the Weyburn storage complex, the Watrous Formation is defined as the ultimate seal. Immediately subjacent to the unconformity surface, and within many of the Paleozoic carbonate beds, diagenesis has resulted in a highly cemented low-permeability zone that forms an effective seal against vertical fluid migration.

The overlying Mesozoic succession includes marine and non-marine sandstones and siltstones but is volumetrically dominated by thick transgressive shales that formed within the Cretaceous foreland basin. Above the Weyburn Field, the thickness of the Mesozoic rocks is 1300–1400 m, but, in the broader study area, it ranges from 675 m in the northeast to 2300 m in the south.

2.2.2 Structure

Basement structures, faults, and intra-sedimentary fracture zones are fundamental structural elements that influenced primary depositional patterns and secondary subsurface processes that ultimately may affect subsurface CO_2 distribution and containment. Faults and fractures are among the most important of these features because they potentially represent direct natural pathways between storage zones and shallower aquifers or the atmosphere. The structural network in the study area was examined through the relationship of basement structures to intra-sedimentary features using approximately 2000 km of 2D seismic data, high-resolution aeromagnetic data, and air photo and Landsat imagery analysis of surface lineaments.

A west-east cross-section extending 157 km through the study area (Figure 2-6) was constructed from four merged seismic lines. Generally, strata are continuous across the region, but in places vertical discontinuities in the seismic images were interpreted as faults. Seismic data indicate that local segments of the subsurface were deformed by anorogenic stresses throughout the Paleozoic and, to a lesser extent, in more recent times. Local structural anomalies, such as small fault zones, are recognized on many of the 2D seismic sections. Nearly all are observable within Paleozoic strata, in contrast to the relatively undeformed Watrous Formation, and to Upper Cretaceous strata which are even less deformed. This suggests that the majority of structural discontinuities are absent above the Midale Beds, although this interpretation may be biased because processing of the seismic data was focused so as to image strata below and immediately above the storage interval, hence reducing the resolution at shallower levels.

The general lack of distinguishable features – and often large spatial separation between seismic survey lines – hinders correlation of faults and fractures

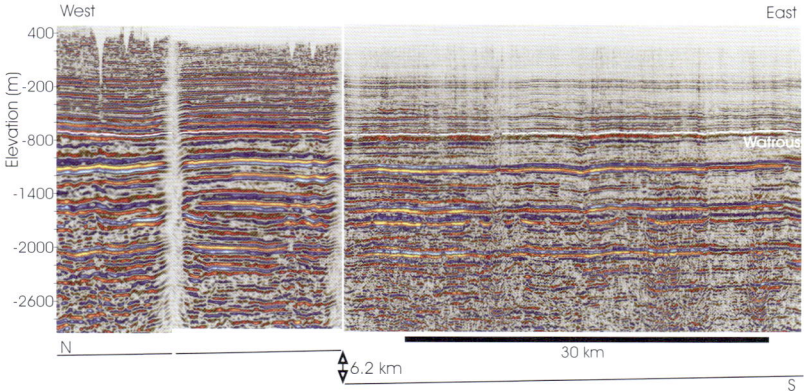

Figure 2-6. *West-east regional seismic line through the project area.*

among seismic profiles. Hence, high-resolution aeromagnetic data were used to help correlate structures among seismic sections and to determine fault and fracture orientations more clearly. The high-resolution aeromagnetic data response is related to the content of magnetic minerals and to the ratio of induced-to-remnant magnetism (Goussev et al., 2004). Commonly, it provides additional information on the orientation of structural features. This integrated seismic–high-resolution aeromagnetic data technique was used to develop a network of interpreted intra-sedimentary faults and fractures for the risk assessment region (Figure 2-7). One linear magnetic anomaly that trends NNW-SSE through the risk assessment region was interpreted as the magnetic signature of a fault identified seismically. This fault is named the Souris River Fault and was included in the geological model for risk assessment.

Subsurface joint and fault zones that are associated with surface structures were identified using remotely-sensed surface lineament mapping – this helps to better interpret seismic and high-resolution aeromagnetic anomalies. This method shows that the study area is characterized by systematic NE-SW- and NW-SE-oriented lineaments and lineament zones and minor N-S and E-W trends (Figure 2-8 and Figure 2-9). Lineaments range in length from <1 km for individual lineaments to >100 km for the longest zones. The NE-SW surficial lineament trend closely parallels the well-documented crustal movement direction (maximum horizontal stress). Structural discontinuities at these lineament locations are inferred to include faults (fractures across which varying amounts of displacement have occurred), flexures, and joint zones (open-mode fracture zones along which displacement has not occurred).

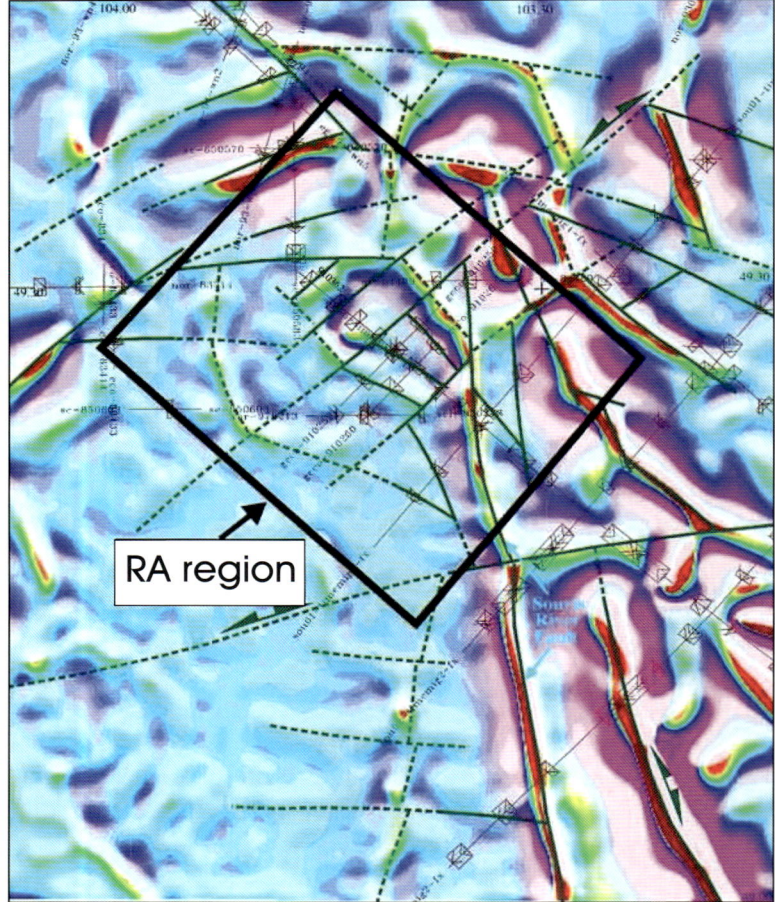

Figure 2-7. *High-resolution aeromagnetic image of the risk assessment region in the project area.*

2.2.3 **Lithostratigraphic Framework**

The regional study comprised a volume of strata about 100 000 km³ with the Weyburn Field at its geographic centre. A major focus in developing the regional lithostratigraphic framework was to provide a consistent database for establishing the hydrostratigraphic framework. Emphasis was placed on identifying and characterizing primary and secondary seals and determining preferential migration pathways, particularly in strata overlying the Mississippian reservoir.

Geological mapping in the regional area was based on more than 3400 wells drilled prior to January 2003. All wells that penetrated Devonian strata

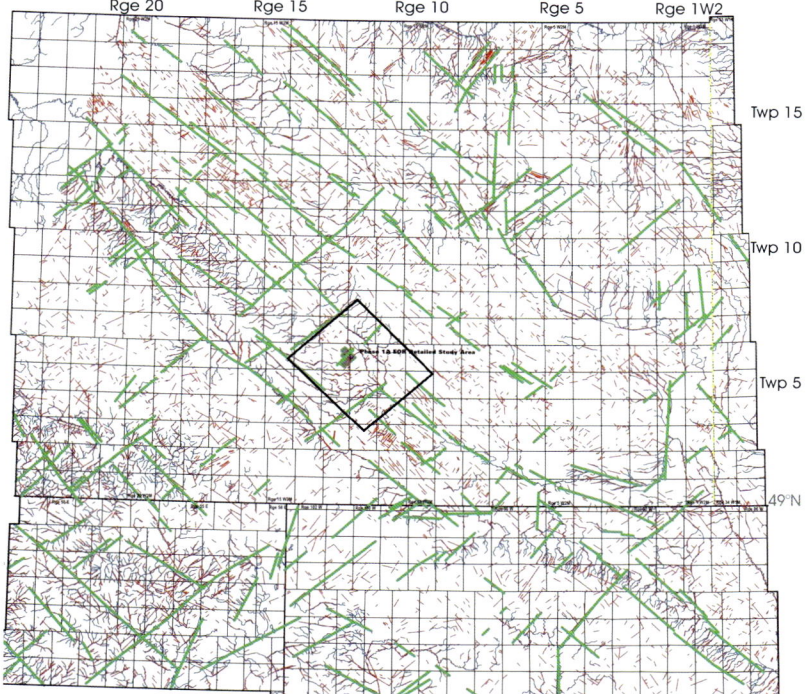

Figure 2-8. *Regional map of main lineaments. Inferred regional lineament zones (heavy green lines), lineaments based on Landsat imagery (thin brown lines), and drainage features based on topographic maps (thin blue lines).*

or deeper were used and, where possible, at least three wells were mapped in each township. Additional wells were used near the Weyburn Field to help delineate the geological units. About 140 individual geological units were mapped, although all units are not present in every well. The geological picks were made primarily through geophysical log analysis and were utilized to generate isopach and structure maps using contouring software (Surfer™). Minor manual editing allowed better definition of salt zero edges and subcrop limits. A limited amount of core was also examined to assist with picks and correlations. To ensure consistency of the stratigraphic correlations, four teams of geologists and geographic information system technicians evaluated specific stratigraphic packages: Precambrian to Silurian, Devonian, Mississippian, and Mesozoic. Cross-sections were constructed, a comprehensive database was compiled, and isopach and structure maps were made of specific rock units.

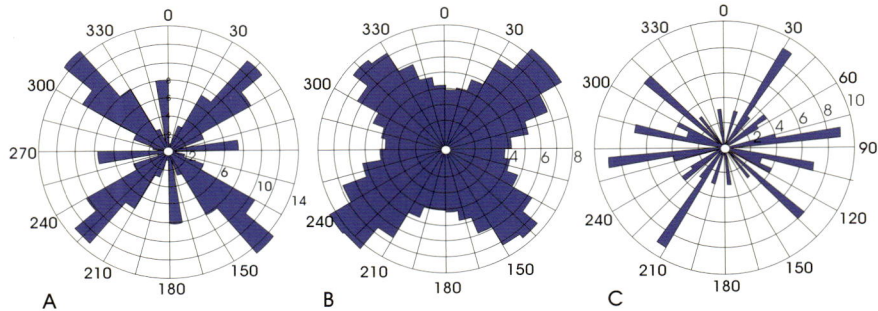

Figure 2-9. *Summary of lineaments mapped in the project area.* **A.** *Air photo and satellite lineaments.* **B.** *Joints in drift and bedrock exposure (Stauffer and Gendzwill, 1987).* **C.** *Fractures in the Weyburn Field reservoir (Bunge, 2000).*

2.2.4 Geological Models

The primary function of the geological models was to provide a framework for other parts of the project requiring geological information (for example, invasion percolation and stochastic engineering modelling). Two different 3D geological models were created. The first (Figure 2-10), called the Phase I model, was constructed over the risk assessment region using GOCAD™ and included all mapped geological, hydrogeological, and geophysical data generated from the larger (200 × 200 km) regional study area. This model described the natural system and included the geometry of the strata, interpreted faults, and many hydraulic and hydrochemical properties associated with those rocks: porosity, permeability, formation water salinity, temperature, hydraulic head, formation pressure, and oil and water saturation. These data were derived from core analyses, laboratory analyses, drill stem tests, geophysical log analyses, and stochastic simulations of permeability data, and supplemented with published literature and historical records.

The lateral extent of the Phase I model was determined using preliminary scoping calculations to ensure coverage that would significantly exceed the expected migration of CO_2 over thousands of years. The vertical extent of the model was somewhat arbitrarily chosen to include strata from about 100 m below the base of the reservoir (Mississippian Tilston Beds) to the ground surface. Horizontal layers included all major hydrostratigraphic flow units, with additional resolution at the reservoir level achieved with the inclusion of the Midale 'Marly' and Midale 'Vuggy' layers, the overlying Midale Evaporite seal, and the underlying Frobisher Beds. Neither the Frobisher Evaporite nor the diagenetically altered zone (strata with reduced permeability) at the top of the Midale Beds was included in order to reduce the processing requirements of the risk assessment.

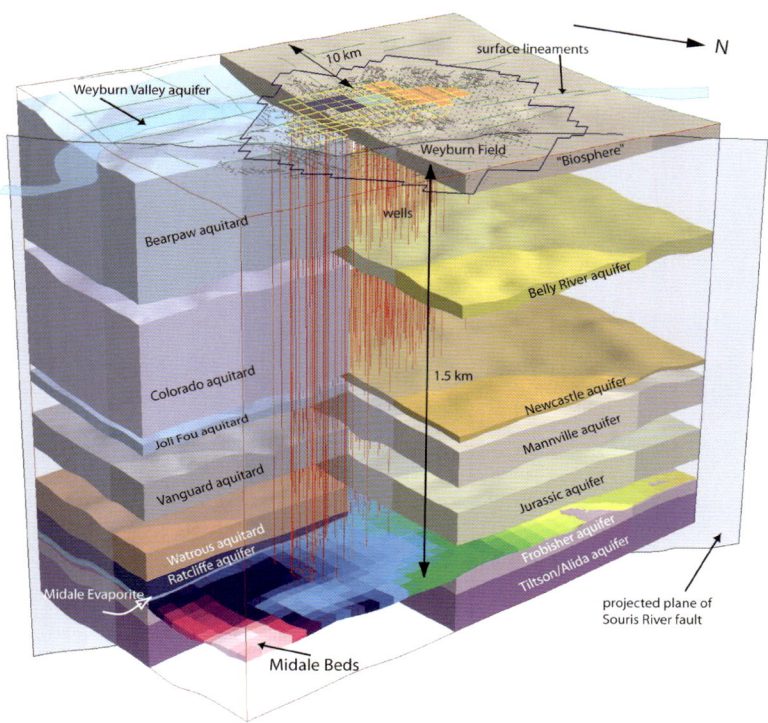

Figure 2-10. *GOCAD™ model of the project area showing major elements of the storage complex. 20× vertical exaggeration.*

The second, updated geological model (Figure 2-11), called the Final Phase model, focused on details of the injection zone and the geological layers immediately above it (see Figure 2-4).

This model covered Twp 5–8, Rge 11–15, W2Mer, as defined by numerical simulations of CO_2 migration above the reservoir level (see discussion below). Data from about 900 wells were used to construct the Final Phase model (including all wells used in the Phase I model). Twenty-nine layers were picked using a consistent set of geophysical log signatures (supplemented with core analyses) with special attention paid to four units missing from the Phase I model (the altered zone at the base of the Lower Watrous Formation, the Oungre Evaporite, the Midale Evaporite, and the Frobisher Evaporite).

The Final Phase model was generated using Petrel™ software. Improvements in data density and computing hardware enabled the implementation of a 50 × 50 m horizontal grid spacing. Special attention was paid to delineating the subcrops of the eight 'truncated' Mississippian units (see Figure 2-4).

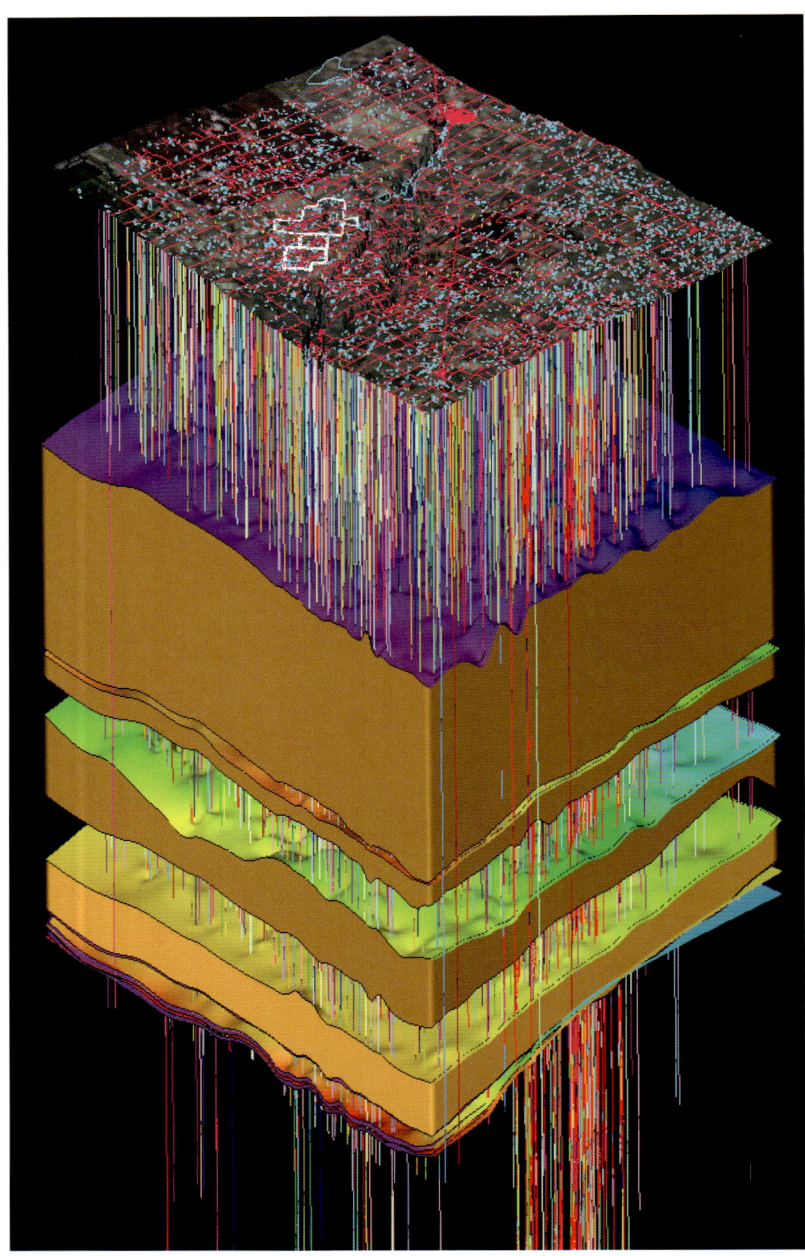

Figure 2-11. *Schematic of the Final Phase geological model showing the digital elevation of the topographic surface and all well penetrations (about 800) used to create the model.*

Geological picks from wells were supplemented with about 2700 'Log Analysis Standard' digital geophysical well log traces for 360 wells, as well as two depth-migrated regional seismic sections.

2.2.5 Geological Mapping
Lower Paleozoic Strata and Salt Dissolution
The Paleozoic succession is about 1350 m thick in the Weyburn area, and many of the deposits exhibit cyclic depositional patterns with relatively porous carbonate intervals capped by non-porous argillaceous carbonates and evaporites.

Evaporites are important because they may be thin and discontinuous bodies or thick and extensive units, both of which influence regional and local fluid flow within the basin. Evaporite beds are effective aquitards, but in parts of the basin they have undergone subsurface dissolution that led to the development

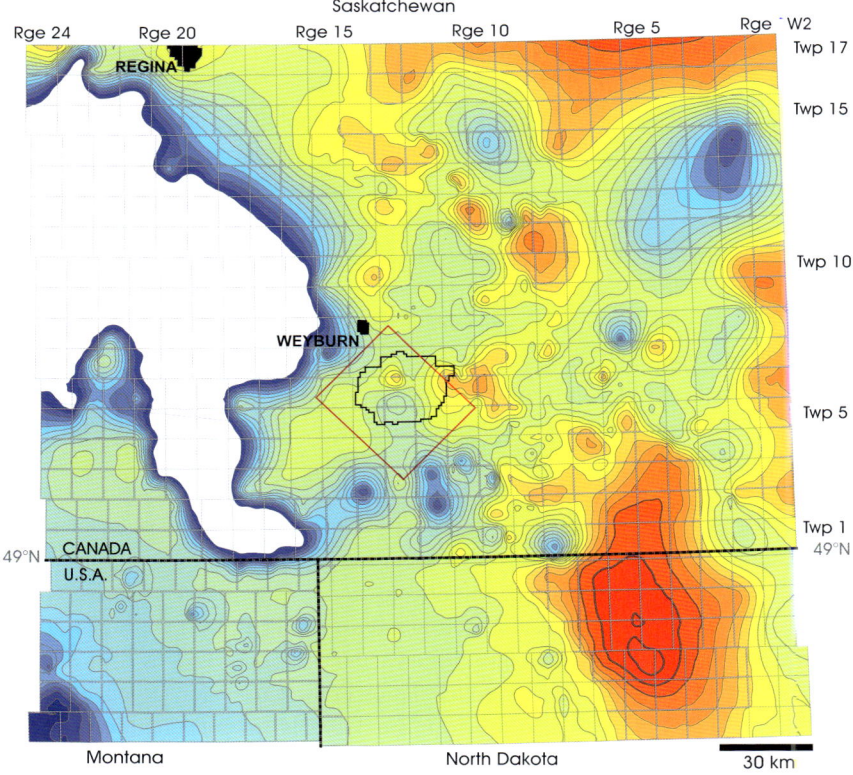

Figure 2-12. *Isopach map of salt, Prairie Evaporite. Contour interval 10 m.*

of collapse structures in overlying beds (De Mille et al., 1964) that impact the integrity of confining strata (Smith and Pullen, 1967; Nemeth et al., 2002).

The distribution of salt dissolution is therefore important for planning future sites for CO_2 storage. In the project area emphasis was placed on mapping evaporite units in Mississippian and older strata to identify areas where evaporite removal has occurred. Twenty-nine evaporite units were mapped in the Paleozoic to Middle Jurassic succession, of which the Middle Devonian Prairie Evaporite is the most extensive (Figure 2-12). The Prairie Evaporite contains thick salt beds mined for potash elsewhere in the basin and forms aquitards that are among the most competent in the basin. It has undergone post-depositional dissolution in both localized areas and large regions and, in the study area, ranges in thickness from >200 m in North Dakota to zero in areas of total dissolution. It is possible to determine the timing of salt dissolution and its impact on stratigraphic continuity by examining variations in thickness of the evaporite beds and of associated overlying and underlying strata (Smith and Pullen, 1967; Nemeth et al., 2002). This, in turn, provides a mechanism to assess the risk of collapse and fracturing of overlying strata that may impact seal integrity or possibly result in cross-formational fluid flow, either of which would reduce the suitability of a formation for long-term storage of CO_2.

Features of collapse structures related to salt dissolution include brecciation and disruption of normal stratigraphic bedding. Recent work using depth-migrated 3D seismic data indicates that, although the disturbed zone is tightly focused above the dissolution area, it may extend vertically hundreds of metres and even potentially to the surface (Mosegaard and Tarantola, 1995). These structures are important because they could provide a conduit for vertical fluid flow across formation boundaries, which is a significant detriment for long-term geological storage of CO_2.

However, no indication of collapse features was observed in strata of the Charles Formation and above, which include the Midale Beds that are laterally equivalent to the reservoir rocks of the Weyburn Field.

The Prairie Evaporite thins by >30 m beneath part of the Weyburn reservoir (for example, in Twp 6–7, Rge 13, W2Mer). Three other Devonian salt beds present elsewhere in the study area are absent at this location, providing further indication that localized salt dissolution occurred in the Weyburn reservoir area. However, anomalous thickening of overlying Late Devonian and earliest Mississippian strata in the same area indicates compensating deposition occurred contemporaneously with dissolution of the underlying Devonian salt. Therefore, the integrity of seals overlying the reservoir was not compromised by earlier dissolution events. In the risk assessment region, there is no thickening of overlying strata that would indicate significant evaporite dissolution following the formation of the Midale Beds.

Mississippian Strata

Mississippian strata include the reservoir beds into which CO_2 is being injected, their laterally equivalent aquifers, and the primary seals for containing the CO_2. Hence, these reservoir formations, and the overlying seals within the Mississippian, form part of the storage complex and are the most important stratigraphic intervals in the Weyburn area. Geological mapping of the Mississippian succession was focused on the following.

1. The regional extent and lateral continuity of reservoir and equivalent beds.
2. The distribution and nature of the containing strata.
3. The provision of sufficient consistency in stratigraphic delineation to permit the resolution of Mississippian hydrostratigraphy into individual aquifers.

The overall Mississippian succession in the study area consists of lithologically similar rocks that form a series of aquifers and thin and variably continuous aquitards. The Mississippian strata were mapped in detail because any natural migration of CO_2 upward or downward out of the Midale Beds will be associated with porous zones in these units.

Mississippian Units (Weyburn and Midale Reservoirs)

Extensive geological mapping of the Weyburn and Midale reservoirs was completed by the CO_2-EOR operators and hence only a summary is presented here.

The Weyburn reservoir occurs within the Midale Beds of the Mississippian Charles Formation and is a carbonate–evaporite cycle deposited in a shallow peritidal environment. The Midale Beds are informally subdivided into a lower limestone layer (the 'Vuggy'), and an upper dolostone layer (the 'Marly'). Each layer has been further subdivided on the basis of depositional and lithological variations for detailed characterization of the reservoir – ultimately aimed at maximizing oil production.

The Vuggy unit was formed mainly in a marine lagoonal environment, where carbonate shoal development produced high-quality reservoir rocks (Burrowes, 2001). At Weyburn, the Vuggy ranges from 10–22 m thick, with the shoal facies being predominantly coated-grain wackestone to grainstone with about 15% vuggy and fenestral porosity. The unit is relatively heterogeneous with permeability in the range of 1–500 mD, but averaging about 20 mD. Inter-shoal deposits also occur in the Vuggy unit and have much poorer reservoir characteristics, with smaller pores and lower permeability. Because shoals are found mainly in the western part of the pool, the planned CO_2-flood rollout is focused on the western portion of the reservoir (Figure 2-3).

Although most early oil production from the Weyburn Field was from the Vuggy unit, current CO_2 injection is mainly into the overlying Marly unit, with the sweep extending down into the Vuggy unit. The Marly unit is a low-permeability microsucrosic dolostone with more homogeneous porosity than the Vuggy, which averages about 26%.

Fractures are present throughout the Midale reservoir and are more common in the Vuggy unit than in the Marly. They have a generally SW-NE orientation (Edmonds and Moroney, 1998). Wegelin (1987) suggested that the inferred regular pattern of fractures in the Midale reservoir may be related to basement features formed during basin development, and that some structural irregularities may also be due to the effects of salt dissolution.

The northern extent of the Midale Beds is sharply defined, as tilted Mississippian layers are truncated by the erosional surface of the sub-Mesozoic unconformity (Figure 2-13).

The contact between the Midale Beds and the unconformity surface, which is directly overlaid by low-permeability siliciclastics of the Triassic Lower Watrous 'Red Beds', is termed the Midale subcrop. This configuration has resulted in the Midale Beds of the northern Williston Basin being among the most hydrocarbon-rich reservoirs of the entire basin, and several other large oilfields are found along the Midale subcrop elsewhere in the basin.

Trapping Units Within the Mississippian Reservoir
The effectiveness of the storage complex is largely dependent on the character, distribution, and extent of bounding seals. The primary Mississippian seals of the Weyburn reservoir include (1) the overlying Midale Evaporite, (2) an altered zone at the subcrop of the Midale Beds, and (3) the underlying Frobisher Evaporite. The relationship of these units to the Midale Beds is shown in Figure 2-4.

Within the Weyburn reservoir, the hydrocarbon (and hence CO_2) trapping mechanism is a combination of stratigraphic, structural, diagenetic, and hydrodynamic components (Wegelin, 1984; Burrowes, 2001). In the Midale Beds, grain size is markedly finer updip towards the subcrop, and irreducible water becomes a factor limiting the movement of hydrocarbons (Burrowes, 2001). In addition, diagenesis within the porous carbonate units immediately subjacent to the unconformity has resulted in micritization that, along with carbonate and anhydrite cement, has effectively occluded most porosity. This zone of alteration is common below the unconformity in many of the Mississippian beds and was mapped as a unit in the Final Phase of the project because of its importance in adding to overall trapping efficiency.

The shallowing-up nature of deposition along the shallow ramp resulted in formation of a very low-permeability evaporite cap, the Midale Evaporite, that serves as an important seal to porous carbonates of the Weyburn Field (see isopach map, Figure 2-14). Similarly, beds of the Ratcliffe cycle and, in turn, Poplar cycle, overlie the Midale Beds but are present only in the southern portion of the

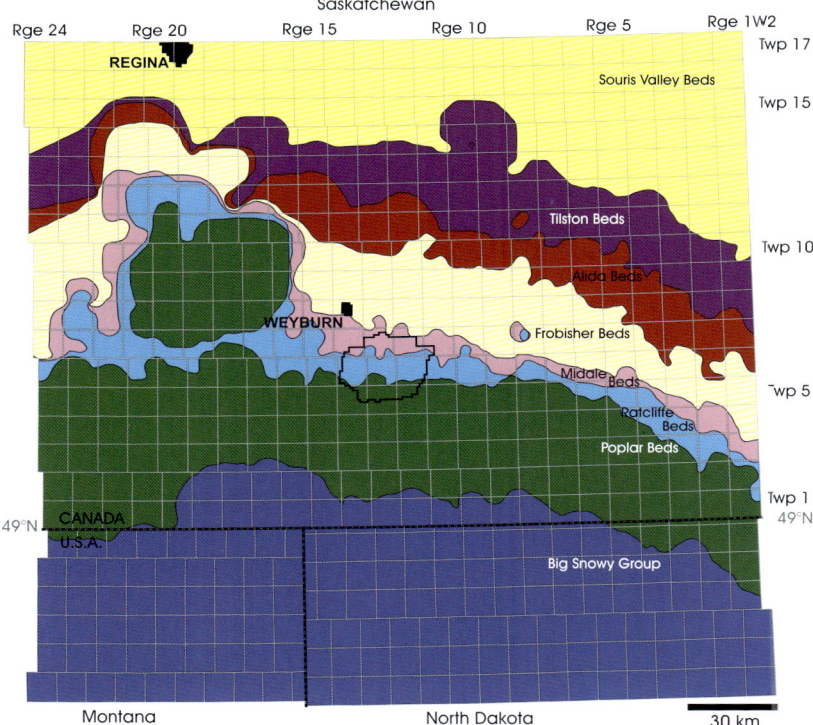

Figure 2-13. *Subcrop of the zero-edge of Mississippian units beneath the sub-Mesozoic unconformity.*

Weyburn reservoir because they are progressively truncated by the sub-Mesozoic unconformity (Figure 2-13).

The Midale Evaporite forms a band of relatively impermeable cap rock that sits conformably on the uppermost unit of the Midale Beds, and its top is defined by a sharp transition to the dolostones of the Ratcliffe Beds (Figure 2-4). On the basis of geophysical logs, the Midale Evaporite is a 2–11 m thick anhydrite layer. Detailed examination of Midale Evaporite core revealed that the unit consists of laminated to massive anhydrite at the base, grading upward into nodular anhydrite with scattered dolostone interbeds.

Fractures are rare in the Midale Evaporite and, where present, are mainly interpreted as having developed contemporaneously with, or shortly after, deposition. There is no evidence of fluid conductance, such as oil staining or precipitation of cements, along fracture surfaces in the Midale Evaporite. Sulphur isotope ratios of primary Midale anhydrites are similar to those of

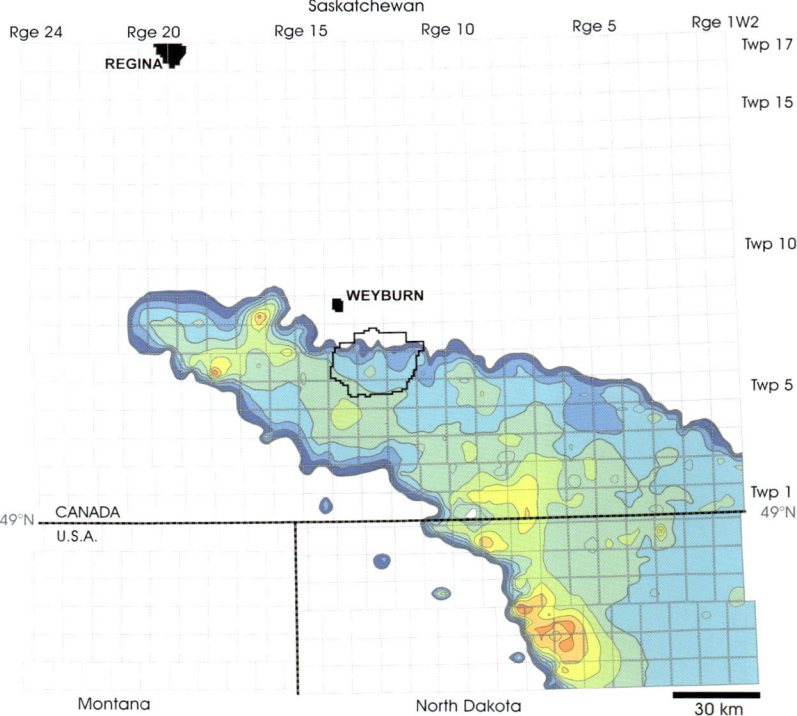

Figure 2-14. *Isopach map of the Midale Evaporite. Contour interval 2 m.*

contemporaneous Mississippian sea water, further supporting the suggestion that these rocks have not experienced significant interaction with post-depositional fluids. Moreover, the Midale Evaporite has clearly been a very effective seal to hydrocarbon migration for the past 50 Ma.

Included as part of the Midale cap rock is the transitional Three Fingers zone at the top of the Midale Beds. This zone contains argillaceous, mottled carbonates that generally have very low permeability.

The Frobisher Evaporite is the cap rock of the underlying Mississippian cycle and forms a bottom seal to the Midale Beds in the northern part of the Weyburn reservoir but is not present beneath the southern part. The Frobisher Evaporite is generally not fractured, although its nodular character makes it slightly prone to small-scale (millimetre to centimetre) fractures (Kemna et al., 2000). In addition, the presence of dolomicrite between the nodules suggests that the unit may be potentially susceptible to fluid migration, because some samples exhibited oil staining.

The diagenetically altered zone occurs along the subcrop region of the Poplar, Ratcliffe, Midale, Frobisher, and Alida beds throughout the Weyburn area and in much of southeastern Saskatchewan. This zone ranges in thickness from 2–10 m, and is immediately subjacent to the sub-Mesozoic unconformity surface at the Weyburn Field. Almost complete porosity reduction has been developed through a complex process of micritization, dolomitization, and anhydritization. Early diagenesis primarily involved micritization and dolomitization, which destroyed much of the original depositional textures of the Midale Beds. Secondary anhydrite cements are present as nodules and veinlets that have effectively reduced porosity (Wittrup and Kyser, 1990).

The Triassic Watrous Formation is the most regionally extensive seal that overlies Mississippian strata throughout the northern portion of the Williston Basin.

Sealing Units in Mesozoic Strata
Mesozoic units are significant in the project area because (1) they contain the most important regional barriers to vertical fluid migration above the Mississippian reservoir and (2) they host aquifers having the greatest transmissivity in the entire stratigraphic section. Shales are the dominant lithology in the Mesozoic succession, in terms of both thickness and areal extent. Mesozoic strata were divided into 38 units and used to define four aquifers and five aquitards between the Midale Beds and the surface (Figure 2-5). Mesozoic aquitards include the Watrous Formation, Vanguard Group, Joli Fou Formation, Colorado Group, and the Bearpaw Formation (see Figure 2-5). Each of these aquitards is regionally extensive and forms significant barriers to vertical fluid migration, with the Colorado Group and Bearpaw Formation, in particular, comprising thick shaly packages.

The Watrous Formation is divided into an upper and lower member, with the Lower Watrous Member forming the most extensive, continuous, thick seal (Figure 2-15). The dominantly non-marine 'red-beds' of the Lower Watrous Member were deposited across the sub-Mesozoic erosional surface and are therefore immediately superjacent to, and forming the trap for, many of the Mississippian oil pools of southeastern Saskatchewan. Informally, the Lower Watrous Member can be locally subdivided into a lower sandstone–siltstone unit and an upper siltstone–mudstone unit, but it generally has a mixed lithological character – grains are of granitic origin with micrite lithoclasts in an argillaceous dolomitic matrix having a clay content rarely exceeding 40%. Anhydrite and dolomicrosparite are the most common secondary cements. Cementation effectively reduces porosity through carbonate replacement and void filling. Clays are mainly interlayered smectite and chlorite with abundant illites and other micas. Porosity in the lowermost section of the Lower Watrous Member can reach 11%,

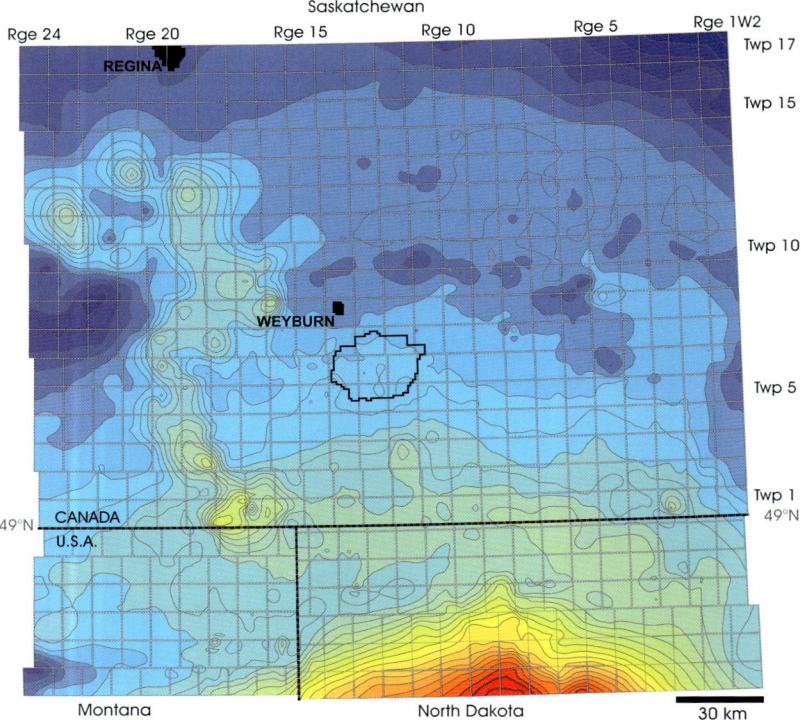

Figure 2-15. *Isopach map of the Watrous Formation.*

although the effective porosity is much less (about 1%). Stratigraphically higher in the Lower Watrous, porosity is even further reduced. No significant fractures were observed in the Lower Watrous Member, although small (1–2 mm) offset microfractures occur in muddier sections. This unit forms one of the most important aquitards in the Weyburn storage complex, ensuring the secure storage of CO_2 and was thus studied in detail.

2.2.6 Historical Natural Seismicity

The Williston Basin lies near the centre of the North American craton, a generally quiescent tectonic region. Southeastern Saskatchewan is in a zone defined as having a low seismic risk (Basham et al., 1982). There is a relatively complete record of seismic activity since the 1950s collected as part of the seismic network of the Geological Survey of Canada. Few earthquakes of note have occurred within the region during the past century. The largest occurred in 1909, about 75 km south of Weyburn, with a magnitude estimated at 5.5 from anecdotal evidence. The return period for such events is unknown but long – of the order of 500 years according to the Gutenburg–Richter graph (Basham et al., 1982). The hypocentre of this earthquake was likely located in the Precambrian basement and it had an estimated nominal fault diameter between 3 km and 7 km and an average displacement about 0.1–0.3 m. These estimates of location and size suggest that no faults would have been created or reactivated in the Weyburn area.

Several natural earthquakes have occurred in historical time within 160 km to the west and south of Weyburn, but none to the north or east. The largest ones are probably controlled by lineament-block boundaries. The smaller ones may be controlled by basement structures or salt removal structures; these typically have magnitudes in the 2–3 range and occur at rates of a few each decade.

According to this information and general knowledge about earthquakes elsewhere, the potential for vertical migration pathways resulting from natural seismicity is low in the project area.

2.3 Hydrogeology

2.3.1 Introduction

Hydrogeological characterization is not normally needed for a CO_2-EOR project because the main focus is on the reservoir. For a CO_2-EOR project to be recognized for CO_2 storage, the operator needs to demonstrate that the CO_2 stays in the reservoir, and that the direction and rate of migration of CO_2 can be predicted should it leak from the storage complex. Flow rates and directions of CO_2 movement are controlled by the hydrogeology of the site. Thus addressing the additional requirements for a CO_2 storage project means extensive work to characterize the hydrogeology of the area. For the project, the objective of the hydrogeological mapping was to characterize the natural hydraulic regime of the sedimentary succession in order to determine the impact of fluid flow on CO_2 storage and to define the necessary boundary conditions for the models used in the risk analysis.

The primary goal was to identify the directions and rates of formation fluid flow through aquifers. Other objectives included the following.
   1. To identify preferential pathways for cross-formational fluid flow.
   2. To assess the competence of low-permeability confining strata (aquitards).
   3. To characterize the hydrochemistry of each aquifer.
   4. To provide hydrogeological data (for example, pressure, temperature, porosity, and permeability) for predictive modelling of CO_2 storage performance as input to environmental risk assessment

2.3.2 Basin-scale Hydrogeology

The basin-scale hydrogeology and hydrochemistry of the Williston Basin was obtained from extensive published literature (Downey, 1984; Downey et al., 1987; Berg et al., 1994; Kent and Christopher, 1994; Busby et al., 1995; DeMis, 1995; Bachu and Hitchon, 1996; Dietrich et al., 1999; Goodway et al., 2010). Previous studies concluded that flow of formation waters generally occurs from the south-southwest to north-northeast across the basin. This cross-basin flow is thought to occur in response to a regional hydraulic gradient created by topographic elevation differences of about 1000 m between exposures of major aquifers at opposite edges of the basin.

Meteoric waters enter aquifers at elevated outcrops along the southern, southwestern, and western basin flanks and are driven towards the basin centre, as suggested by very fresh formation waters with meteoric isotopic signatures (Rostron and Holmden, 2000) found in the basal aquifers (Hanor, 1994). In the deepest portions of the basin, meteoric waters mix with brines (Hitchon, 1996) and the resulting mixtures are displaced updip from the basin centre on to the northeastern basin flank (Downey et al., 1987). This cross-basin flow system seems to have originated relatively recently, with the current boundary conditions set up in the Eocene in association with the onset of the Laramide Orogeny. In some areas of the basin, the effects of hydrodynamic forces have been observed as tilted oil-water contacts in oil pools and as brine inflows that threaten subsurface potash mines (Wegelin, 1987; DeMis, 1995).

2.3.3 Hydrogeology of the 'Biosphere'

Introduction

Protection of potable water resources is of paramount importance when assessing potential storage sites. Thus, characterization of the hydrogeology of the strata above a CO_2 storage site is required to understand the distribution, hydraulic properties, flow directions, flow rates, water quality, and pre-existing usage of near-surface aquifers. This understanding enables reliable predictions to be made of possible interactions that might occur with deeper aquifers units.

In general practice in Saskatchewan, 'shallow' aquifers refer to groundwater resources between the top of the Cretaceous bedrock (Bearpaw Formation or Pierre Shale) and the ground surface. In the framework of the Weyburn CO_2 project, the 'zone' between the top of the Bearpaw Formation/Pierre Shale and the ground surface is referred to as the 'biosphere'. The biosphere consists mainly of unconsolidated glacial deposits – in contrast to the underlying geosphere, which comprises consolidated rocks.

Three major hydrogeological investigations were undertaken to understand fluid flow in the biosphere. A regional hydrogeological investigation covered the same area as the regional geology study. A more detailed investigation of the hydrogeology was carried out in a 3 × 3 township block over the study area (Meneley, 1983). This included detailed mapping of existing geological, hydrogeological, and remote sensing data, along with field work (geophysics and drilling) to delineate an important local shallow aquifer. The third hydrogeological investigation was an extensive shallow-well groundwater sampling program. Water sampling activities were repeated several times, and are a key component of the monitoring efforts.

The importance of undertaking a detailed investigation of the shallow hydrogeology of the project site cannot be overstated. Understanding the shallow hydrogeology and background hydrochemistry is of critical importance for several reasons.

1. Information on the baseline state of the shallow aquifers is essential in view of their possible connections with deep aquifers.
2. Determination of the hydraulics of shallow groundwater systems assists in understanding and predicting their behaviour.
3. Correct representation of the shallow groundwater system is needed in monitoring and verification plans, as well as in risk assessment.

Data Sources and Methods

Information on groundwater quality and water level data were obtained from several sources, including Saskatchewan Research Council, Saskatchewan Watershed Authority, North Dakota State Water Commission, Montana Bureau of Mines and Geology, United States Geological Survey and various consulting reports.

The geological section was subdivided into hydrostratigraphic units (Figure 2-16), and the four major aquifers in the study area are delineated as follows (Figure 2-17).

1. Bedrock aquifers in the Late Cretaceous–Tertiary unit.
2. Drift aquifers in the Empress Formation.
3. Undifferentiated Quaternary aquifers of glacial till and stratified deposits.
4. Shallow surficial aquifers.

A brief description of each of these major aquifers is given below.

Late Cretaceous–Tertiary Aquifers

Sandstones and siltstones of the Late Cretaceous–Tertiary Eastend-to-Ravenscrag succession lie immediately below the biosphere but are difficult to distinguish and so are considered together as an undifferentiated Late Cretaceous–Tertiary unit (Figure 2-16). This unit is present in the southern part of the regional study area (Figure 2-17), and is a complex system containing aquifers and aquitards that vary significantly in size, thickness, and hydraulic conductivity.

Flow in these aquifers is controlled by the regional topography and by the Estevan Valley aquifer system which acts as a drain. In the western portion of the study area, flow is westward towards the Missouri and Yellowstone channels. North of the Estevan channel, flow is directed southward toward the channel, whereas south of the Estevan channel flow is northward. On a more local scale, flow is thought to be controlled by present-day valleys with discharge in the

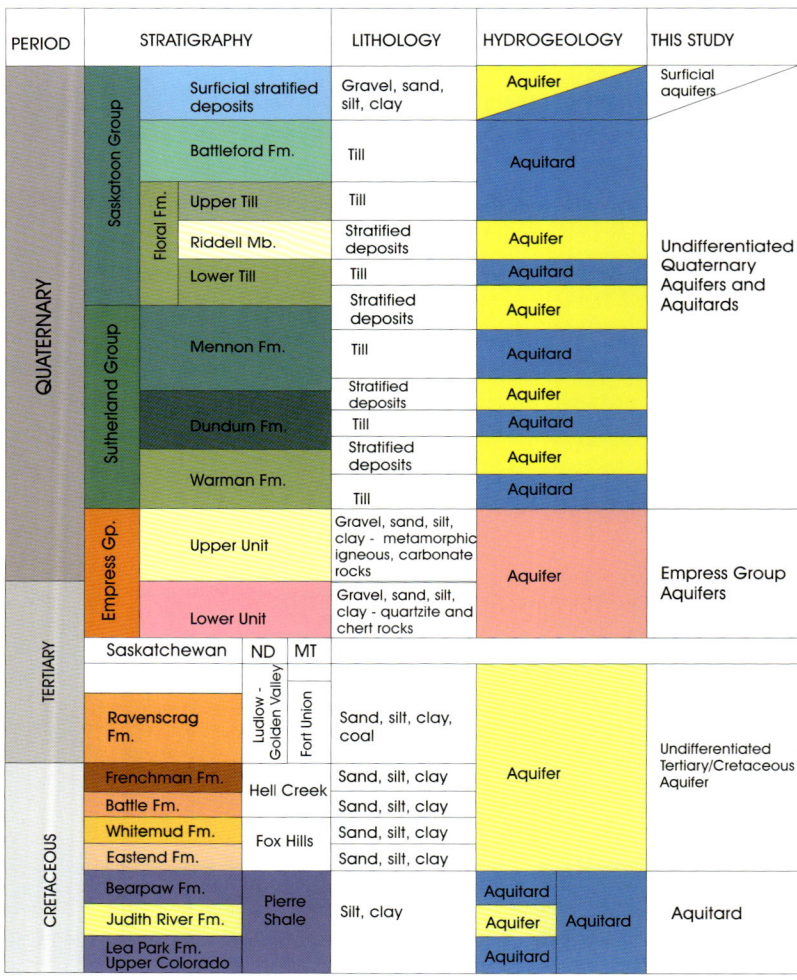

Figure 2-16. *Stratigraphy, lithology, and hydrogeology of shallow aquifers in the project area.*

form of springs and seeps. Water quality in this aquifer varies widely, with about 75% of the water having salinity in the 500–3000 mg/L range and chemical composition predominantly of the Na–HCO$_3$ type, with lesser amounts of Na– SO$_4$, Na–Cl, Ca/Mg–HCO$_3$ or Ca/Mg–SO$_4$ types.

Empress Formation Aquifers

Empress Formation aquifers are the most important aquifers in the biosphere. In southern Saskatchewan, Empress Formation sediments often fill valleys incised into the underlying bedrock to form buried-valley aquifer systems, such as the Weyburn and Estevan Valley aquifers.

The Estevan Valley aquifer system consists of the interconnection of the Estevan, Missouri, Yellowstone, and North channels (Figure 2-17). This is the most extensive shallow aquifer system in the region, with channels 2–4 km wide filled with up to 80 m of Empress Formation sediments (Van Stempvoort and Simpson, 1994). Recharge to the Estevan Valley aquifer occurs over most of the aquifer and is the result of downward flow through the overlying confining layer. In addition, there is lateral flow from the Late Cretaceous–Tertiary aquifer into the Estevan Valley aquifer system. Typical water chemistries in the Estevan Valley aquifer are of the Na–HCO$_3$ type with a salinity of 1750 ± 490 mg/L and characterized by low SO$_4$ contents.

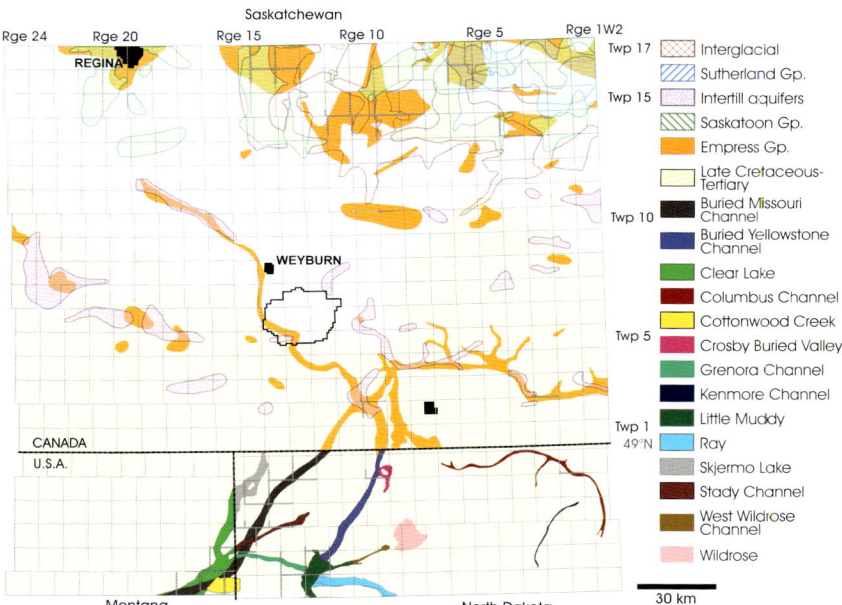

Figure 2-17. *Map of shallow aquifers in the project area.*

The Weyburn Valley aquifer is another buried-valley type aquifer and is part of the Weyburn channel. The Weyburn channel was incised into bedrock, either into silts and clays of the Bearpaw Formation/Pierre Shale or into silts and sands of the Eastend to Ravenscrag formations. The channel is filled with 10–60 m (average 30 m) of sediments of the Empress Formation, and is overlain by 30–80 m of drift, mainly till. The Weyburn valley aquifer yields water of the Na–SO_4/Cl type with a salinity of 2500 ± 300 mg/L. Higher salinity is found where the bedrock is Upper Cretaceous marine and brackish rocks. Lower salinity reflects meteoric water recharge, presumably along fracture lineaments. This aquifer is a significant source of water for the City of Weyburn.

The continuity and location of the North Channel and Weyburn Valley in the vicinity of the Weyburn Field are not fully constrained (Meneley, 1983) and so detailed mapping, geophysics, and a drilling program were undertaken to define the limits of this aquifer. These included an electrical resistivity tomography survey and twelve test holes (including six monitoring wells) to define the extent and hydraulic continuity of the Weyburn Valley aquifer with the Estevan Valley aquifer. Results indicated the presence of a groundwater divide west of the Weyburn Field and a transmissivity barrier in the southern part of the aquifer.

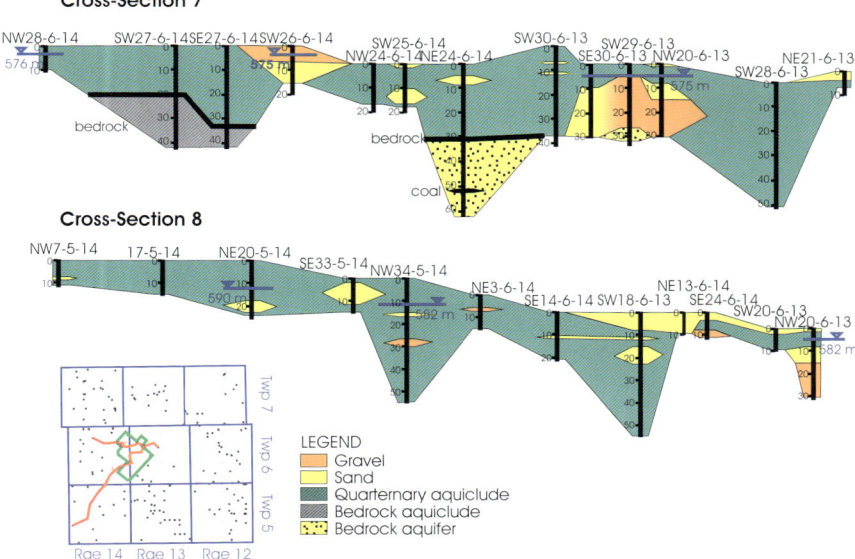

Figure 2-18. *Two typical near-surface lithological cross-sections from the Weyburn area showing the distribution of shallow Quaternary aquifers in the vicinity of the Weyburn Field (inset map).*

The Tableland–Hitchcock aquifer is another significant aquifer of the buried-valley type, incised into Eastend and Ravenscrag formation rocks. It fills a channel 4–5 km wide, 10–30 m thick, and covered by 30–40 m of drift. This valley is not continuous across the study area and is separated by a hydraulic barrier (Rge 9, Twp 2, Sec 16, 17, 20 and 21; Van Stempvoort and Simpson, 1994). The Tableland aquifer extends from the U.S. border to the barrier, and the Hitchcock aquifer occurs between the barrier and the Estevan Valley. Water in the Tableland and Hitchcock aquifers is generally of the Na–HCO_3 or Na–SO_4 type, with salinity in the 1200–3835 mg/L range.

Inter-till Aquifers
Quaternary drift deposits contain shallow aquifers in isolated pockets of stratified granular sediment within unsorted and unstratified glacial till. At the local (farm) scale, these are the most important aquifers for potable water supply for small communities and farms in the area, and hence the most important for potential impact by CO_2 operations. These aquifers occur at various depths and vary in extent (Figure 2-18) and the quality of the water is variable but commonly of the Ca/Mg–SO_4/HCO_3 type with salinity typically in the range 1000–3000 mg/L.

Surficial Aquifers
The uppermost aquifers in the biosphere occur in surficial sands and gravels in recent alluvial, glacio-fluvial, and fluvial deposits. Little is known about the surficial aquifers in Saskatchewan, but published records indicate limited areal extent and depth, with typical water of the Ca/Mg–HCO_3 type with a low salinity (<1000 mg/L).

2.3.4 Deep Hydrogeology

Data Collection and Interval Testing
Sufficient data are required to map the regional flow field – in both the injection zone and the strata above and below – in order to understand the flow system and identify potential pathways of fluid escape. Given typical data densities in sedimentary basins, it is necessary to examine as large an area around the site as possible, in order to capture maximum information from available data.

A hydrogeological database of more than 5300 fluid pressures and formation temperatures and more than 8500 formation water analyses was assembled from public and private data sources. This was supplemented with petrophysical data from core analyses in key aquifers and more than 100 privately collected wellhead water samples. The geological framework for the hydrogeological study was re-mapped as described above. Hydraulic and hydrochemical data were assigned to their respective aquifers and then tested against structural controls provided by the geological mapping. Where aquifer nomenclature was incomplete or vague, data were assigned to individual aquifers through a series of 'interval tests'. The detailed formation picks and the intervals from pressure data were used to compare the bounding surfaces of the individual geological units to the imaginary surfaces defined by the listed drill stem test intervals. In this manner, each pressure measurement was correctly assigned to its corresponding aquifer. Data from multiple aquifers were not used in the study.

Data Preparation
The hydrogeological data were screened to remove inaccurate or non-representative samples using both automated and manual techniques. Only pressure data with full Horner-type extrapolations were used. Each pressure measurement was screened for production-influenced pressure drawdown using a quantitative index. The index accounted for radial proximity of a drill stem test to production or injection wells and the duration of production or injection (after Toth and Corbet, 1986; Bachu and Hitchon, 1996). Chemical data were also screened to remove non-representative data (for example, mixtures, drilling mud filtrates, spent acid-fracture fluids, and injection waters) using automated techniques (Hitchon, 1996) supplemented by manual inspection.

Pressure data were used for pressure-depth profiles, and converted into potentiometric surfaces and hydraulic cross-sections. Variable-density flow effects were present in many aquifers, and procedures for identifying and evaluating these effects are discussed separately below. The end product of the mapping of hydraulic heads was the creation of a 'water driving force' field for each aquifer. In aquifers with no density-related flow effects, the water driving force is simply the negative gradient of its hydraulic head field. For aquifers with significant density-related flow components, the water driving force is a density-corrected net driving force for fluids in the aquifer. Calculation of the water driving force is detailed below. Hydrochemical data were mapped as individual ions, ion ratios, and salinity. In this manner, all geological and hydrogeological data were synthesized into a regional picture of fluid flow in the study area.

Formation water fluxes and velocities were estimated by combining the mapped water driving force with permeabilities derived from drill stem tests and porosities measured on cores. Conditional geostatistical simulation of regional aquifer permeability fields was used to supplement univariate statistical summaries of the hydraulic property data. Stochastic images of aquifer permeability were produced that honoured the permeability data at the data locations and reproduced the spatial statistics between the scattered permeability data in four key aquifers: the Mannville, the Midale, the Jurassic, and the Newcastle. Knowledge of the permeability field structure is needed to resolve high-conductivity pathways that may act as macro-scale conduits for preferential flow and lateral CO_2 transport. The selection criterion for a single permeability field realization was a minimum difference between observed and model-calculated steady-state hydraulic heads.

Driving force calculations were used to predict the directions of lateral migration of separate-phase (supercritical or gaseous) CO_2 in key aquifers. Densities of CO_2 at aquifer conditions were estimated using a tabulated solution for an equation of state for CO_2. The in-situ CO_2 density was then combined with the water driving force to calculate the trajectory of stringers of separate-phase CO_2 migrating through each aquifer.

Hydrostratigraphy

After all of the pressure and chemical data were processed, a hydrostratigraphic framework consisting of 18 major aquifers and 13 major aquitards was developed (Figure 2-5). For practical purposes and similar behaviour, the subsurface was divided into three hydrostratigraphic sections: Deep Paleozoic, Mississippian, and Mesozoic.

Deep Paleozoic Aquifers and Aquitards

The hydrostratigraphic subdivision of pre-Mississippian strata has been elucidated in several studies (for example, Hitchon, 1996; Iampen 2003). Several aquifers are mapped within what had previously been considered thick confining aquitard units (for example, Busby et al., 1995; Dietrich et al., 1999; Downey, 1984, 1986) or undifferentiated aquifer systems (Bachu and Hitchon, 1996).

Mississippian Aquifers and Aquitards

All previous regional hydrogeological studies in the Williston Basin mapped Mississippian strata as a single aquifer system (for example, Downey, 1984; Downey et al., 1987; Hannon, 1987; Toop, 1992; Berg et al., 1994; DeMis, 1995; Toop and Toth, 1995a, 1995b; Bachu and Hitchon, 1996). Although Mississippian strata may appear to act as one large aquifer system at the basin scale because of broadly similar chemistry and hydraulic heads, there are significant differences in these parameters at the local scale. Separating out the individual aquifers and aquitards has previously proved an intractable problem because of (1) the lack of regionally continuous aquitards, (2) sparse, often older (hence, with poorer definition) geophysical logs, and (3) an inconsistently defined geological framework for the hydrogeological units. As a result, the geological framework was re-mapped for this project using a newly derived, consistent set of geophysical well log picks, examination of cores, and a newly developed geological model. For the hydrogeological portion of this study, the Mississippian strata were divided into seven aquifers based on stratigraphic subdivisions (Figure 2-5).

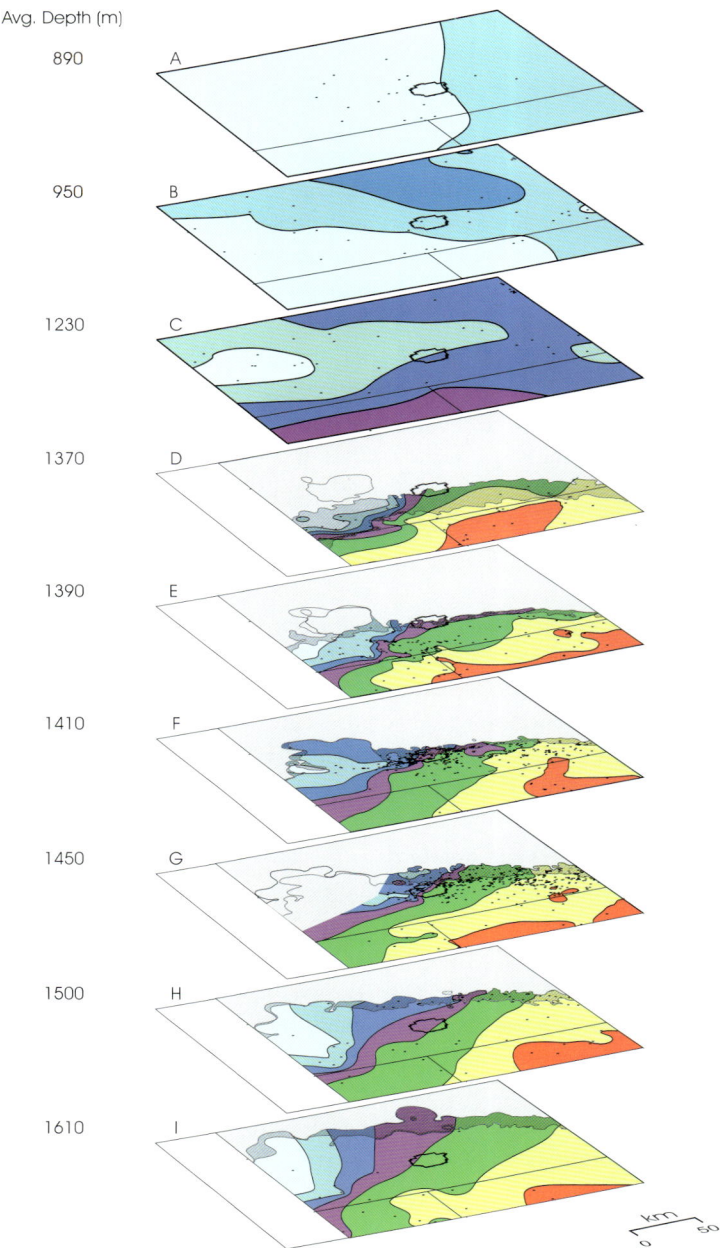

Figure 2-19. *Salinity (mg/L) in selected Mesozoic and Mississippian aquifers.* **A.** *Newcastle,* **B.** *Mannville,* **C.** *Jurassic,* **D.** *Poplar,* **E.** *Ratcliffe,* **F.** *Midale,* **G.** *Frobisher,* **H.** *Alida and* **I.** *Tilston. Colour legend shown on Figure 2-20.*

Mesozoic Aquifers and Aquitards

The Mesozoic aquifer system overlies the Mississippian system (Figure 2-5) and continues to the base of either 'shallow groundwater', or the 'lower limit of potable water' or the 'biosphere' (as termed in this project). For this study, the base of the biosphere was defined as the deepest water-bearing unit that is currently being used, or could potentially be used, for human consumption. In the Weyburn area, the top of the shales of the Bearpaw Formation (Pierre Shale equivalent) was assumed to be the base of the biosphere and hence the top of the deep hydrogeological units.

2.3.5 **Results and Discussion**
Formation Water Chemistry

Geochemical mapping revealed large variations in formation water chemistry (both in composition and salinity) within and among aquifers in the study area. These variations are evident in salinity patterns shown from typical aquifers in each of the three main aquifer groups: pre-Mississippian Deep Paleozoic (not shown) and Mississippian and Mesozoic (Figure 2-19). Vertical variations in salinity are visible in a typical salinity cross-section (Figure 2-20).

Deep Paleozoic Aquifers

Salinity patterns for deep Paleozoic formation waters are similar in overall character. Fresh waters (salinity <5 g/L) are found in the west and north of the map

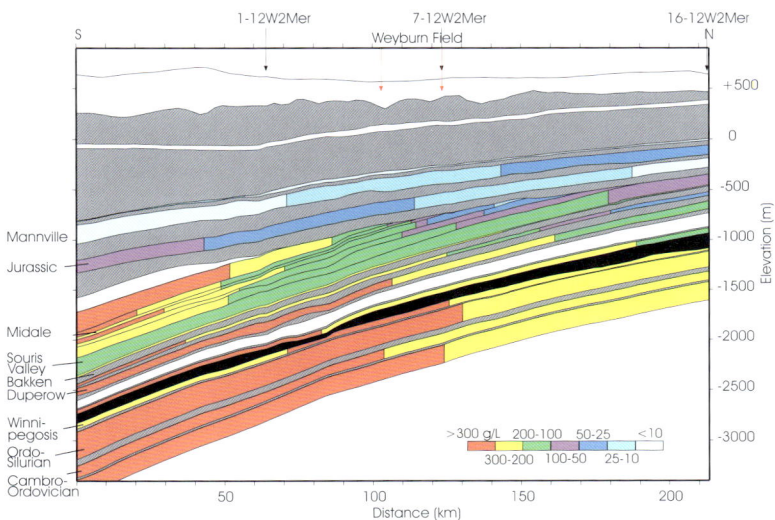

Figure 2-20. *North-south salinity (mg/L) cross-section through the project area.*

area, Na–Cl brines with intermediate salinity (100–300 g/L) occupy the central area, and Ca–Na–Cl brines (salinity >300 g/L) occur in the south and east areas. In general, there is no clear correlation between salinity and depth in these aquifers.

Within each aquifer, salinity systematically increases toward the deeper part of the basin in the east-southeast (red areas in Figure 2-19). The degree of penetration of fresh meteoric waters (Ca–SO$_4$ type) decreases with depth. These meteoric waters migrated considerable distances from their presumed points of entry along the south and western flanks of the Williston Basin. Brines (salinity >300 g/L; Ca–Na–Cl type waters) in the deep Paleozoic aquifers reflect an origin from seawater concentrated through evaporation (Hanor, 1994). Intermediate salinity waters (10–100 g/L) constitute various mixtures and are best characterized as Na–SO$_4$ type waters.

Mississippian Aquifers
Salinity of formation waters in Mississippian aquifers has a similar range and distribution to those in the underlying Paleozoic aquifers. Of particular interest is the steep salinity gradient (from <50 g/L to >150 g/L) within the Midale and Frobisher aquifers across the Weyburn Field (see Figure 2-19F), from invading fresher waters (from the west transition) to Na–SO$_4$ type waters. Although waters in Mississippian aquifers have similar chemical compositions and origins, the spatial distribution of water types varies between aquifers, indicating differences in aquifer hydraulics.

Mesozoic Aquifers
Formation waters in Mesozoic aquifers are markedly different from those in underlying aquifers. Salinity is much lower, with typical values in the 10–50 g/L range, and a maximum of 85 g/L. There is no consistent spatial pattern of salinity within, or between, the Mesozoic aquifers. The dominant water type in all Mesozoic aquifers has a Na–HCO$_3$ composition, in sharp contrast to the Paleozoic aquifers, in which this water type does not occur. These characteristics suggest an active flow regime and formation waters of predominantly meteoric origin. The most relevant difference between the Mesozoic aquifers and the underlying Mississippian aquifers is the absence of brines (salinity >100 g/L) in any of the Mesozoic aquifers within the study area.

Driving Forces and Fluid Flow Directions
Each aquifer has its own unique flow characteristics, although there are similarities. Consequently, a discussion of aquifer flow characteristics focuses on a representative Mississippian aquifer (Midale Beds) into which CO$_2$ is being injected (Figure 2-21) and a Mesozoic aquifer (Mannville aquifer; Figure 2-22).

Flow patterns of formation waters are normally interpreted from gradients of equivalent freshwater hydraulic head, with flow directed toward lower hydraulic head values (Figure 2-22). However, the presence of brines in Mississippian and pre-Mississippian aquifers – with their potential for significant density-related flow effects – complicates the process of determining flow directions.

For aquifers with significant variations in formation water density, a parallel approach to mapping equivalent freshwater hydraulic heads and calculating point estimates of density-corrected driving forces was used. Density-related flow effects become significant where (1) the aquifer slope increases, (2) density (salinity) increases, (3) the gradient of equivalent freshwater hydraulic head decreases, or (4) any combination of these factors (Bachu, 1995; Alkalali and Rostron, 2003).

A density-corrected water driving force was calculated over a regular grid of points in each aquifer by first mapping the temperature field in each aquifer

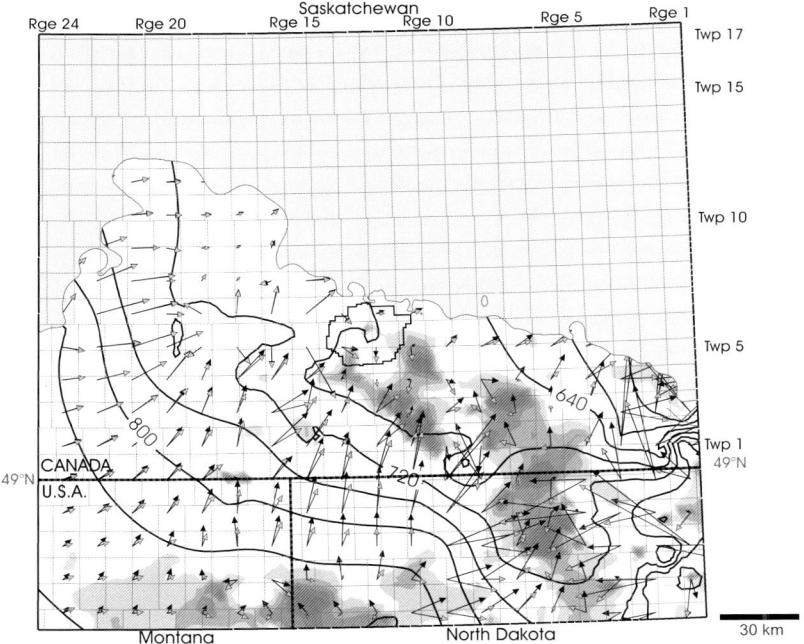

Figure 2-21. *Potentiometric surface (m above sea level) in the Midale aquifer with flow directions and magnitudes, inferred from the gradient of equivalent freshwater hydraulic heads (black vectors). Density-corrected water driving force vectors (grey) diverge from these where density related flow effects are significant (shaded areas).*

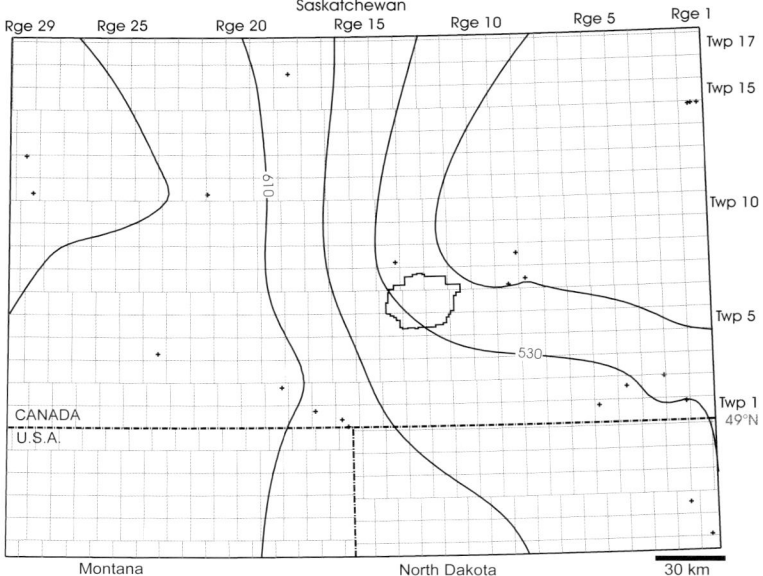

Figure 2-22. *Fresh water hydraulic heads (m above sea level) in the Mannville aquifer.*

and then converting that temperature and the measured salinity to a density distribution at ambient temperature and pressure. The water driving force was calculated by adding a buoyancy correction to the gradient of the equivalent freshwater hydraulic head. Flow directions are indicated by the direction of the water driving force vectors, and an indication of the magnitude of the driving force is given by the length of the vector. Figure 2-21 shows the significant density effects in the Midale aquifer and can be compared to those of the Mannville aquifer, where density effects are insignificant (Figure 2-22).

Water driving force analysis indicates that the deep brines are dynamic, in agreement with Downey (1986), although the non-uniform pattern of brine movement has not been shown previously. In many cases, local areas of intense density-driven flow correspond to complete reversals of flow, opposing the local hydraulic gradient. The result is either an effective stagnation of flow or a complete reversal and down-dip sinking of brines (see Figure 2-21). As a result, down-dip movement of dense brines can only help retard any potential up-dip migration of buoyant CO_2.

Within Mississippian aquifers (for example, the Midale aquifer, Figure 2-21), hydraulic head distributions are similar to those in the underlying aquifers. Near the Weyburn Field, the average direction of fluid flow in the Midale aquifer is diverted to the east-southeast due to the hydraulic barrier created by the Watrous aquitard at the Midale subcrop. Hydraulic gradients are generally comparable to those of underlying aquifers, except for local areas in the east, with gradients of up to 20 m/km. Locally, high gradients in equivalent freshwater hydraulic heads indicate the existence of barriers to lateral flow. Density-corrected driving force vectors within the Midale aquifer indicate that flow directions are correctly represented by equivalent freshwater hydraulic heads in the west due to low salinity in that portion of the aquifer (Figure 2-21). Water driving forces are deflected down-dip in the southern half of the Weyburn Field, as increasingly dense brines and a southward-increasing structural gradient become important controls on flow. The steep salinity gradient across the Weyburn Field in the Midale aquifer marks the approximate position of this transition to buoyancy-dominated flows. Other Mississippian aquifers show similar fluid flow patterns.

In Mesozoic aquifers, average water-flow directions are more variable between aquifers. Magnitudes of hydraulic gradients are also lower than those observed in the underlying aquifer groups. Flow directions in the Jurassic aquifer are directed north-northeast, similar to those in underlying Mississippian aquifers. Within the Mannville aquifer, hydraulic heads drive fluid flow in a generally west-to-east direction (Figure 2-22). In the Newcastle aquifer, formation waters flow south-to-north, following the elongate orientation of the permeable Newcastle sandstone, a pattern previously observed by Hannon (1987).

Vertical Pressure Gradients
Vertical pressure profiles indicate the presence or absence of a vertical component to the hydraulic gradient, which cannot be quantified from potentiometric surface maps (for example, Toth, 1978). Two pressure-depth profiles near the Weyburn Field (Figure 2-23) reveal important elements of the hydraulic regime through the entire stratigraphic section.

There are four key observations.
1. There is no significant vertical component to flow in the pre-Mississippian and Mississippian aquifers, because all pressure data fall along the nominal hydrostatic gradient (Figure 2-23). A hydrostatic vertical pressure gradient indicates a lack of vertical flow, with formation water flow almost entirely horizontal in these units.
2. In the Weyburn Field there are slight overpressures at depths of 1260–1510 m (Figure 2-23). The overpressures could indicate (1) minor isolation of that portion of the aquifer, (2) pressurization effects due to water injection, or (3) energy loss in the water phase, due to lateral flow through the aquifer. In any case, these data fall along a single hydrostatic gradient, indicating that water flow through the reservoir is predominantly horizontal.

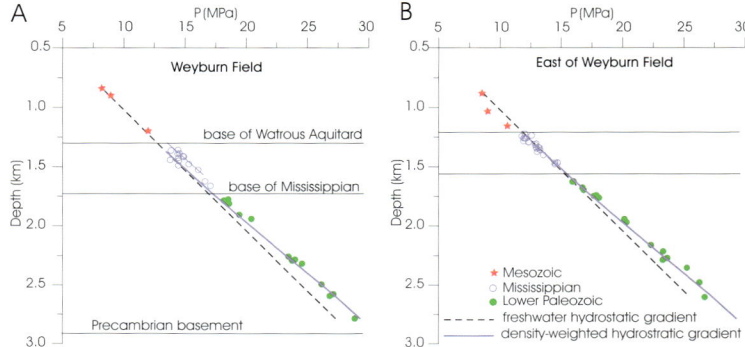

Figure 2-23. *Composite pressure-depth profiles.* **A.** *Near the Weyburn Field. Pressure data fall along the nominal hydrostatic gradient despite some overpressures observed in the upper Mississippian data.* **B.** *East of the Weyburn Field, where pressure data fall along the nominal hydrostatic gradient, except for data from the Mesozoic Jurassic and Mannville aquifers, which are underpressured and fit a super-hydrostatic gradient.*

3. East of the Weyburn Field, pressure-depth plots show a change in slope or an apparent break between the Midale aquifer and the Watrous aquitard. Because of the low permeability of the Watrous aquitard, the most likely interpretation of the break in the pressure-depth profile is a hydraulic discontinuity between the Midale aquifer and the Jurassic aquifer.
4. Mesozoic aquifers are normally pressured, but may be underpressured (Figure 2-23, at depths <1250 m). East of the risk assessment region, there is thus a super-hydrostatic vertical pressure gradient between the Jurassic and Mannville aquifers. Where pressure configurations such as these are present, it is possible that upward flow may occur locally in the shallower aquifers overlying the storage zone in areas where intervening aquitards have relatively high vertical permeability.

Permeabilities, Water, and CO_2 Flow Directions

Formation water fluxes and velocities were estimated by combining the mapped water driving forces with a calculated formation permeability field. The permeability fields were created using conditional geostatistical simulation (Deutsch and Journel, 1998) of regional aquifer permeability fields supplemented with univariate statistical summaries of the hydraulic property data. Multiple stochastic images of aquifer permeability were produced that honoured the permeability data at the data locations and reproduced the spatial statistics between the scattered permeability data in four key aquifers. A single realization of permeability distribution was chosen to model the flow velocity within selected aquifers. The

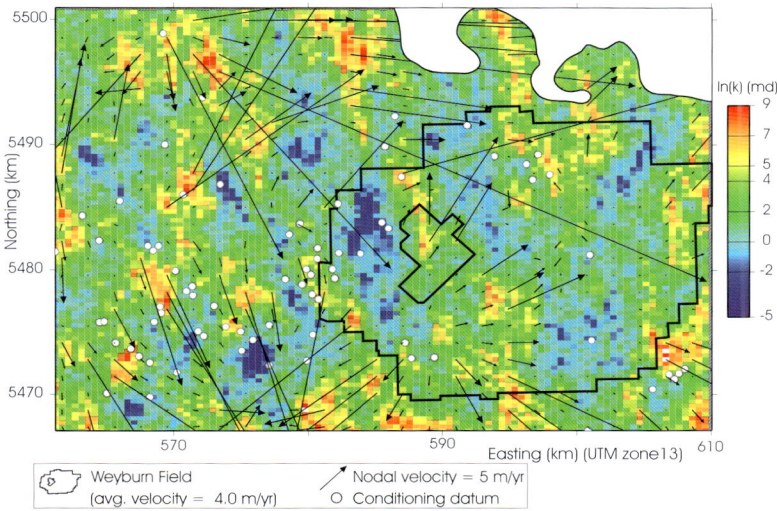

Figure 2-24. *A single realization of the Midale aquifer permeability field over the Weyburn near-field area with calculated formation water flow velocities. Vectors represent calculated point water flow velocities at grid nodes, assuming spatially constant porosity. Vector scaling is linear and a reference vector is shown.*

selected realizations exhibited the smallest difference between observed and steady-state hydraulic heads calculated by the model, based on one set of specified head boundary conditions. Once the hydraulic conductivity field for each aquifer was calculated, the grid node hydraulic gradient and hydraulic conductivities were used to calculate the flux, which was divided by an average value for porosity, to determine water velocity at a given node.

Simulation of the permeability distribution in the Midale aquifer was focused on the near-field area around the injection site (Figure 2-24).

The permeability distribution displays a slight N-S anisotropy that may potentially reflect the influence of large-scale fractures or other structural trends. The Midale aquifer has an average permeability of about 35 mD (after correction for sample bias due to clustered preferential sampling of reservoir-quality rock). Using an average porosity fraction of 0.14, a representative estimate of bulk water flow velocity through the Midale reservoir near the Weyburn Field is 0.80 m/a. In contrast, Jurassic, Mannville, and Viking aquifers have connected sandstone bodies with permeabilities exceeding 10 D, as a consequence of which these aquifers have an active hydrodynamic regime with estimated formation water velocities of 4, 9, and 2.5 m/a, respectively.

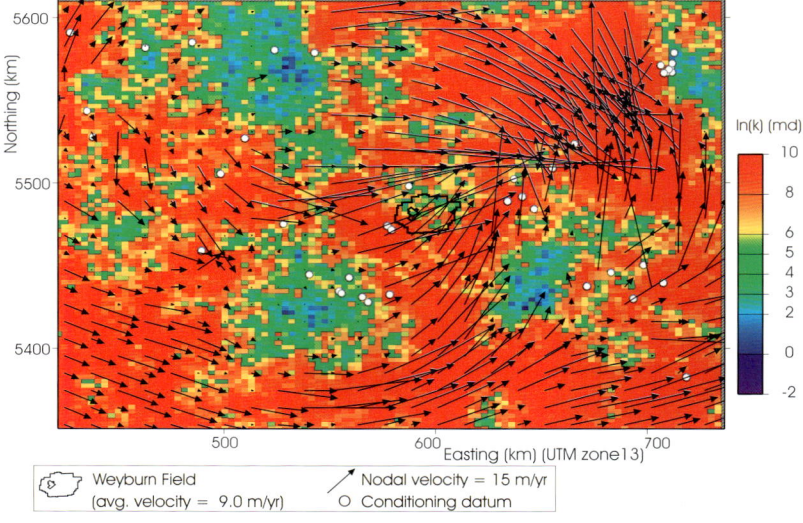

Figure 2-25. *A single realization of the Mannville aquifer permeability field over the entire map area and calculated formation water flow velocities. Vector scaling is linear, and a reference vector is shown.*

Velocities for the Mannville aquifer in the Weyburn Field are shown in Figure 2-25. A sample realization of the permeability structure of the Mannville aquifer reveals large channel structures interpreted to reflect relatively thick, extremely permeable sandstone, which is consistent with results from geological mapping. Calculated point velocities in these channels approach 50 m/a.

Conclusions

Hydrogeological conditions in the study area appear very favorable for CO_2 storage in the Midale (and other) aquifers. Regional mapping of the deep hydrogeology demonstrates the following.

1. The Watrous Formation is a strong regional aquitard that separates the deep hydrogeological system (including the Midale aquifer) from a shallow (1000–300 m depth) hydrogeological system characterized by (1) lower salinity, faster flowing formation waters and (2) higher permeability aquifers. There is no evidence for regional flow of formation waters from the Midale aquifer across the Watrous aquitard into the upper aquifers within the project area, and therefore the Watrous aquitard should serve as an excellent primary seal for CO_2 injected into the Midale reservoir at the Weyburn Field.

2. Low flow velocities (<1 m/a) and favourable (horizontal) flow directions in the Midale aquifer prevent formation water flow from acting as an effective transport agent, thus hydrodynamically trapping injected CO_2.
3. Vertical pressure profiles indicate negligible vertical flow in the Midale aquifer at the Weyburn Field, which is consistent with the interpretation that flow is slowly moving (<1 m/a) horizontally along the subcrop through the relatively high-permeability aquifer.
4. The steep salinity gradient (salinity <50 g/L to >150 g/L) across the CO_2 injection area must be accounted for in geochemical modelling and risk assessment.
5. The deep hydrogeological regime beneath the reservoir may be neglected in order to simplify the system models developed to support risk assessment.
6. High flow velocities in overlying aquifers (1–10 m/a) are important input parameters to the system model for scenario analysis of any CO_2 leakage into overlying zones.

2.3.6 Lessons Learned

1. Classical hydrogeological mapping techniques, coupled with novel analytical methods, demonstrate that the Weyburn site is suitable for CO_2 storage.
2. Regional hydrogeological and hydrochemical mapping must be conducted on formations above and below the injection zone in an area extending beyond the limits of the CO_2-EOR project – both to ensure sufficient data for mapping and to understand the directions and rates of fluid flow in and around the injection site.
3. Procedures and results from this work may help to identify and characterize other sites for CO_2 storage in sedimentary basins worldwide. However, geological information in the northern Williston Basin is likely more abundant and readily available than that typically encountered in most areas considered for geological storage of CO_2.
4. Hydrogeological characterization justified limiting the system model and risk assessment studies to the Mississippian and Mesozoic aquifer units because there is little, if any, contribution to fluid flow from the underlying Lower Paleozoic aquifers.
5. Additional work is required on petrophysical characterization of low-permeability clastic aquitards and their effectiveness as seals to vertical CO_2 migration. This may entail active monitoring of these aquitards using instrumented boreholes.

6. Integrating fault and hydrogeological data may be useful in estimating the hydraulic influence of fractures in seals and aquitards to help in determining long-term monitoring programs and predict potential leakage areas.
7. Continued assessment of existing hydrogeological data is needed to further understand fluid flow in aquifers overlying the Weyburn reservoir.

2.4 Containment
2.4.1 Hydraulic Properties of Seals
Introduction

Existing and emerging regulations for CO_2 storage, such as the European CCS Directive (European Commission, 2009) and the federal regulations for Class VI wells under the U.S. Environmental Protection Agency Underground Injection Control Program (U.S. Environmental Protection Agency, 2010), require proof of containment of CO_2 in the storage complex. The primary controls on trapping of CO_2 in the subsurface depend on the hydraulic properties of the layers of rock overlying the injection zone, the so-called cap rocks or 'seals'. Containing CO_2 in the subsurface may depend on more than one seal. In the context of the project, the seals are referred to as follows.

PRIMARY SEAL: Midale Evaporite, directly above the CO_2 injection zone.

ULTIMATE SEAL: Watrous Formation, a regionally extensive aquitard.

OTHER SEALS: four low-permeability Mesozoic aquitards (Vanguard, Joli Fou, Colorado Group, and Bearpaw) above the Watrous Formation and Quaternary low permeability aquitards in the biosphere.

Investigating the hydraulic properties of these rock units would not normally be conducted for primary hydrocarbon production, or for CO_2-EOR activities, thus leaving a significant data gap in proving long-term containment of CO_2. The following sections summarize the methods and results of characterization efforts on various seals above the reservoir in the project area.

Primary Seal Properties – Midale Evaporite

The sealing properties of the Midale Evaporite were investigated using three different approaches, two using direct hydraulic testing, and the third using indirect measurements. 'Conventional' core analysis techniques were used on representative samples, while gas-liquid breakthrough experiments were conducted at reservoir conditions on samples using three different gases (nitrogen, supercritical CO_2, and methane). These hydraulic methods were supplemented with indirect measures of seal integrity using geological, petrographic, and isotope geochemistry techniques.

Gas permeabilities were measured on core plugs cut from the Midale Evaporite and Midale Evaporite/Midale Marly M1 Unit transition zone in three wells

(14-1, 12-11, and B8-23-6-11 W2Mer). Measured porosity was between 0.05 and 0.15 with an average of 0.10. Gas permeability ranged from 0.03–1.82 mD, with an average of 0.7 mD. Grain densities were in the 2.84–2.96 g/cm^3 range.

Measured porosities were very low (between 0.002 and 0.015, and generally ≤ 0.005) and liquid permeabilities were from about 7–5000 nanodarcys. Measured higher permeabilities were possibly due to generation of micro cracks during handling or laboratory testing. The breakthrough pressure to supercritical CO_2 ranged from about 5–11 MPa, and from about 2–30+ MPa when using nitrogen or methane as the test gas. There was good agreement between breakthrough pressure and interfacial tension (IFT) of the testing fluids. It was concluded that the Midale Evaporite is a high quality seal and that the Midale Beds can therefore be used as a CO_2 storage site after well abandonment.

Geological and geochemical investigations of the Mississippian seals of the Midale reservoir included the overlying Midale Evaporite and the altered zone at the subcrop of the Midale Beds. Even though fractures are present throughout the Midale reservoir, very few fractures were observed in the Midale Evaporite and none suggested any evidence of fluid movement, such as oil staining or precipitation of cements along fracture surfaces. Sulphur isotope ratios of primary Midale Evaporite anhydrites are similar to those of contemporaneous Mississippian seawater, which further supports the suggestion that these rocks did not experience significant interaction with post-depositional fluids. Moreover, the Midale Evaporite has clearly been a very effective seal to hydrocarbon migration for the past 50 Ma.

Studies were also made of the diagenetically altered zone that occurs along the subcrop region of the Poplar, Ratcliffe, Midale, Frobisher, and Alida beds (Figure 2-13). This zone ranges in thickness from 2–10 m and is immediately subjacent to the sub-Mesozoic unconformity at the Weyburn Field. It is a zone of almost complete porosity reduction that developed through a complex process of micritization, dolomitization, and anhydritization. Early diagenesis involving micritization and dolomitization destroyed much of the original depositional textures (and porosity) of the Midale Beds. Secondary anhydrite cements are present as nodules and veinlets that have further reduced porosity. Petrographic and fluid inclusion studies showed that most diagenetic minerals in the altered zone predate migration of oil into the Midale Beds, except for some metasomatic anhydrites that formed contemporaneously with oil migration during the Late Cretaceous or Early Tertiary (Osadetz et al., 1998).

Ultimate Seal Properties – Watrous Formation
The Lower Watrous Member forms the ultimate seal for containment in the project area and is an important trap for many of the Mississippian oil pools of southeastern Saskatchewan. Informally, it can be subdivided locally into a lower

sandstone–siltstone unit and a more uniform upper siltstone–mudstone unit. Despite its 'shale-like' appearance on geophysical well logs, the Watrous Formation has a mixed lithological character. Grains are of granitic origin with micrite lithoclasts in an argillaceous dolomitic matrix, with a clay content rarely exceeding 40%. Anhydrite and dolomicrosparite are the most common secondary cements, and they effectively reduce porosity through carbonate replacement and void filling. No significant fractures were observed in the lower Watrous Formation, although minor small (1–2 mm) offset microfractures are present in the muddier units. All the facies have carbonate and/or anhydritic cement, suggesting that the lower Watrous Formation in the project area has the potential to act as an effective seal to cross-formational fluid flow.

The Lower Watrous Member is characterized by a mixed lithology with about six major components (Table 2-1). This mineral assemblage is fairly constant throughout the formation, as verified by X-ray diffraction (XRD) and petrographic study. Mercury porosity was measured on four samples from the Watrous Formation in the well 21/05-16-07-12 W2Mer. Samples were selected in the most porous siltstones to assume a 'worst case' for CO_2 storage, that is, facies where the maximum porosity would be expected. Table 2-2 shows that the total porosity for the siltstones is about 0.1, ignoring the low value in one sample (YLN03-002). Test results on this sample indicated that the pores are very small

Mineral	Mean (%) (semi-quantitative)	Average Bulk (%)
Quartz	27	10.6
Dolomite	15	40.2
Calcite	9	23.3
Microcline	9	–
K-Feldspar	–	3.1
Albite	–	3.4
Plagioclase (An_{25}–An_{30})	9	–
Anhydrite	21	13.3
Micas/Illite	6	–
Kaolinite	–	0.5
Muscovite	–	4.6
Chlorite	2	–
Ankerite	–	0.4
Siderite	–	0.1
Hematite	1	–
Zeolites (Analcime?)	1	–

Table 2-1. *Percentage of minerals in the Watrous Formation.*

Sample No.	Depth (m)	Total Mercury Porosity	'Free' Mercury Porosity	Calculated Permeability (mD)
YLN03-002	1350.08	0.0311	0.0105	0.61
YLN03-005	1343.50	0.1124	0.0145	64.17
YLN03-007	1335.30	0.0927	0.0145	9.07
YLN03-009	1328.96	0.0823	0.0135	4.80

Table 2-2. *Mercury porosity measured on plugs from 21/05-16-7-12 W2Mer and permeability calculated using the Marshall model.*

and are not accessible to mercury. The pore radius may have a bimodal (0.01 to about 4 µm) or monomodal (0.2 µm) distribution with a 'free' porosity range of 0.01–6 µm. The permeability was calculated from the mercury porosity measurements by the Marshall model.

Porosity values were also calculated from well logs, which accounts for the complex matrix and permeability extrapolated from the values calculated on plugs from mercury porosity data (Table 2-3).

Cores were also sampled for porosity and permeability measurements in three other wells (Table 2-4). Helium porosity values are relatively high (0.099–0.174), but permeability is generally low (0.02–1.87 mD, disregarding sample WRA10).

In order to obtain a single 'average' value of porosity and permeability for the Watrous Formation for numerical modelling and risk assessment, all measured data and their associated scale of measurement were reviewed. Most frequently, values of porosity ranging between 0.04 and 0.06 are common, with corresponding estimated permeability values of 0.1–1 mD. Based on these results, it was decided to use 0.04 for large-scale effective porosity and 0.8 mD for permeability for the Watrous Formation. These are conservative values because, as mentioned above, they reflect measurements obtained from the observed higher porosity sections of cores. Stratigraphically higher in the Lower Watrous Formation, porosity is even further reduced.

Other Seal Properties

Above the Watrous Formation are four low permeability Mesozoic aquitards (Vanguard, Joli Fou, Colorado Group, and Bearpaw) that serve as secondary seals in the project area (Figure 2-10 and Figure 2-11). Coring of targeted units with missing hydraulic data was done in conjunction with the drilling of a well by Cenovus. A wireline method was used (to reduce rig time) and the core was collected in aluminium sleeves to minimize material loss, particularly in the

Well ID	Phi (minimum) (%)	Phi (maximum) (%)	Phi (average) (%)	K minimum (mD)	K maximum (mD)	K average (mD)
9-10 W2Mer	~0	15.4*	2.92	0.04	274.3*	0.41
9-9 W2Mer	~0	15.6*	2.72	0.01	305.58	0.33

*single value due to the uppermost dolomitic beds at the boundary between lower and upper Watrous.

Table 2-3. *Porosity and permeability of the Lower Watrous Formation, estimated from well logs.*

Well ID	Sample No.	Top depth (m)	Kmaximum (mD)	K90 (mD)	Kvertical (mD)	Phi
21/06-08-006-13 W2Mer	FD 15	1366.90	0.26	0.19	0.13	0.138
21/06-08-006-13 W2Mer	FD 16	1368.85	0.10	0.09	0.02	0.129
21/06-08-006-13 W2Mer	FD 17	1370.52	0.04	0.04	0.00	0.110
21/06-08-006-13 W2Mer	FD 18	1372.45	0.10	0.09	0.03	0.132
21/06-08-006-13 W2Mer	FD 19	1374.00	0.04	0.03	0.00	0.990
101/12-09-007-13 W2Mer	WRA 9	1316.19	1.87	1.82	0.35	0.129
101/14-17-006-13 W2Mer	WRA 10	1370.84	34.0	33.9	21.70	0.174

Table 2-4. *Porosity and permeability of the Lower Watrous in other wells.*

dominantly clastic upper formations. This method worked well for the upper formations, but results were very disappointing for two of the lower formations. Cored intervals were as follows.

Lea Park Formation – cut 12.2 m, recovered 9.5 m (78%).
Viking Formation – cut 14.4 m, recovered 14 m (97%).
Mannville Group – cut 12.6 m, recovered 11 m (87%).
Gravelbourg Formation – cut 1.6 m (three attempts, two unsuccessful), recovered 0.9 m (56%).
Lower Watrous/Poplar Beds – cut 14.2 m, recovered 14 m (99%).
Ratcliffe Beds – cut 2 m, recovered 2 m (100%).

Twenty-nine samples were preserved (by sealing in wax) for later geomechanical and petrographic analysis.

LEA PARK FORMATION: Samples from the Lea Park Formation (Colorado Group aquitard) from 602.3–611.2 m were analysed for hydraulic properties, using a new system for measuring the permeability and CO_2 breakthrough pressure in shales. Porosities were measured, using high-pressure mercury injection capillary pressure tests on two samples, with values of 0.18 and 0.25. Permeability results for the Lea Park shale ranged from 14–35 nanodarcys. The CO_2 breakthrough pressure for the Lea Park shale was 0.02 MPa, indicating that the Lea Park shale will not withstand large pressures before allowing CO_2 to flow through it. However, because the permeability is extremely low, the flow rate would be low. In other words, the low permeability of the Lea Park shale will be the controlling factor in terms of the rate of potential CO_2 leakage through it.

PIERRE SHALE AND BEARPAW FORMATION: Other hydraulic conductivity data on Cretaceous aquitards in and near the project area are scarce. A hydraulic conductivity of 1.4×10^{-12} m/s was found for the Pierre aquitard (van der Kamp et al., 1986). This was based on a Horslev analysis of the latter part of a rise of the water level in a monitor well which had been completed beneath the Weyburn Valley aquifer near the City of Weyburn. Studies of the Pierre Shale in South Dakota indicate local hydraulic conductivities in the range of 10^{-13} to 10^{-14} m/s, but with a regional hydraulic conductivity of the order of 2×10^{-9} m/s (Bredehoeft et al., 1983; Neuzil et al., 1984; Neuzil, 1993). The difference between these local and regional hydraulic conductivities was attributed to the presence of widely spaced fractures and fracture zones (for example, see Neuzil et al., 1984). Peterson (1954) reported hydraulic conductivities in the range of 3×10^{-8} to 3×10^{-12} m/s for the Bearpaw Formation in the vicinity of the Gardiner Dam. Finally, constant-head tests on undisturbed samples from the Snakebite Member of the Bearpaw Formation 140 km south of Saskatoon yielded hydraulic conductivities in the range of 10^{-11} to 10^{-12} m/s (Cey et al., 2001).

GLACIAL TILLS: The Weyburn area is blanketed by a succession of Late Pleistocene glacial till deposits up to about 30 m thick (Christiansen, 1992). The till belongs to the Battleford Formation of the Saskatoon Group. Tills sampled as part of the project varied slightly in grain size and degree of sorting, and contain laterally discontinuous sandier interlayers. Geochemical, mineralogical, palynological, and sedimentological evidence strongly indicates that the till in the Weyburn area is from a single till sheet. The till succession in this area is the uppermost barrier to any potentially migrating, deeply sourced CO_2.

Hydraulic properties of glacial tills were not determined as part of the project. Only four measurements of hydraulic conductivities of till aquitards are reported in the study area (Table 2-5). A larger review of available data on the hydraulic properties in the Prairie Provinces was provided by Maathuis and Thorleifson (2000), who concluded that weathered and fractured till may have a vertical hydraulic conductivity in the range of 10^{-7} to 10^{-9} m/s, whereas thick unfractured till units may have a hydraulic conductivity in the range 10^{-10} to 10^{-11} m/s. These same authors indicate that the porosity of tills may vary from 0.25–0.40, and that it may have a specific storage coefficient of the order of 10^{-4} m^{-1}.

2.4.2 Fracture Mapping Using Seismic AVOA Analysis

Currently, there are few practical means, outside of extensive and costly core sampling, to detect the presence of fractures within the sealing units of potential geological storage sites. However, fracture mapping using seismic amplitude analysis [amplitude versus offset and azimuth (AVOA) analysis], provides a means of identifying potential subvertical fracture zones over large areas, based on their associated azimuthal seismic anisotropy (for example, Rüger and Tsvankin, 1997; Gray et al., 2002; Jenner, 2002; Hall and Kendall, 2003; Al-Marzoug et al., 2006).

Method

AVOA analysis was applied to 2001 3D dynamite seismic data from the Phase 1A area of the Weyburn Field, focusing on the reservoir cap rock and the directly overlying primary seal (Midale Evaporite and Ratcliffe Beds) and Watrous Formation, respectively (Figure 2-26). Anisotropic modelling was used to assess how well the anisotropy of the thin cap rock can be separated from anisotropy of the reservoir unit. Details of the analysis and modelling methodologies can be found in Duxbury et al. (in press).

The Midale Evaporite and Ratcliffe Beds are treated as a composite cap rock in the AVOA analysis because they have similar bulk geophysical properties (see Figure 2-27) and are not seismically resolvable as individual units. The cap rock thins from 25–30 m in the southern parts of the Phase 1A area to 8 m in the north as the Ratcliffe Beds are truncated by the sub-Mesozoic unconformity.

Unit	Hydraulic Property	Method/Comment
Estevan Valley aquifer system	$T \approx 1170$ m^2/d* $S = 6 \times 10^{-4}$* $K_v \approx 8 \times 10^{-9}$ m/s* $L = 37$ km*	Vandenberg-type curve analysis of 29-d pump test (van der Kamp, 1985)
Estevan Valley aquifer system	$T = 720$ m^2/d* $S = 2.9 \times 10^{-4}$* $K_v = 3 \times 10^{-10}$ m/s* $L \approx 48$ km*	Numerical model: best fit to measured drawdown for 29-d pump test (van der Kamp, 1985)
Tableland Aquifer confining layer	2×10^{-7} to $<1 \times 10^{-9}$ m/s	Hydraulic response test (proprietary report)
Weyburn Valley aquifer	10^{-9} to 10^{-10} m/s	Calculated vertical hydraulic conductivity of the confining layer (van der Kamp et al., 1986)

*Note: L = leakage length (m) = $(Tc)^{\frac{1}{2}} = (T b'/K'_v)^{\frac{1}{2}}$, where c = vertical resistance (d), b' = thickness confining layer (m) and K'_v = vertical hydraulic conductivity confining layer (m/s)

Table 2-5. *Literature values of hydraulic properties of till aquitards in the project area.*

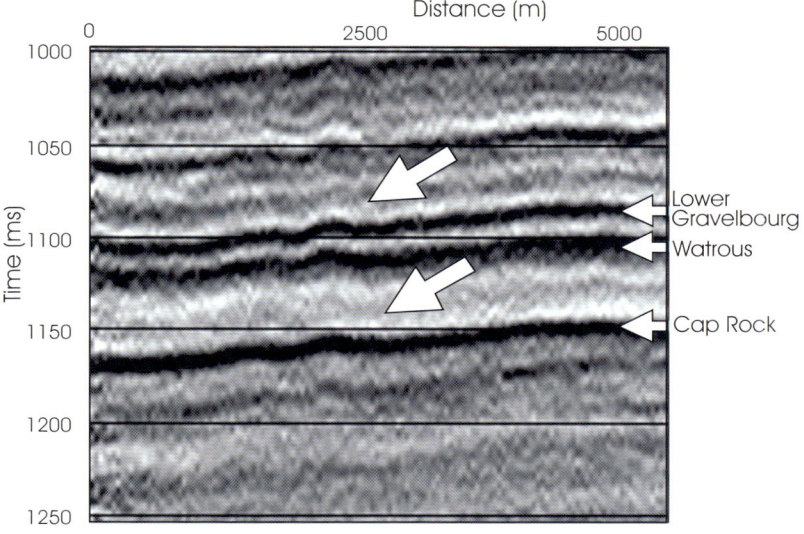

Figure 2-26. *Stack image highlighting the zones of interest. Arrows denote the location of highs in the reflection horizons, which are interpreted as resulting from salt dissolution.*

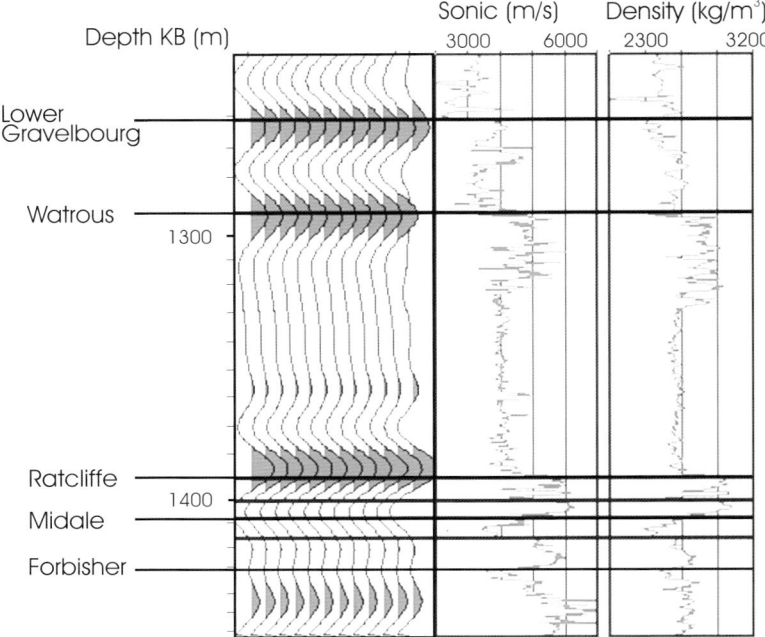

Figure 2-27. *Representative sonic and density logs for the reservoir (Midale Marly and Midale Vuggy) and the overlying seals (Midale Evaporite, Ratcliffe Beds, and Lower Watrous Member). Calculated seismic reflection response is shown on the left. AVOA analysis was performed on reflections from the top of the Watrous and Ratcliffe Beds.*

Fractures are rare within the Midale Evaporite and are interpreted as having developed contemporaneously with deposition or shortly thereafter, possibly in association with dissolution of underlying salt layers (for example, Bunge, 2000).

The Watrous Formation forms an extensive seal in the Weyburn area. The Lower Watrous Member has an average thickness of about 65 m. No significant fractures were observed within this unit, although microfractures (1–2 mm offset) have been observed in muddier layers. The Upper Watrous Member comprises a stacked succession of anhydrite, dolomitic limestone, and shales.

Modelling

The limited thickness of the cap rock introduces the likelihood that interference of seismic reflections from the underlying reservoir (known to be fractured, for example, Bunge, 2000) may contribute to observed anisotropy at the cap rock. Anisotropy in the overburden may also influence the cap rock response. These

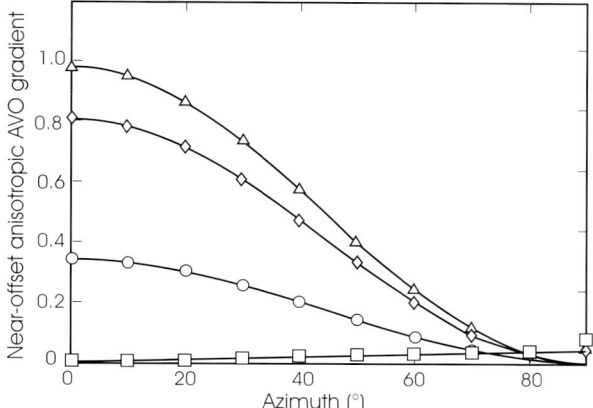

Figure 2-28. *Near-offset AVO (amplitude versus offset) gradient response for the top of the cap rock. Curves are shown for anisotropy in a given layer: 20 m thick cap rock (diamond), 75 m Watrous (circle), 50 m overburden (square), and both cap rock and Watrous layers (triangle). Curves are plotted according to the most positive AVO gradient direction. Response is shown for a crack density of 0.1, equivalent to the average lower Vuggy reservoir.*

possible effects were explored through modelling, to determine the expected magnitude of anisotropy from the cap rock and to assess the feasibility of differentiating anisotropy from the cap rock from that of either the overburden or underlying reservoir.

The effects of anisotropy above the cap rock (overburden and Watrous layers) are summarized in Figure 2-28. In all scenarios, the largest synthetic AVOA response as determined at the cap rock horizon is associated with the presence of anisotropy in the cap rock. However, anisotropy in the Watrous layer can make a significant contribution to the cap rock response, whereas anisotropy in the shallower overburden layer has little effect.

The effect of anisotropy in the underlying reservoir layer is shown in Figure 2-29. Where the cap rock is thicker than 15 m, the effects of an underlying anisotropic reservoir is relatively small, and anisotropy in the cap rock should be distinguished on the basis of magnitude. For cap rock thicknesses <15 m, it would be difficult to differentiate anisotropy in the reservoir from the overlying cap rock on the basis of magnitude.

Results

Anisotropy results obtained for the seismic reflections from the Upper Watrous Member and cap rock are shown in Figure 2-30. The Watrous horizon (Figure

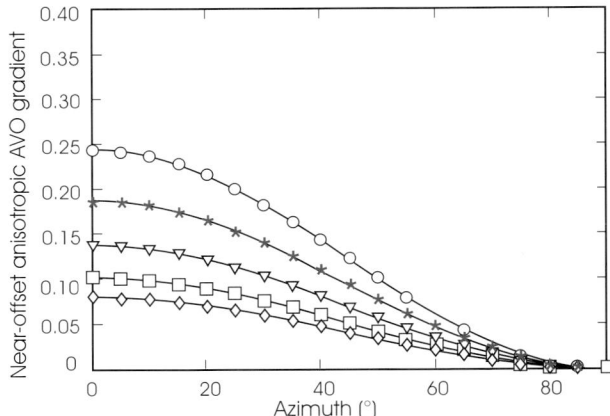

Figure 2-29. *Near-offset AVO gradient response assuming an isotropic cap rock above an anisotropic reservoir. Interference effects from the underlying reservoir for cap rock layer thicknesses 10–30 m, and reservoir thickness 10 m. Cap rock thickness is indicated using symbols: 10 m circle, 15 m asterisk, 20 m triangle, 25 m square, and 30 m diamond, and plotted according to the azimuth of the most positive AVO gradient. Increasing anisotropy from the reservoir is observed with decreasing cap rock thickness.*

2-30, left column) generally shows little substantial anisotropy in contrast to the cap rock horizon (Figure 2-30, right column) which is characterized by isolated zones of significant anisotropy. Overall, uncertainties are higher for the cap rock horizon than for the Upper Watrous Member, presumably due to the reduced thickness and increased stratigraphic complexity of the cap rock.

To isolate areas where the measured anisotropy is significant, the inversion results have been winnowed based on (1) magnitude of anisotropy, (2) calculated uncertainties, (3) model correlation, (4) root mean square difference of observed and modelled amplitudes, and (5) adequate azimuth and offset coverage. Plots of anisotropy after winnowing (Figure 2-30C; Figure 2-30F) show a few areas with clusters of vectors having consistent orientations. The isolated singular vectors are likely exceptions, and it is difficult to assess their significance.

Figure 2-30. Left column. *AVOA inversion results for the Watrous horizon. The Upper Watrous Member shows low levels of anisotropy when compared with the cap rock horizon. A cluster of high magnitude anisotropy with spatially consistent orientation is observed in (C). These features occur on the flanks of salt dissolution structures observed in seismic sections.* **A.** *Normalized near-offset AVOA magnitude of anisotropy. Tick marks indicate the orientation of anisotropy.* **B.** *Per cent uncertainty in the magnitude of anisotropy.* **C.** *Residual anomalies after masking represent points with the highest model correlation, lowest uncertainty, and above average anisotropy*

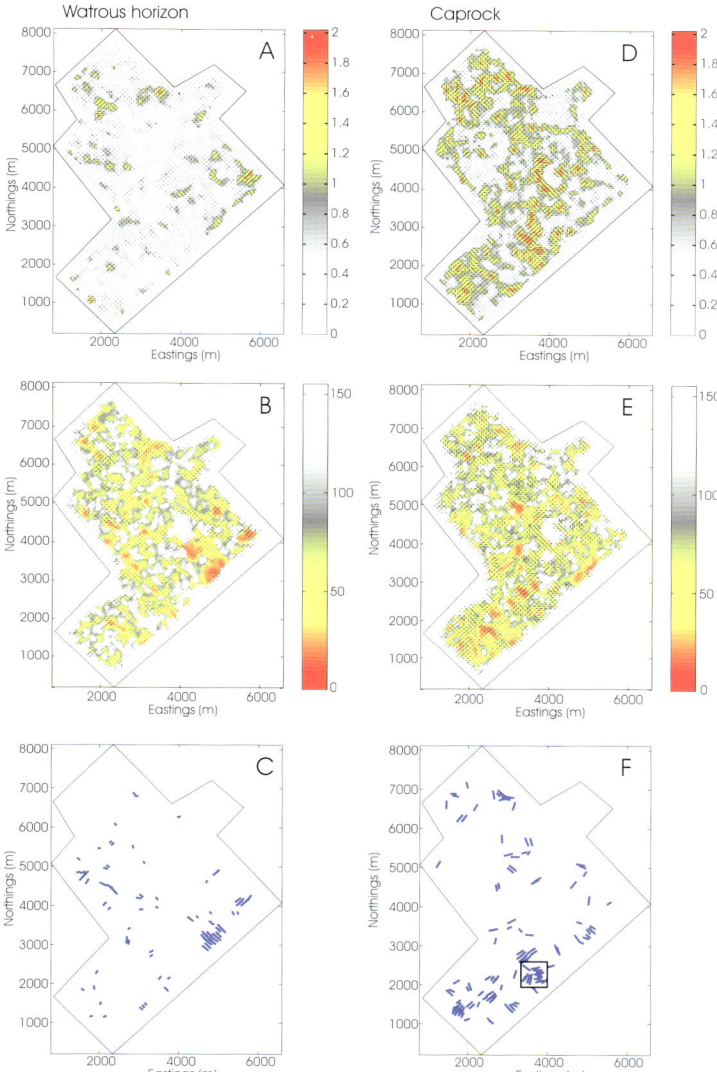

magnitude. **Right column.** *AVOA inversion results for the cap rock horizon. A seismic section through the boxed area in (D) is shown in Figure 2-26. Clustered anisotropic zones appear to be correlated with the flanks of salt dissolution structures.* **D.** *Normalized near-offset AVOA magnitude of anisotropy. Tick marks indicate the orientation of anisotropy.* **E.** *Per cent uncertainty in the magnitude of anisotropy. Lower per cent uncertainty is observed over anomalies with consistent orientation.* **F.** *Anisotropy orientations after winnowing procedure preserving points with the highest model correlation, lowest uncertainty, and above average anisotropy magnitude.*

Anisotropy orientation distributions are shown for the cap rock and Upper Watrous Member in Figure 2-31. Distributions for both horizons (Figure 2-31A and Figure 2-31B, respectively) show a prominence of NE and NW orientations, parallel and normal to the direction of regional maximum horizontal stress. However, the NW orientation is predominant for the Upper Watrous Member, in contrast to the cap rock horizon where the NE orientation is dominant. Comparison of the orientation distributions for the cap rock with the orientation of open vertical fractures determined for core from the underlying reservoir (Figure 2-31C) shows a very good correspondence.

Interpretation of Results
The observed zones of azimuthal seismic anisotropy are consistent with the presence of subvertical zones of increased elastic compliance. Potential causes include (1) lattice-preferred orientation of constituent minerals (anhydrite, dolomite), (2) strain-related mineral fabrics, (3) preferential closure of randomly oriented crack-like voids in a non-isotropic stress field, and (4) the presence of crack-like voids having a predominant orientation. Of these possible causes, the first three are considered to be less likely (Duxbury et al., in press). Preferential closure of cracks striking near-orthogonal to the direction of maximum horizontal stress would result in the observed seismic anisotropy being aligned with the NE-SW maximum horizontal stress direction at Weyburn, consistent with the observed predominant orientations of the observed seismic anisotropy in the cap rock (30°–50°; Figure 2-31A).

However, natural vertical fractures with a predominant NE-SW orientation have been documented within the Weyburn reservoir from core analysis (for example, Beliveau et al., 1991). A more detailed study on a small subset of cores from the Weyburn Field (Bunge, 2000) found prominent open fracture orientations of 40°±5°, 105°±7° and 148°±11° at the reservoir level (Figure 2-31C). Less information is available for the cap rock, but several generations of natural vertical fractures have been observed there as well, although much less frequently. The orientation distributions of seismic anisotropy of both the cap rock and the Upper Watrous Member show a generally good correspondence with the fracture orientations (NE-SW and NW-SE) determined by Bunge (2000), although the dominant seismic anisotropy orientations for the cap rock and Watrous horizon are orthogonal (NE-SW versus NW-SE, respectively). This correspondence generally supports the interpretation of the seismic anisotropy results in terms of fractures. Further, the AVOA-determined orientations are consistent with the microseismic shear-wave splitting results of Verdon et al. (2011) for the zone above the reservoir.

Of the several generations of natural fractures identified in cores from the carbonates of the Weyburn reservoir, late-stage open fractures have been attributed to salt dissolution within the underlying evaporites (for example, Bunge, 2000). Some of the more prominent anisotropic zones mapped at the cap rock and Watrous horizons correspond to salt dissolution structures in the Watrous and overlying Lower Gravelbourg formations (see Figure 2-26). However, it should be noted that there may be some inaccuracy in the locations of the seismically identified anisotropic zones, as the AVOA analysis was performed on unmigrated data. Salt dissolution in underlying layers causes a sag in overlying beds, which can result in fracturing. This effect may be enhanced on the flanks of structural highs which are interpreted as over-thickened layers resulting from the differential timing of underlying salt dissolution.

Seismic anisotropy analysis is capable of identifying broad zones of fracturing, but it cannot determine the scale (microscopic versus macroscopic) of the individual cracks, or whether there is associated permeability. Other corroborating means are required to test for hydraulic communication across the cap rock. It was concluded that none of the natural fracture systems within the Midale Evaporite have the capacity to transmit fluids, because open fractures are discontinuous and others are healed by later diagenetic anhydrite cements. The trapping of hydrocarbons within the reservoir over geological time qualitatively supports this conclusion. However, the 50 years of enhanced oil recovery in the field could be the source of these zones of inferred vertical fracturing. Jenner (2002) noted that shear failure of the Midale Evaporite during EOR operations within the field is highly unlikely, due to the strength of anhydrite, but also noted that microfracturing could occur at lower stresses when the rock is stressed beyond its damage threshold. This process would inherently exploit pre-existing zones of weakness, such as natural fracturing due to salt dissolution.

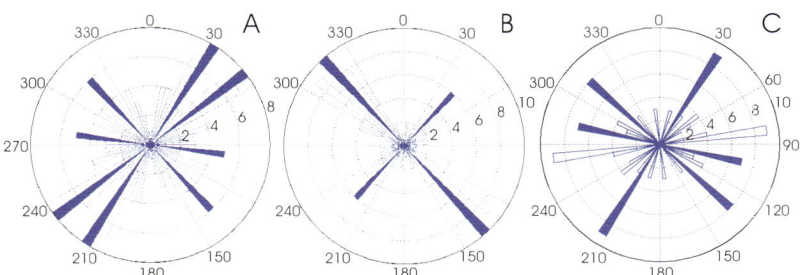

Figure 2-31. *Anisotropy orientation distribution after winnowing for:* **A.** *Cap rock results (Figure 2-30F), and* **B.** *Watrous results (Figure 2-30C).* **C.** *Fracture orientation distribution as determined by Bunge (2000) for oriented cores and downhole televiewer records from the reservoir. Bunge notes that in (C) open fracture directions are roughly 30°, 315°, and 285°, while the healed fracture directions are 80°.*

Microseismic monitoring in the Weyburn Field has located very-small-magnitude seismic events within and directly above the reservoir since the start of CO_2 injection (Verdon et al., 2011), which may indicate small-scale reactivation of pre-existing fractures.

Conclusions

Seismic AVOA analysis applied to 3D seismic data from the Weyburn Field has identified zones of azimuthal anisotropy. These indicate the limited presence of subvertical zones of increased elastic compliance within a composite cap rock layer comprising the Midale Evaporite and Ratcliffe Beds, and to a lesser extent within the Watrous Formation. Assessment of the significance of these results leads to the following conclusions.

1. The predominant orientations of the observed anisotropy are parallel and normal to the direction of maximum horizontal stress (NE-SW) and agree closely with previous fracture studies on core samples from the underlying reservoir. This generally supports interpretation of the seismic anisotropy results in terms of broad zones of vertical fracturing, although the scale of the individual fractures or their hydraulic properties are unconstrained by the seismic data.
2. The regional seal (Watrous Formation) has relatively few inferred fracture zones, whereas the reservoir cap rock has several isolated areas of interpreted fracture-related anisotropy. The seismic data are unable to resolve whether these inferred fracture zones in the cap rock are exclusive to the Midale Evaporite or the Ratcliffe Beds.
3. Modelling results indicate that anisotropy in the cap rock (combined Midale Evaporite and Ratcliffe Beds) should be adequately distinguished from that of the underlying reservoir (known to be fractured) in regions where the cap rock thickness >15 m. This covers the southern part of the monitoring area.
4. Anisotropic zones are observed to correlate spatially in some cases with salt dissolution structures in the cap rock and overlying beds, as interpreted from 3D seismic cross-sections. However, it should be noted that there may be some inaccuracy in the locations of the seismically identified anisotropic zones, as the AVOA analysis was performed on unmigrated data.
5. The interpreted fracture zones could represent hydraulically conductive zones or, alternatively, damage zones that could be become conductive if high enough pressures are achieved during ongoing CO_2 injection. Other independent means are required to further assess these possibilities. In either case, these zones likely warrant attention in ongoing monitoring operations.

2.5 Natural Analogues

Natural accumulations of CO_2 occur within Devonian carbonates and Cambrian siliciclastics about 400 km west of the project site on the western flank of the Williston Basin in southwestern Saskatchewan (Figure 2-32). Some reservoirs in this area contain inert gases with >80% CO_2 and have sustained flow rates of up to 425 000 m^3/d. Nearly 2 Mt of CO_2 are estimated to occur in this area (Lane, 1987).

The naturally occurring CO_2 and associated N_2 and He (Lee, 1962) are generally trapped in the Devonian Duperow Formation (Figure 2-33) within a succession of thin cycles of carbonates capped by evaporite units. The cycles resulted from deposition within shallow, periodically restricted waters along a carbonate platform, a geological environment similar to that which produced the Midale Beds.

The age of these accumulations has been estimated by age-dating Tertiary intrusions at the Bearpaw and Little Rocky Mountains in Montana. The CO_2 found in Devonian rocks of southwestern Saskatchewan was likely generated during intrusion of hot alkaline magma into Lower Paleozoic carbonates in

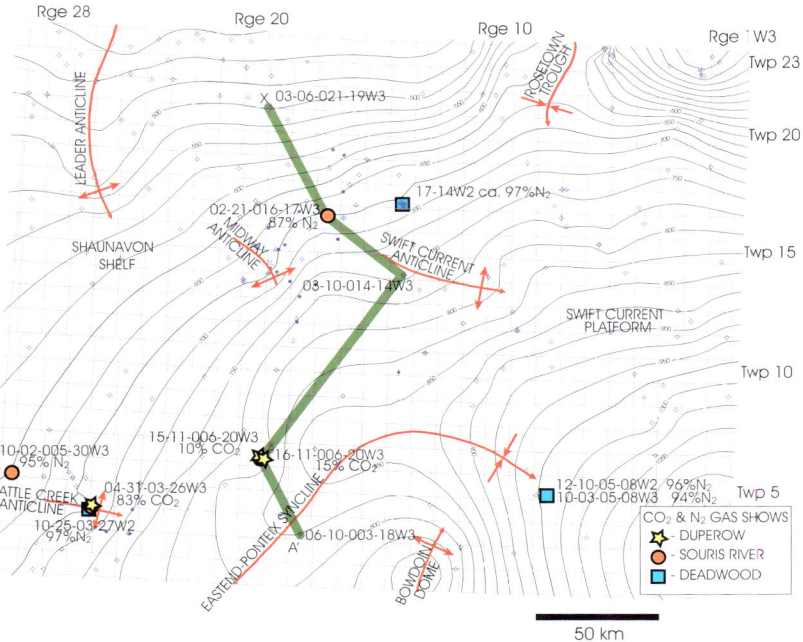

Figure 2-32. *Structure map of the Duperow Formation showing the location of stratigraphic units in which CO_2 and N_2 have been found. Contour interval 25 m.*

Montana about 50 Ma ago. The CO_2 migrated about 100 km north, up-dip, to where it became trapped in local structures formed previously through dissolution of underlying salt layers. The CO_2-bearing intervals occur at depths of 1675–2050 m below surface, within Devonian carbonates that are lithologically and genetically similar to carbonates encountered at the Weyburn site.

Natural gas produced from shallow wells (penetrating only Mesozoic strata) in the same location as the deeper natural CO_2 contains only a trace or non-detectable amounts of CO_2. This suggests that leakage from the underlying natural CO_2 reservoirs into shallower beds in southwestern Saskatchewan has not been significant during the past 50 Ma (Lake and Whittaker, 2006).

The Duperow Formation is a legitimate analogue for the Midale Beds at the Weyburn CO_2 injection site, based on similar lithofacies, whole-rock composition, mineral composition, and porosity. Analyses included XRD, X-ray fluorescence (XRF), and electron microprobe analysis of major minerals as well as back-scattered electron and X-ray imaging to identify trace phases and silicate minerals. LPNORM analysis was used to quantify modal concentrations of minerals species.

Results showed that the distribution of porosity in the Midale Vuggy unit is virtually identical to that of the Duperow Formation, but the Marly unit has significantly higher porosity. Analysis revealed that the major mineral phases at both sites are calcite, dolomite, and anhydrite, with accessory pyrite, fluorite, and celestine. The composition of dolomite and calcite is similar in both localities. Silicate minerals comprise <3 wt. % on a normative basis in the Duperow Formation, with quartz the only phase identifiable using X-ray diffraction. Silicate minerals rich in alkaline earths, such as plagioclase feldspar, appear in normative calculations, but have not been identified in thin section. Quartz and K-feldspar are the only silicate minerals identified in thin section in the Duperow Formation and Midale Beds. Although the Midale Beds contain significantly higher amounts of K-feldspar, it is unlikely to participate in carbonate mineral precipitation due to the absence of alkaline earths. Hence, physical and solution trapping are likely to be the primary trapping mechanisms at both sites. These accumulations, coupled with the striking similarity between these lithologies, provide support for the long-term security of CO_2 injection at Weyburn.

Conclusions
1. Mineral trapping of CO_2 is largely dependent on the dissolution of alkaline earth silicate minerals to provide the cations necessary for formation of carbonates. The absence of such minerals precludes long-term mineral trapping at either site. Although the content of silicate minerals is higher in the Midale Beds, this will not enhance the

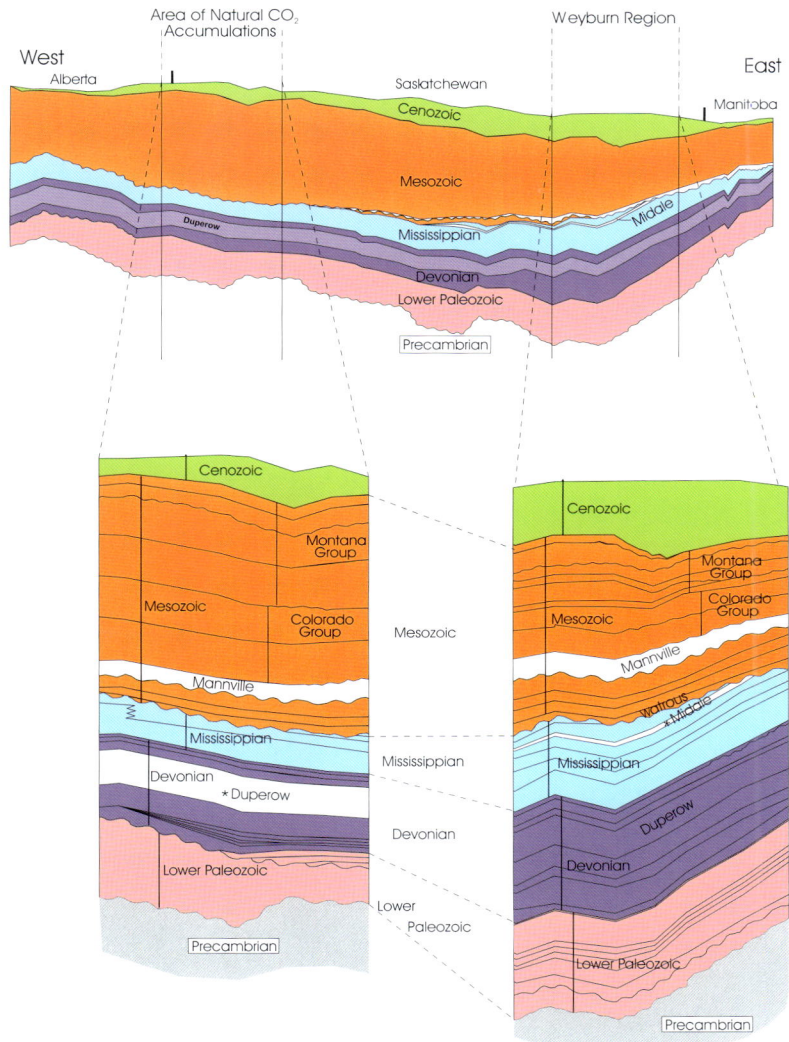

Figure 2-33. *Comparison of the geological setting between the natural CO_2 sites in southwestern Saskatchewan and the Weyburn injection site. The top diagram depicts the relatively continuous strata across the intervening distance between the two areas. The detailed geological columns indicate the broad similarity of geological setting in both areas – where the CO_2 is contained in Paleozoic carbonate reservoirs capped by anhydrite layers and in turn overlain by approximately 1500 m of Mesozoic shales, siltstones, and sandstones. At the Weyburn site, CO_2 is being injected into the Mississippian Midale Beds, whereas in southwestern Saskatchewan the CO_2 occurs naturally, mainly in the Devonian Duperow Formation.*

potential for mineral trapping over that of the Duperow Formation because of the nature of the silicate minerals present.
2. Reactive transport models developed for the Weyburn project should also be applicable to the Duperow lithologies, given the similarity of mineral constituents, mineral chemistry, and the whole rock between the two areas.

2.6 Geomechanics

2.6.1 Introduction

Emerging regulations for CO_2 storage are expected to require that the operator demonstrate an understanding of the potential implications of pressure buildup for containing CO_2 within the site. This includes understanding the potential impact on faults, induced fractures, induced seismicity, and abandoned wells. Geomechanical characterization of the site – with emphasis on the injection zone, its primary seal, modelling, and monitoring – is an important step in developing this understanding.

2.6.2 In-situ Stress Regime

The magnitude and orientation of in-situ stress influences the pressure at which tensile or shear fractures may develop, and existing fractures may re-open (in tension) or reactivate (in shear). Therefore, it is standard practice to compile and interpret available stress-related data to assist in the operational planning of subsurface fluid injection projects, and it is recommended that new data be obtained to fill gaps in the existing data.

Suitable data and estimates of the magnitude and orientation of in-situ stress were compiled and, although the data are quite varied, some general conclusions may be summarized as follows.

1. At the depth of the Midale Beds, the vertical stress gradient is about 24 kPa/m. In the Midale Field, Shell Canada suggested that minimum horizontal stress gradients in pressure-depleted zones of the Midale Beds may be in the range 13–14 kPa/m (McLellan et al., 1992). However, values in undepleted zones of the reservoir may be somewhat higher (with 18 kPa/m suggested as an upper limit), given that pressure depletion is known to reduce the magnitude of horizontal stress.
2. No direct measurements of maximum horizontal stress gradients have been made. Based largely on regional in-situ stress studies (see Figure 2-34) the Weyburn Field is believed to be located near the transition from a transpressional (strike-slip) stress regime to an extensional (normal fault) stress regime. This implies that the maximum horizontal stress is similar in magnitude to, or slightly greater than, the vertical stress. A likely upper limit of 28 kPa/m was proposed by project researchers.

3. The azimuth of the maximum horizontal stress is roughly northeast-southwest (Figure 2-34), which is consistent with the tectonic history of the Western Canada Sedimentary Basin (this orientation is roughly normal to the trend of the Rocky Mountains).
4. During well integrity testing of a Weyburn well in 2011, small-scale hydraulic fracture tests (mini-frac tests) were conducted at three depths in the Watrous Formation. Minimum horizontal in-situ stress gradients interpreted from two of these tests were approximately 16 kPa/m, while the gradient interpreted for the third test was approximately 21 kPa/m.

2.6.3 Rock Mechanical Properties

Elastic properties of rocks are important because they control the magnitude of rock deformation and stress changes that occur in response to changes in pressure or temperature. Rock strength properties are important because they dictate the threshold at which rock failure may occur if pressure- or temperature-induced stress changes become large. Further, the frequency, orientation,

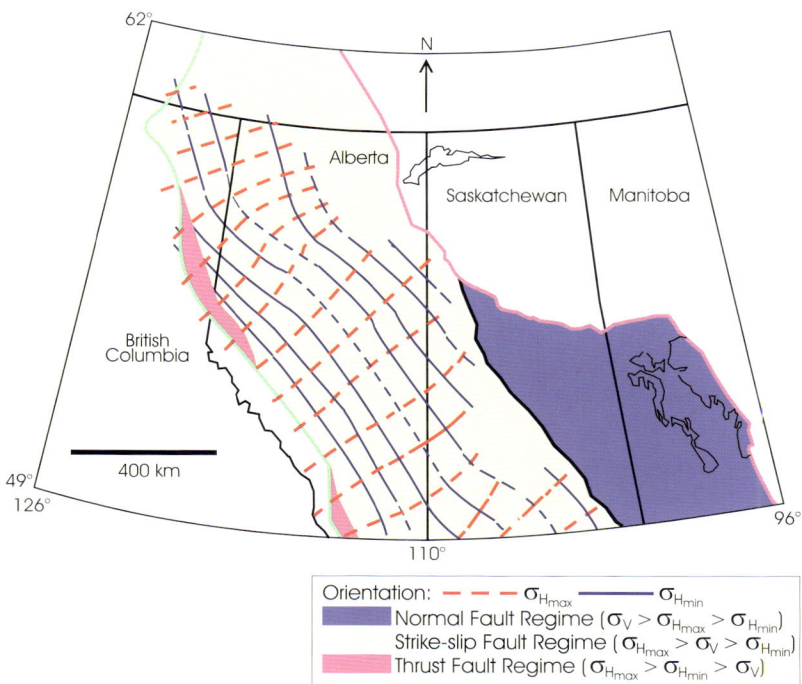

Figure 2-34. *Stress regimes and horizontal stress orientations interpreted for the Western Canada Sedimentary Basin (after Bell and Babcock, 1986; Bell et al., 1994).*

and attributes of natural fractures are also important because they can influence the large-scale deformability and strength of rock formations.

An extensive study of rock mechanical properties included (1) a literature-based component, which generally focused on cap rock lithologies, such as shale, but which also included published laboratory testing results on reservoir cores (Vuggy unit) taken from the Midale Field (McLellan, 1996), (2) a log-based component, which involved the calculation of dynamic elastic properties, using bulk density and full-wave (dipole shear imager) sonic log data from five wells in the Weyburn area, and (3) laboratory testing on four anhydrite cores from the Midale Evaporite and the Three Fingers zone of the Midale Beds. After the mechanical properties of these cores were measured, they were inspected to assess the presence and characteristics of natural fractures. Some small fractures were observed, but all had healed. As such, they were not deemed to be likely flow paths or preferential surfaces for mechanical failure.

The elastic properties were estimated for all stratigraphic units from the base of the Frobisher Beds to the top of the Mannville Group. The rationale was to include the injection zone, adjacent formations, and overburden units that could potentially be exposed to CO_2 in the event of leakage. In the case of upward leakage, it was assumed that the Mannville Group would serve as a conduit for lateral flow, hence limiting the potential exposure of post-Mannville rocks to CO_2. For projects where the monitoring of ground surface deformation is being considered as a means of tracking the CO_2 plume, elastic properties should be estimated for the entire overburden succession (perhaps based on a simplified stratigraphic model).

Rock strength properties were interpreted from the base of the Frobisher Evaporite to the top of the Midale Anhydrite. The rationale for analysing this subset of the stratigraphic section was based on the assumption that injection-induced stress changes would be far below the failure strengths of all units, except those within or immediately adjacent to the injection zone. The static elastic properties and Mohr–Coulomb strength properties estimated for these most critical units are listed in Table 2-6. Two scenarios are presented for elastic properties, namely 'soft' and 'stiff'. The stiff properties are the dynamic elastic properties calculated from well log data, and are an upper bound estimate of elastic properties. The soft properties were calculated based on the following.

1. Published data which demonstrate that static elastic properties can be significantly lower than dynamic values (by about 30%).
2. An empirical equation which accounts for the effects of natural fractures on formation-scale elastic properties, as quantified using the Geological Strength Index (Hoek et al., 2002).

Stratigraphic Unit	Soft properties			Stiff properties			Strength properties	
	Young's modulus (GPa)	Poisson's ratio	Shear modulus (GPa)	Young's modulus (GPa)	Poisson's ratio	Shear modulus (GPa)	Friction Angle (°)	Cohesion (MPa)
Midale Evaporite	22.7	0.26	9.0	61.3	0.28	23.9	44	18.2
Marly	10.3	0.29	4.0	33.4	0.29	12.9	40	3.5
Upper Vuggy	18.3	0.31	7.0	54.8	0.31	20.9		
Lower Vuggy	15.4	0.29	6.0	48.7	0.28	19.0	47	3.5
Frobisher Evaporite	15.8	0.31	6.0	51.6	0.29	20.0	44	18.2

Table 2-6. *Rock mechanical properties interpreted for the Midale Beds and adjacent strata.*

Subsequent to the work carried out on the project, it has become apparent that the soft properties are more representative of actual conditions at Weyburn.

In conclusion, in-situ stresses and rock mechanical properties are important parameters because they affect the geomechanical response of an injection site and hence play a role in defining the limits of injection pressures. If initially unavailable, data, and samples which enable in-situ stress and mechanical measurements to be made, should be collected during site characterization activities. Key data types include geophysical logs (oriented caliper, bulk density, full-wave sonic, bore hole imager), small-scale hydraulic fracture tests (micro-frac or mini-frac), and core samples for use in laboratory testing programs (for example, triaxial compression).

2.7 Geochemistry

Geochemical characterization is another key factor differentiating the regulation of CO_2-EOR and regulation for CO_2 storage. There are two main reasons for this: (1) the different potential for degradation of containment features (both geological seals and well integrity), and (2) the added contribution that dissolution and mineralization may have to the long-term trapping of CO_2 in the subsurface. Emerging regulations for CO_2 storage are therefore formulating specific requirements to ensure that relevant geochemical issues are adequately understood, in particular their potential impact on the integrity of new and existing wells. Examples of specific requirements include the following.

1. The requirement to gather baseline geochemical data in aquifers above confining zone(s).
2. The requirement to model and demonstrate an adequate understanding of site-specific dissolution and mineralization rates.
3. The requirement to perform periodic monitoring of groundwater quality and geochemical changes above the confining zone(s) that may stem from the CO_2 injection operation.
4. The requirement to perform periodic corrosion monitoring of CO_2 injection wells.
5. The requirement to use well construction materials, cements, and cement additives that are compatible with the CO_2 stream, carbonic acid, formation fluids, or other fluids which the materials may come in contact with over the lifetime of the project.
6. The requirement to determine which abandoned wells within the storage complex have been plugged with materials compatible with the CO_2 stream, carbonic acid and formation fluids.
7. The requirement to plug and abandon wells within the storage complex with materials compatible with the CO_2 stream, carbonic acid, and formation fluids.

8. The requirement to perform corrective actions on wells using materials compatible with the CO_2 stream, carbonic acid, and formation fluids, where appropriate.

Information on the chemical composition of formation waters is an important parameter for site characterization, and is a key requirement in many of the newly proposed regulations for CO_2 storage projects (European Commission, 2009; U.S. Environmental Protection Agency, 2010; DNV, 2012). Geochemical data are required for two main purposes: (1) as input into processes such as water–rock interaction and cement–rock interaction, and (2) to identify and trace migrating formation fluids.

2.8 Recommendations

Some research results from the Weyburn-Midale Monitoring and Storage Project are relevant to the CO_2-EOR operations at these fields, and are therefore site-specific. However, the design, scope, and processes of the research are applicable to other CO_2-EOR operations as well as to CO_2 storage. Therefore, the following recommendations have wide relevance. They are listed under three broad headings: study area, geology, and hydrogeology.

2.8.1 Study Area

Several vertical and areal scales are required in order to cover all possibilities for secure storage of CO_2. The depth of the vertical scales and the size of the areas will depend on the data available. When determining the scales, consideration should be given to the fact that different disciplines may need different scales. The largest scale should be regional in nature and the smallest probably a risk assessment region close to the injection site.

2.8.2 Geology

Geological characterization of the study area is critical to many aspects of CO_2 storage, not least in defining the storage complex. The size of the area for regional geological mapping should be constrained by data distribution in the stratigraphic unit with the minimum information. This will commonly be the deepest zero-flow boundary (often the Precambrian basement). All strata from this boundary to the surface should be mapped, although not all in the same detail. The storage complex requires the maximum detail. Generally, as much detail as possible should be used at the beginning, because at that stage it will not be known if 'lumping' will be required later. Better to lump later than split later!

The geological framework will include standard features such as maps of stratigraphic surfaces and bed thickness, as well as cross-sections. These will commonly be developed from well logs and seismic information. Fundamental structural elements such as basement structures, faults, fracture zones, joints and other lineaments should receive particular attention as possible conduits for fluid flow (natural or induced). Remotely sensed surface lineament mapping may be useful in this regard. In the Weyburn area, structures resulting from salt dissolution are important. In other areas the issue could be faults associated with salt domes. The important point is to be aware of the geological history of the area and how this may impact the rocks and contained fluids, as well as the ways in which these may be perturbed by the injection of CO_2.

Numerical models, used to calculate flow parameters in reservoirs, are critical to evaluating CO_2 storage sites. Geological models should be constructed mainly with a view to their application to numerical models, rather than to illustrate complex geological situations. To this end, it is important to use techniques that treat such parameters as permeability and porosity with the characteristics required by the numerical models.

2.8.3 Hydrogeology

Hydrogeological characterization of a study area is as important as geological characterization, and it starts with the geological framework. The objective is to identify the direction and rate of fluid flow, both of the natural situation and of the effects due to perturbation by injected CO_2. All vertical and areal scales mapped as part of the geological characterization should have equivalent hydrogeological characterization.

There are two parts to hydrogeological characterization: hydrostratigraphy and hydrochemistry. The former considers the properties of aquifers, aquitards and aquicludes, and the latter the properties of formation waters. The main sources of data for hydrogeological characterization are (1) the geological framework, (2) measurements of such properties of the rocks as permeability and porosity, obtained from drill stem tests, core analysis, and pump tests, and (3) analysis of major, minor, and trace elements, and of stable isotopes in formation water.

Hydrogeological information is commonly presented as maps and cross-sections of hydraulic head and as pressure-depth plots. Formation water data may be displayed as both maps and cross-sections, commonly using only salinity data. Other parameters which may be useful include distribution maps of saturation with respect to anhydrite, if anhydrite plugging is suspected. Variable density formation waters are common in many sedimentary basins, in which case hydraulic head data should be converted to 'water driving force' data, so that the direction and flux of water flow can be determined.

Identifying seals and determining their integrity is of paramount importance. The sort of information that may be of assistance includes, but is not limited to, (1) core analysis of shales and other rocks, to determine permeability and porosity, (2) vertical pressure gradient plots, (3) steep gradients in permeability and salinity distribution maps, (4) fracture mapping, using seismic AVOA analysis, and (5) comparison with natural analogues such as nearby reservoirs with high-CO_2 natural gases.

Falling under the general classification of containment is the identification of what may be termed the primary seal (which is directly above the injection zone) and ultimate seal (which should be a regionally extensive aquitard) – although at Sleipner there is only a single seal. These aquitards should receive special attention including, but not limited to, determination of (1) their permeability and porosity using a variety of methods, (2) their bulk chemical composition and mineralogy, and (3) other physical properties, such as Young's modulus, Poisson's ratio and shear modulus. It would be of value to supplement information on these properties of seals with a study of the regional stress field, to determine the nature and magnitude of the horizontal stress.

In summary, effectively all the characterization techniques that were used in the Weyburn-Midale Monitoring and Storage Project are applicable to the storage of CO_2. Other storage sites may require additional techniques.

3

STORAGE PERFORMANCE PREDICTIONS

Summary

The long-term isolation of CO_2 at geological storage sites was evaluated using reactive transport models and invasion percolation models. Reactive transport models explicitly represent transfer processes among fluids and the rock mass. They were used to elucidate the impact of reaction-dependent changes in reservoir porosity and permeability, with respect to the integrity of the reservoir and seals. In CO_2 storage operations, it is possible to account for all CO_2–oil–formation water–rock/mineral interactions.

Invasion percolation models are significantly different from traditional Darcy flow simulations, using capillary pressures rather than permeability and viscosity. They provide two important advantages: faster computation and less loss of geological detail due to averaging or upscaling. Both vertical and lateral breach points were identified as a function of threshold pressure.

The CO_2 storage capacity of a reservoir varies with changes in both pore volume and the mass partitioning of emplaced CO_2 among distinct physical (hydrodynamic, structural, residual) and chemical (hydrocarbon, aqueous, and mineral) trapping mechanisms. These were identified during the study with respect to both CO_2-EOR and post-EOR phases.

The ultimate CO_2 storage capacity of a given reservoir and seal system is physically dependent upon its structural, compositional, transport, and hydrological characteristics. Most important among these are the structural closure volume, mineralogy, aqueous salinity, porosity, permeability, ambient flow fields, and pressure-temperature conditions.

Continuous minimization of discrepancies between predicted and measured results through refined modelling and monitoring (that is, successful history matching) is essential for long-term forecasting.

Recommendations arising out this research begin on page 115.

Authors: J.W. Johnson (Schlumberger Doll Research, Cambridge, MA, USA), and **B.J. Rostron** (University of Alberta, Edmonton, Alberta).

3.1 Introduction

Advanced computational methods, such as reactive transport modelling and compositional reservoir simulation, are well suited to predicting the long-term isolation performance of geological CO_2 storage sites. Specifically, these modelling capabilities are used to confirm (1) that there is sufficient CO_2 injectivity and ultimate capacity in the target reservoir, and (2) that in-situ chemical and stress perturbations imposed by CO_2 injection will not change reservoir integrity or degrade overlying seals. This chapter focuses on the methods and key outcomes of the modelling studies as related to potential CO_2 migration, storage capacity, and dynamic CO_2 mass partitioning among the various physical and chemical trapping mechanisms.

3.1.1 Context

The goal of CO_2 storage is to securely isolate the injected CO_2 from sensitive overlying environments such as aquifers with potable water, soils, and the atmosphere. There are four components to modelling this process.
1. Capacity of CO_2 (incremental and total).
2. Footprint (by area and volume).
3. Containment (efficacy of natural seals).
4. Risk (uncertainty bounds on capacity, footprint, and containment forecasts).

In principle, the feasibility of transitioning a CO_2-EOR operation into a long-term CO_2 storage site hinges on demonstrating that acceptable values of these four components will be achieved. In this context, 'acceptable' values are those that ensure compliance with all regulations within the relevant jurisdiction.

Reactive transport modelling is now preferred over traditional compositional reservoir simulation because of its explicit representation of fluid–rock mass transfer processes, and the impact these have on changes in the integrity of the reservoir and seals (Figure 3-1). Because this impact is far greater during CO_2-EOR operations than other oil recovery methods, it is necessary to account for CO_2–oil–formation water–rock interactions.

3.1.2 Objectives

The key objectives of performance prediction for a CO_2 storage project are as follows.
1. To demonstrate anticipated performance (capacity, footprint, and containment).
2. To use simulated performance dynamics to design an effective monitoring program which will document these forecasts.
3. To ensure protection of health, safety, and the environment.

4. To support optimization and risk management of the CO_2 injection operations in full compliance with applicable regulations.

3.1.3 Modelling to Support Risk Management

The difference between modelling and monitoring requirements for CO_2-EOR and CO_2 storage projects is the fundamental shift in motive from optimizing oil recovery to assuring CO_2 containment. In particular, modelling and monitoring programs for CO_2 storage projects must predict and verify that injected CO_2 will be contained in the storage complex, and that displaced fluids will not pose a risk to human health or the environment. Additional requirements for modelling include the following.

1. Dynamic models must account for geomechanical and geochemical processes and demonstrate their coupled impact on reservoir and seal properties.
2. Modelling must be able to assess CO_2 and displaced formation fluid migration within the storage complex, which could degrade containment efficacy.

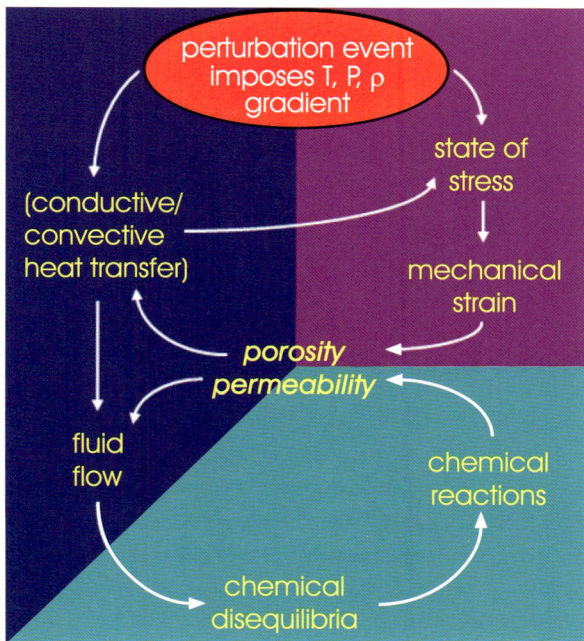

Figure 3-1. *Example of a set of integrated subsurface processes that may be resolved by using reactive transport modelling. Modified from Norton (1984).*

3. Modelling must assess lateral and vertical extents of the region within which stress perturbations may be sufficient to drive CO_2 or formation fluids out of the storage complex.
4. The long-term (decades to millennia) migration (footprint) and mass partitioning of emplaced CO_2 must be simulated.

Monitoring programs must be able to verify key output parameters, and specifically to detect the following.
1. Indicators of migration along potential escape pathways for CO_2 or displaced formation fluids that have been identified through modelling.
2. Trends that may cause pressure to reach levels sufficient to cause initiation or propagation of fractures, re-activation of faults, induced seismicity, or other negative impacts.
3. Trends that may cause pressure at a 'vulnerable point' (for example, an improperly abandoned well, or a fault crossing the confining seal).
4. Geochemical degradation of the cap rock or well integrity, where this could provide a significant risk of loss of containment.
5. Indicators of CO_2 or displaced formation fluids in shallower or deeper zones, indicating that these fluids have migrated from the injection zone.

Modelling and monitoring programs must proceed in parallel, with continuous inter-program feedback to ensure that the monitoring program is designed in accordance with the latest modelling results, and vice versa.

3.1.4 Post-injection Performance Validation

In terms of regulatory compliance, CO_2-EOR projects are generally allowed to terminate most modelling and monitoring activities as soon as CO_2 injection stops, and once all wells have been properly abandoned. In contrast, CO_2 storage projects – whether initially associated with EOR or not – will be required to continue these activities well into the post-injection phase. The objectives of post-closure modelling and monitoring will generally be the same as during active CO_2 injection, albeit with a potential for reduced frequency of modelling updates and discrete monitoring activities. The main focus during this phase is on providing evidence to support the following objectives.
1. Modelling and monitoring provide continued assurance that the injected CO_2 is contained within the storage complex (that is, there are no detectable leaks beyond the allowable limits).
2. The storage complex is evolving towards long-term stability, and is understood in sufficient detail to permit accurate assessment of future changes.

3. There is negligible risk of future CO_2 leakage, or formation fluid displacement, which could have a negative impact on human health or the environment.

3.2 CO_2 Migration

3.2.1 Introduction

From the point of view of project design and risk assessment, accurate uncertainty-bound forecasts of CO_2 migration out of the storage complex rank high on the list of essential data to be determined. These forecasts are needed (1) to establish lease boundary requirements, (2) to construct effective reservoir, overburden, and near-surface monitoring programs, and (3) to demonstrate to stakeholders and the public that the long-term fate of injected CO_2 is understood within acceptable limits. Here, we present the methods and results of several project studies focused on predicting CO_2 migration, within and outside the Midale reservoir, during both CO_2-EOR and CO_2 storage phases.

3.2.2 Designed (Intra-reservoir) Migration

Accurate modelling of CO_2 migration within the reservoir is required to design an effective reservoir-monitoring program. For CO_2-EOR projects, there is commonly ample data to design an effective monitoring program. In contrast, accurate forecasting of ultimate CO_2 migration within laterally unconfined CO_2 storage sites is subject to significantly increased uncertainty.

During Phase 1 of the Weyburn project, the compositional reservoir simulator GEM® was used to predict CO_2 migration within the reservoir. The initial approach was to generate fine-grid simulations of single-pattern EOR-storage performance for three injection strategies. Traditional upscaling techniques were then used to produce coarse-grid analogs and, finally, full-field 75-well pattern simulations that were designed to capture overall injection, production, and fluid distributions within each single pattern, but not the details for each individual well.

During the Final Phase of the Weyburn project, a more recent version of GEM®, including fluid–rock mass transfer (reactive transport) and mechanical dispersion processes, was used to reassess CO_2 migration for one of the original fine-grid single patterns. This version allowed improved boundary conditions, and the original simulation was extended out to 2010.

The initial simulations were validated (history matched) against both laboratory and field observations. At the laboratory scale, successful simulation history matching was obtained for a series of CO_2-coreflood experiments conducted using different oil samples. At the field scale, the fluid production history was successfully matched for each of the three single patterns that formed the basis of subsequent 75-well pattern simulations.

This full-field model was then used to predict CO_2 migration, storage capacity, and mass partitioning during CO_2-EOR, where current and planned operational strategies were followed as closely as possible in an initial 'EOR base case' scenario. Also investigated were an alternative EOR case and two distinct post-injection CO_2 storage scenarios, which focused on promoting additional CO_2 storage.

Here, we illustrate the use of compositional reservoir simulation and reactive transport modelling to forecast CO_2 migration during and beyond EOR operations. First, the approach used to carry out such simulations for a single pattern is reviewed, as well as the results obtained. Emphasis is placed on fundamental considerations associated with achieving accurate representation of reservoir properties, spatial data distribution, boundary conditions, and integrated processes. We then outline the use of traditional upscaling techniques to extend single pattern analyses into full-field simulations that encompass multiple patterns, and present the results obtained from the 75-well pattern simulation at Weyburn.

3.2.3 Single Pattern Simulations

Spatial Domains and Property Distributions

Input data were based primarily on Phase 1A models and data sets, with updated production data. In order to minimize uncertainties associated with boundary well allocation fractions, a half-pattern buffer was used around the well pattern. Initial upscaling from the Phase 1A model to the fine-grid, single-pattern models involved coarsening the vertical description to nine layers representing different flow units within the upper Marly and lower Vuggy zones, while the areal gridblock size remained at 57.4 m × 57.4 m.

Porosity was upscaled by simple arithmetic averaging, and absolute permeability was upscaled in a stepwise fashion. Permeability values were converted from air to liquid permeability. Horizontal permeability was calculated as the weighted arithmetic mean of its constituent layer permeability measurements, and vertical permeability was calculated as a weighted harmonic mean. Because fine-scale laminations and baffles reduce vertical permeability, vertical permeability values were modified to achieve consistency with a vertical-to-horizontal permeability ratio of 0.1 (as estimated for the Midale reservoir by Encana). A final significant correction to permeability values was implemented to account for the effect of known fracturing.

All fine-grid single-pattern simulations provided a good history match of Phase 1 fluid production, and good agreement with the observed production and pressure data through 2010.

3.2.4 Base-case Models

Based on the successful match of single-pattern histories, the original fine-grid simulations were extended until a total of 25 years of post-waterflood operation was simulated. Operational guidelines recommended by Encana for Phase 1 were used for the Final Phase predictive CO_2-flood period.

Post-2010 simulations also included a series of gas-oil ratio (GOR)-control runs where GOR shut-in limits of 4000, 6000, and 8000 were imposed on producers. In addition, the predictive CO_2-EOR simulations were carried out to 2070 (60 years). Selected time-lapse snapshots of simulated CO_2 migration in Pattern 1 are shown in Figure 3-2. In these (and subsequent) CO_2 distribution plots, CO_2 content is shown as global mole fraction, which includes CO_2 in the oil, gas and water phases. The time-lapse results from both simulations show progressive migration of CO_2 through the pattern during several decades of ongoing CO_2-EOR. When reasonably well-constrained, dynamic models of this kind can be used effectively to design sampling/imaging locations and frequencies for geochemical and geophysical monitoring programs.

Effect of Mechanical Dispersion

Diffusion and mechanical dispersion are two primary mechanisms that lead to mixing and dissipation of the miscible slug (CO_2-diluted oil).

Mechanical dispersion is an important solvent-mixing process caused by variations of flow velocity in the reservoir. It results in a dilution of the solvent at the advancing edge of flow. The most important variables influencing dispersion are (1) the magnitude and distribution of permeability (including anisotropy), (2) fractures, or other reservoir heterogeneity, (3) molecular diffusion, and (4) reservoir shape or flow system aspect ratio.

There is considerable uncertainty in the mechanical dispersion coefficient (or dispersivity), because it varies over a wide range of values at the field scale. Because mechanical dispersion depends on fluid velocity, it is clear that variations in both permeability and porosity play a significant role. At the field scale, even reservoir formations that are considered to be homogeneous have zones of different porosity and permeability. As a result, the dispersion parameter may be a few orders of magnitude higher in the field than the value obtained in the laboratory.

Two sets of dispersivity values were assumed for the Weyburn project, one based on experimental data (macro-scale mixing parameters) and the other on length scale correlations (field-scale mixing). Simulations were also done using three different sets of mechanical dispersion coefficients. Figure 3-3 shows comparison between the two extreme cases for these two mechanical dispersion scenarios for the fine- grid model. Clearly, the effect of mechanical dispersion

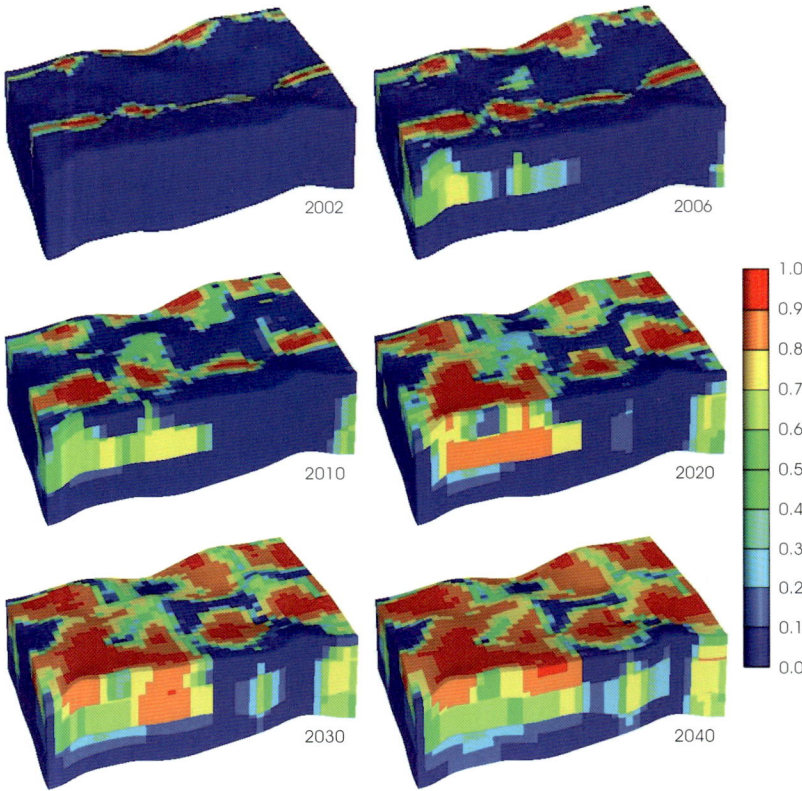

Figure 3-2. Distribution of CO_2 (as a global mole fraction) in Pattern 1 from 2002 (after 2 years of EOR) to 2040 (after 40 years). Producing wells were shut in when GOR exceeded 6000.

on CO_2 migration is significant. These results illustrate that supercritical and dissolved CO_2 migrates to a greater extent when the dispersion is larger, highlighting the importance of further refining the precision of dispersivity values used in the simulations. These effects using the coarse-grid model are shown in Figure 3-3.

Upscaling from Fine-grid to Coarse-grid Simulations

Full-field reservoir simulations require a second tier of upscaling. The objective is to reproduce important features of the fine-grid model, while simultaneously upscaling pseudo-functions, such as absolute and relative permeability. The three coarse-grid simulations obtained are then used as building blocks for the

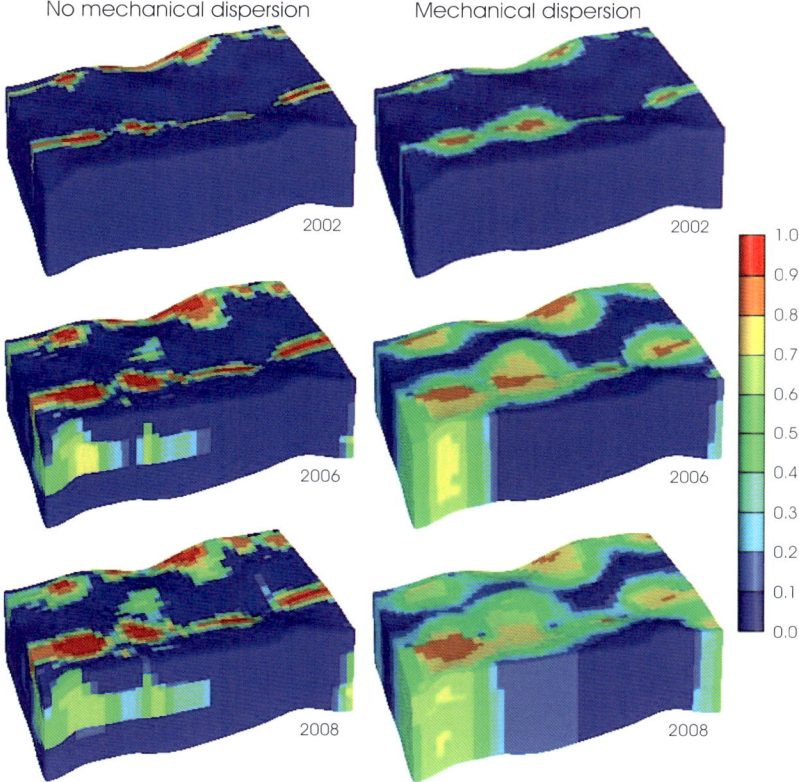

Figure 3-3. *Impact of mechanical dispersion on the distribution of the global mole fraction of CO_2 in the coarse-grid model for 2002, 2006, and 2008.*

75-well pattern model. For the coarse-grid, the block size was 133.9 m × 133.9 m, with only two vertical layers, one each for the Marly and Vuggy zones. The second-tier upscaling method was validated by reasonable agreement with results from the fine-grid and coarse-grid simulations for Patterns 1 to 3.

Time-lapse CO_2 migration in the Marly and Vuggy zones of Pattern 1, as predicted by the fine-grid and coarse-grid models, is shown in Figure 3-4. As is readily apparent, the coarse-grid simulations – which, by design, do in fact accurately capture the overall injection, production, and fluid distributions within the pattern – do not accurately represent these parameters on a well-by-well or local-area basis. As a result, the upscaling is detrimental in the context of predicting single-pattern CO_2 migration, and thus for designing local monitoring strategies.

Full-field Simulations

Upscaling problems were overcome as follows: First, by using the Schlumberger GeoQuest program FloGrid® to translate the geological model into a format for use in a GEM® simulation. Although the geological grid was oriented N-S, the simulation grid was oriented NE-SW in order to take well geometry into account. Second, the Computer Modelling Group program Modelbuilder® was used to vertically upscale the 15 flow units representing the Marly and Vuggy zones into two layers, while maintaining the same horizontal resolution.

The complications associated with a 75-pattern simulation required a number of simplifications, as follows.

1. Constant oil properties were used, even though they vary across the Weyburn Field.
2. Two sets of relative permeability curves were used, one for the upper Marly zone and one for the lower Vuggy zone.
3. Pseudo-well fluid production and injection rates were kept constant, to the extent feasible.
4. All water-alternating-gas (WAG) patterns alternated CO_2 and water injection simultaneously.

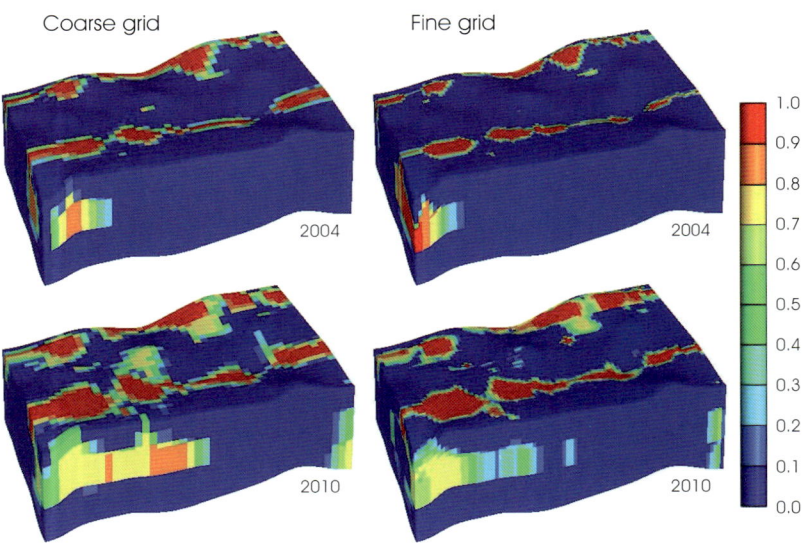

Figure 3-4. *Comparison of the distribution of CO_2 global mole fraction between the coarse-grid and fine-grid models, for the years 2004 and 2010.*

All of these assumptions are justified, because the 75-pattern simulation produced an acceptable history match. Further, the simulation was designed principally to provide the final CO_2 distribution and mass partitioning at the end of EOR, for the risk assessment of potential CO_2 leakage and migration.

3.2.5 **CO_2-EOR Scenarios**

Using the 75-pattern model, initial predictive simulations were carried out to estimate CO_2 migration, storage capacity, and mass partitioning during the CO_2-EOR project. Twenty-five years of CO_2-EOR were simulated for each pattern, with CO_2 injection starting in September 2000 (Phase 1A area) and terminating in 2033 (last roll-out patterns). Encana's planned guidelines for CO_2-EOR development provided a set of operational constraints that were imposed on the simulations to the extent possible. The constraints were as follows.

1. To implement the anticipated CO_2 injection strategy in each currently active and projected pattern.
2. To implement the projected pattern roll-out plan.
3. To achieve a time-integrated target hydrocarbon pore volume of about 50–60%.
4. To assure a maximum CO_2 supply rate of 2.69 Mm^3/d (total supply of 20 Mt CO_2).
5. To enforce a recycle plan of about 60% total CO_2 injection.
6. To enforce reservoir pressure constraints: >18 MPa (for miscibility); <27 MPa (fracture pressure).
7. To implement a depletion plan: water flush depleted patterns to liberate some CO_2 for recycle.

The hydrocarbon pore volume constraint was achieved by adjusting initial water saturation to an extent dependent upon CO_2 injection strategy. In the base case scenario, the aggregate CO_2 injected from 2000–2033 was 45.9% of the hydrocarbon pore volume. However, these full-field models of CO_2 migration are of little utility in designing monitoring programs.

An alternative CO_2-EOR scenario was also simulated. Here, CO_2 injection rates were increased by 30% from base-case values, with the intent of promoting additional CO_2 storage during active EOR. In this alternative scenario, aggregate CO_2 injected into the 75 patterns during CO_2-EOR from 2000–2033 was roughly 30% larger than in the base case, as expected. Again, while it was useful (in fact necessary) to compare full-field storage performance predictions for various operating strategies, the practical utility of such models is limited to that specific application.

3.2.6 Post-EOR-CO_2 Storage Scenarios

The effect of various post-EOR operational strategies on long-term CO_2 migration was investigated through a series of 75-pattern simulations. In these models, the time-frame was extended beyond EOR completion in 2033 out to 2055. Thus, the overall simulation time-frame was 100 years (1955–2055). Two fundamental scenarios were investigated. In the first, injection was continued until reservoir pressure reached 29.5 MPa (fracture limit for a 22.61 MPa/km gradient that relates minimum principal stress to overburden weight with a safety margin of 90%). In the second, GOR was constrained not to exceed 1500. Despite the limitations noted above, it is clear from the plots that these post-EOR operations will exploit most of the 75-pattern area for long-term storage.

3.2.7 Impact of CO_2–Oil–Formation Water–Rock Mass Transfer

The evolution of the transport properties of the reservoir and seals – principally, porosity and permeability (due to CO_2–(oil)–formation water–rock mass transfer), can significantly influence long-term CO_2 migration. The magnitude of these changes is dependent on reservoir mineralogy, formation water composition, CO_2 impurities, and the relative importance of fracture and matrix transport, with the greatest effect anticipated for fracture-dominated carbonate reservoirs. For water-wet hydrocarbon reservoirs, the impact is largely independent of the presence of an oil phase, and thus is common to both CO_2-EOR and CO_2 storage sites.

Most compositional reservoir simulators do not model this type of interaction, but these effects must be accounted for in the context of CO_2 storage modelling. Its inclusion facilitates improved history matching in terms of achieving realistic parameter tuning. Accurate modelling of these changes requires detailed characterization of the mineralogy of the reservoir and the seal, of formation water and injection (make-up) water composition, and of the amount of impurities in the injected CO_2.

3.3 Potential CO_2 Migration Above the Reservoir

The requirement to demonstrate the ability of a site to retain CO_2 over long time periods has been formally defined in some jurisdictions, for example in the EU CCS Directive (European Commission, 2009) and in the new rules for Class VI wells by the U.S. Environmental Protection Agency (2010). This requirement is best assessed through numerical simulations, given the long and forward time requirements for predictions, and has been one of the primary objectives of the Weyburn project since its inception.

Three main approaches were applied: (1) deterministic and stochastic models were used to assess the fate of CO_2 within the geosphere and abandoned wellbores, (2) probabilistic risk assessment techniques were used to investigate the potential application of these methods for geological storage projects, and (3) invasion-percolation simulations of CO_2 movement above the injection zone were carried out, including determination of the potential effects of more than 4000 wellbores in the risk assessment area.

The first two techniques were reported during Phase 1 and are not considered here. Results from the invasion-percolation simulations are reported below.

3.4 Invasion Percolation Modelling
3.4.1 Introduction

Assessment of the migration of CO_2 beyond the reservoir is one of the primary objectives of the project. In Phase 1, numerical simulations of fluid flow in the 10 × 10 km risk assessment region were combined with semi-quantitative estimates of wellbore leakage, to complete an assessment of the ability of the Weyburn reservoir to securely store CO_2 for an extensive period of time. At the time, that approach was the best-practice, but questions regarding boundary conditions, geological heterogeneity, and treatment of wellbores merited additional investigations in the Final Phase.

The CO_2 storage site and five overlying aquitards are penetrated by more than 4000 wells in the 40 × 50 km study area. The high well density and regional scale present a major challenge for flow modelling. Conventional simulation methods require relatively coarse grids, which lead to a loss of geological detail due to averaging or upscaling, and hence the inability to represent fine detail in areas with high well concentrations. However, a capillary-dominated flow assumption (Cardona et al., 2002) allows for cellular automata modelling. This allows use of flow rates typical of migration, and provides a better approximation of non-continuum flow physics (capillarity, rather than viscosity flow), as well as very high resolution. This method is known as invasion percolation.

3.4.2 Technology

Invasion percolation models are significantly different in behaviour from traditional Darcy flow simulations, in that the simulation is dependent upon threshold and capillary pressures, rather than permeability and viscosity.

The buoyancy of CO_2 promotes gravity segregation, causing the migrating plume to be sensitive to the topography of seals that overlie aquifers (Singh et al., 2010). This results in discrete fill-spill pathways of migrating and pooling gas. Where pooling occurs, invasion percolation determines vertical and lateral breach points as a function of threshold pressure. The limiting condition for invasion percolation is a capillary force that exceeds the viscous force by a ratio of 10 000 to 1. The key relationship is the capillary number equation, in which the capillary number (C_a) is expressed as a function of the viscosity of the more viscous phase (μ) (formation water), the flux of the viscous phase (q), and the interfacial tension between phases (γ). The dominant term is the flux rate. For typical migration rates (less than centimetres-per-second) invasion percolation is a reasonable approximation. For example, hydrocarbon migration typically has a capillary number of 10^{-10} to 10^{-14} (England et al., 1987). Under these conditions, high-resolution invasion percolation simulators yield fast numerical solutions (Cardona et al., 2002). For CO_2 migration, the viscosity of the formation water (0.6–0.8 mPas) and interfacial tension between the formation water and CO_2 (25–35 mN/m) constrain the maximum percolation flux rate for the Weyburn reservoir as approximately 4 mm/s ($C_a < 10^{-4}$). Regional migration is unlikely to break this condition.

The Young–Laplace equation describes the governing physics of invasion percolation. A capillary pressure occurs at the interface between two fluids, for example, CO_2 and formation water. The pressure is a function of the surface tension at the interface. A popular expression of the Young–Laplace equation relates the gravitationally stable column height of an oil or gas trap to the capillary pressure (for example, Hobson, 1954). If the capillary pressure of the CO_2 column exceeds the threshold pressure of the seal, the CO_2 will breach the trap and migrate vertically. If the threshold pressure of the seal is sufficient to backfill

Hydrostratigraphic Unit	Depth to top (m)	Thickness (m)	Permeability (mD)	Porosity	Flow Velocity (m/a)
Till aquifer	0	54	15	0.2	n/a
Bearpaw aquitard	54	359	0.189	0.05	–
Belly River aquifer	413	60	15	0.2	n/a
Colorado aquitard	473	399	0.189	0.05	–
Newcastle aquifer	872	27	19.5	0.24	n/a
Joli Fou aquitard	899	38	0.0063	0.05	–
Mannville aquifer	937	135	3772	0.25	9.3
Vanguard aquitard	1072	100	0.0063	0.05	–
Jurassic aquifer	1172	124	567	0.16	4.0
Watrous aquitard	1296	97	0.8	0.04	–
Poplar/Ratcliffe aquifer	1393	26	1	0.06	–
Midale aquitard	1419	7	0.00001	0.003	–
Midale Marly	1426	13	79	0.14	0.8
Midale Vuggy	1439	11	74	0.125	0.8
Frobisher aquifer	1450	66	8	0.15	0.8
Alida aquifer	1516	156	5	0.12	0.8

Table 3-1. *Mean parameters for the main hydrostratigraphic units in the Weyburn model area.*

the trap, and exceeds the spill-point pressure, the CO_2 spills laterally, continuing the migration pathway. In this manner, invasion percolation approximates the flow physics of migration. The balancing of capillary pressures (fluid) and threshold pressures (rock) is thought to be an accurate representation of the governing physics of multiphase migration. Simulations match the generally observed phenomena of discrete migration pathways, pooling, breaching, and lateral spill.

Here, it is assumed that the CO_2 is distinct from the formation water (immiscible and insoluble). It has been shown that CO_2 is partially soluble in formation water over the range 1–5 wt. % with respect to formation water (Hassanzadeh et al., 2007). However this aspect was not addressed by the migration simulator in this study.

3.4.3 Model Description

Modelled Area and Volume

The regional study area is 40 × 50 km, and the geological data were based on the Petrel model. Eleven aquifers and aquitards were included in the flow modelling (brief details provided in Table 3-1).

Treatment of Wellbores

Because wells have the highest potential risk of CO_2 movement in the geosphere, the problem was addressed in two steps. First, an estimate was made of the range of effective hydraulic properties associated with a single well, and second, the model was populated, as closely as possible, with the number, position, and effective permeability increase associated with each well.

Single Wellbore Permeability-Equivalency

Significant column heights (and high capillary pressures) are required to overcome the threshold pressures associated with shale aquitards in order for vertical CO_2 migration to take place. As a result, migration pathways commonly breach shales at pooling locations associated with fractures, fault zones, and wells. Wellbores are inevitably associated with some geomechanical damage and local fracturing of the country rock (Rawlings et al., 1993), especially in weak strata. However, the attributes of well-related fractures are difficult to quantify – the extent of fracture penetration, characteristic fracture aperture, vertical and lateral extent of fracture networks, fracture density, and so forth. Given the significance of this poorly constrained variable, the modelling approach was to use a range of apertures that may be characteristic of well-related fracture networks. These are shown in Table 3-2. These permeability scenarios are hypothetical because no empirical data for wells and near-well aquitard conditions are available to test them.

Source Terms for CO_2
Source terms for CO_2 modelling were specified in two ways using (1) all 1011 wells located above the storage site, and (2) a reduced set of 760 wells that penetrate the Midale Evaporite south of the subcrop pinchout of the Ratcliffe Formation. To approximate a reasonable source term for possible migration pathways, a square area 15 × 15 km over the current injection zone was designated as the potential source area of leaking wells above the Watrous Formation.

The source term for CO_2 in the 'Attic Space Scenario' (described below) is defined as all well locations within the 75-well pattern area, based on Phase 1 reservoir simulations for 2055.

It is important to note that these source scenarios are not intended to create a 'worst case scenario' of hundreds to thousands of leaking wells. A key modelling assumption is that any one of these wells might potentially provide a leakage pathway, and so it becomes necessary to establish the spectrum for all potential pathways.

As such, CO_2 source terms do not discriminate between the likelihood of any individual well leaking. In cases with many source terms, these allow consideration of all potential migration pathways, just as these many wells provide for every possible starting point for migration above the current storage site. Attic space scenarios with 760 wells in the region of the Ratcliffe aquifer address questions of sources of CO_2 below the Watrous Formation.

Permeability Interpolation from Sparse Regional Data
Hydrodynamic simulations of CO_2 movement require the specification of horizontal permeability values throughout the simulation domain. Regional horizontal permeability distributions were previously established for major aquifers within the Williston Basin, by stochastic inversion modelling of the hydraulic head response that matched the sparse calibration data for the region (Khan et al., 2007). Because the regional permeability fields were calculated over

Fracture	Slightly Leaky	Moderately Leaky	Very Leaky
Fracture half-width (μ)	1	3	8
Permeability (mD)	0.05	0.5	5
Pth CO_2–H_2O (kPa)	100	41	16
Pth Hg–Air (kPa)	1100	450	175
CO_2 column (m)	16	6.5	2.5
Pth = threshold pressure.			

Table 3-2. *Scenarios for fracture-related flow associated with well paths through aquitards.*

a larger area than the model area, the permeability fields were interpolated from the regional distribution by using a minimum curvature algorithm.

Well logs and regional observations indicate that the aquifers also contain vertical heterogeneities, but there are insufficient data to constrain the model. While vertical heterogeneity, such as thin shale layers, may increase the trapping potential of an aquifer, as observed at Sleipner (Hermanrud et al., 2010), the structural control on regional migration is thought to be the topography of the cap rock interface between aquifer and aquitard. As such, the general behaviour of the model is assumed to be unaffected by absence of intra-aquifer vertical heterogeneity.

3.4.4 Results

Hydrostatic Case

The pilot study for regional CO_2 migration assumed a hydrostatic setting. This allowed for a fast first approximation of the model behaviour, and established the baseline against which hydrodynamic simulations are compared. This scenario used the 15 × 15 km source term for each aquifer. The resulting simulation provides an estimate of the path of migration of CO_2 should CO_2 ever reach that aquifer.

Results from the hydrostatic case study (Figure 3-5) reveal a predominantly northerly migration pattern for each aquifer, reflecting the gentle regional southerly dip. These results can also be used to identify potential well-breach locations: areas where CO_2 could accumulate beneath each cap rock and potentially build enough column height to migrate vertically upward. Wells associated with pools of trapped CO_2 were assessed for column heights and ranked as possible breach locations, with 62 wells identified for all aquifers. The location of these wells varied for each aquifer, with a weak northeast-southwest trend for all wells, and riskier wells more to the northwest for shallower aquifers. A small cluster of wells in the southeast corner of the storage site also appears to present a risk at the deeper aquifer level. With respect to risk potential, the higher the threshold pressure associated with a well, the greater the column height required for a CO_2 breach. Cap rocks above oilfields typically retain columns measured in hundreds of metres. Thus it is a notable characteristic of Mesozoic aquifers of the Williston Basin that the traps have small column heights (generally <15 m). These very small columns have a very low risk for leakage, because the smaller capillary pressures associated with such pools require wide aperture fractures within the shales for a breach to occur. Thus, migration pathways are highly unlikely to leak, due to low column heights and low saturations.

In summary, hydrostatic simulations indicate that the aquitards represent a very low risk for leakage due to the lack of topography on the surfaces between the aquifers and aquitards. Pressure gradients between the aquifers may also

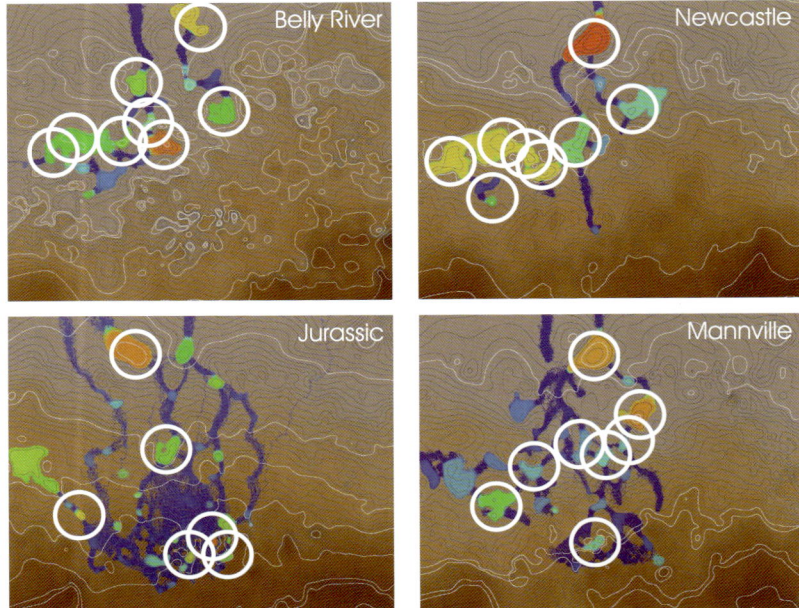

Figure 3-5. *Migration pathways for the regional aquifers and column heights for pools (• low saturation, • 8–10 m, • 6–8 m, • 3–6 m). Circles indicate flagged wells associated with pools.*

affect the seals. This aspect is tested in the hydrodynamic modelling of the next scenario.

Hydrodynamic Case

Hydrodynamic migration modelling includes the effects of dynamic pressure fields in the flow domain. A dynamic component to flow can be imparted by pressure variations caused by geological heterogeneities or density variations that modify the flow field, as expressed in deviated migration pathways and tilted hydrocarbon-water contacts.

Modelling conditions were effectively identical to the hydrostatic scenario, except that the Belly River aquifer was removed from the mesh, due to an absence of pressure and permeability data, and regional permeability and pressure distributions for three deeper Mesozoic aquifers (Newcastle, Mannville, and Jurassic) were included.

ONE MICRON SCENARIO: This hydrodynamic scenario assumes a fracture aperture of 1 micron (the slightly leaky case). All CO_2 is retained within the

Jurassic aquifer (Figure 3-6) for migration above the Watrous Formation, based on potential well penetrations of the Watrous aquitard in the source area. The overlying Vanguard aquitard will not breach under these conditions. Pools within the Jurassic have insufficient column height to overcome the threshold pressure of a one micron fracture pathway.

The migration trend within the Jurassic aquifer is to the northeast as a result of a hydrodynamic influence on the migrating CO_2. A comparison with the hydrostatic situation indicates that the hydrodynamic influence is significant, deviating the migration trend by about 45°. A cluster of closures over the storage site result in about 20 pools that would potentially retain about 1.3 Mt of migrating CO_2. Migration pathways beyond the storage area, to the north, are notable for the absence of pools and significant fill-spill trapping.

THREE MICRON SCENARIO: The second scenario assumes a fracture aperture of 3 microns (the moderately leaky case). In this case, CO_2 is partially retained within the Jurassic aquifer, but breaches the Vanguard and migrates into the Mannville aquifer (Figure 3-7). The simulation indicates that the overlying Joli Fou aquitard will not be breached under these conditions. There is no CO_2 migration into the Newcastle aquifer. Pools within the Jurassic aquifer are reduced in size (compared to the previous model case), with the 10 largest pools retaining only 100 kt of CO_2. As such, the breaching of the Vanguard aquitard reduces the secondary storage potential of the Jurassic aquifer by a factor of three to four. However, the numerous large pools within the Mannville increase the overall storage capacity of the geosphere to >2.5 Mt, with the 18 largest pools within the Mannville accounting for 1.7 Mt of potential storage.

The hydrodynamic influence within the Jurassic aquifer is the same, with migration deviated to the northeast. For the Mannville aquifer, the regional pressure gradient is to the northeast (blue arrow, Figure 3-7), and closer in alignment to the northerly migration direction and related topographic influence under hydrostatic conditions. A notable change is the loss of a westerly hydrostatic fill-spill pathway within the Mannville aquifer. While most pooling locations within the Mannville aquifer occur directly over the storage site, a small number of pools also occur to the north and northeast.

EIGHT MICRON SCENARIO: The third hydrodynamic scenario assumed a fracture aperture of 8 microns (the very leaky case). Little CO_2 is retained within the Jurassic aquifer, breaching both the Vanguard and Joli Fou aquitards at pooling locations, and migrating into the Newcastle aquifer (Figure 3-8). Simulations indicate that the overlying Colorado aquitard also breaches under these conditions. There is little retention of CO_2 in the Newcastle aquifer at around 60 kt within two pools. In the deeper aquifers, pools within the Jurassic aquifer are slightly reduced from the previous scenario (that is, moderately leaky), retaining about 340 kt of CO_2. However, the potential capacity of the Mannville increases

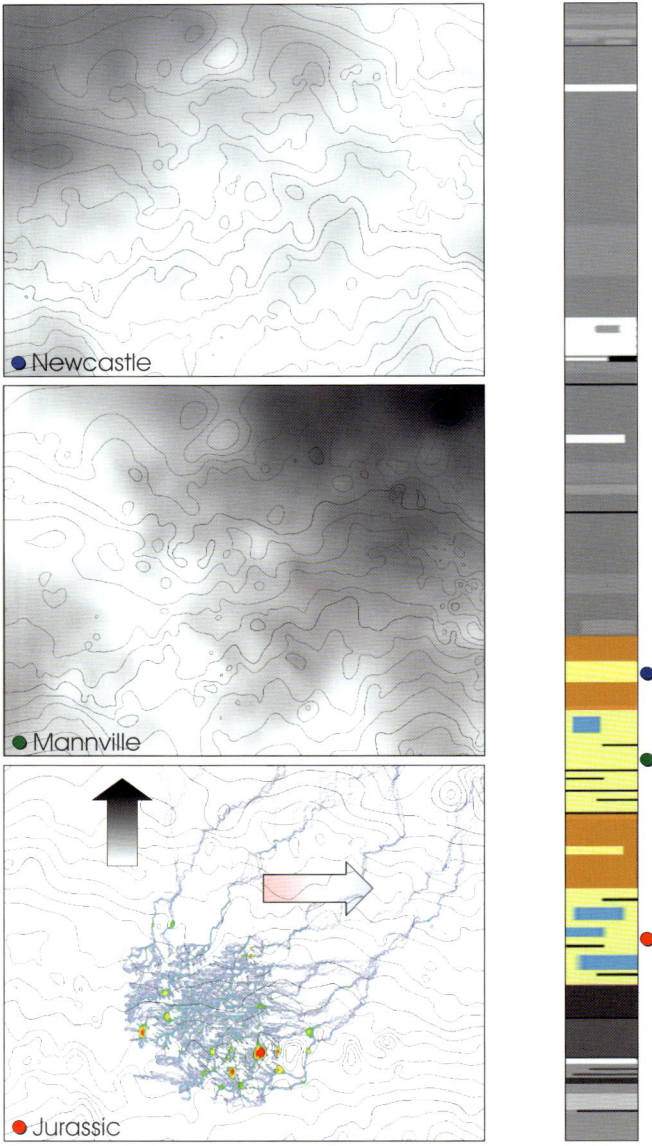

Figure 3-6. *Hydrodynamic results for the 1 micron scenario. The coloured arrow represents the regional overpressure trend (decreases to the east). The grey-scale arrow represents the northerly direction of hydrostatic migration as a result of regional dip. Migration pathways appear as light blue, pools are spectral (red-yellow-green), aquifers with no CO_2 migration appear in grey scale. The stratigraphic column on the right indicates the aquifer level.*

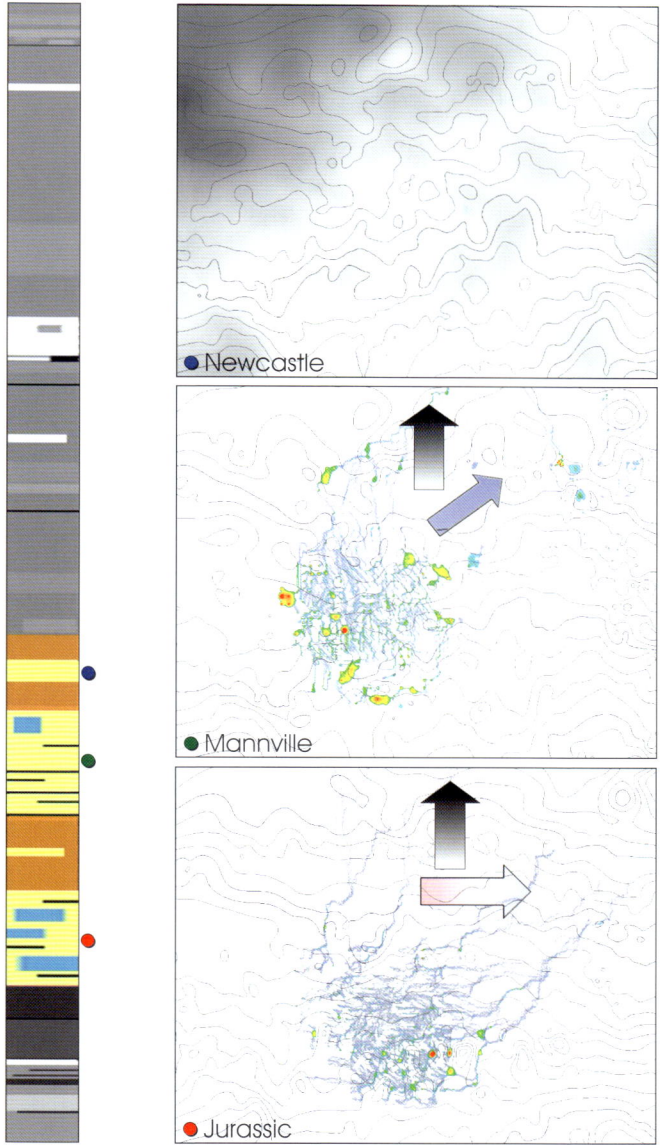

Figure 3-7. *Hydrodynamic results for the 3 micron scenario. Coloured arrows indicate regional overpressure trends (decreases to the northeast for the Mannville, to the east for the Jurassic). Grey arrows represent the northerly topographic influence for hydrostatic migration. Migration pathways appear as light blue, pools are spectral (red-yellow-green).*

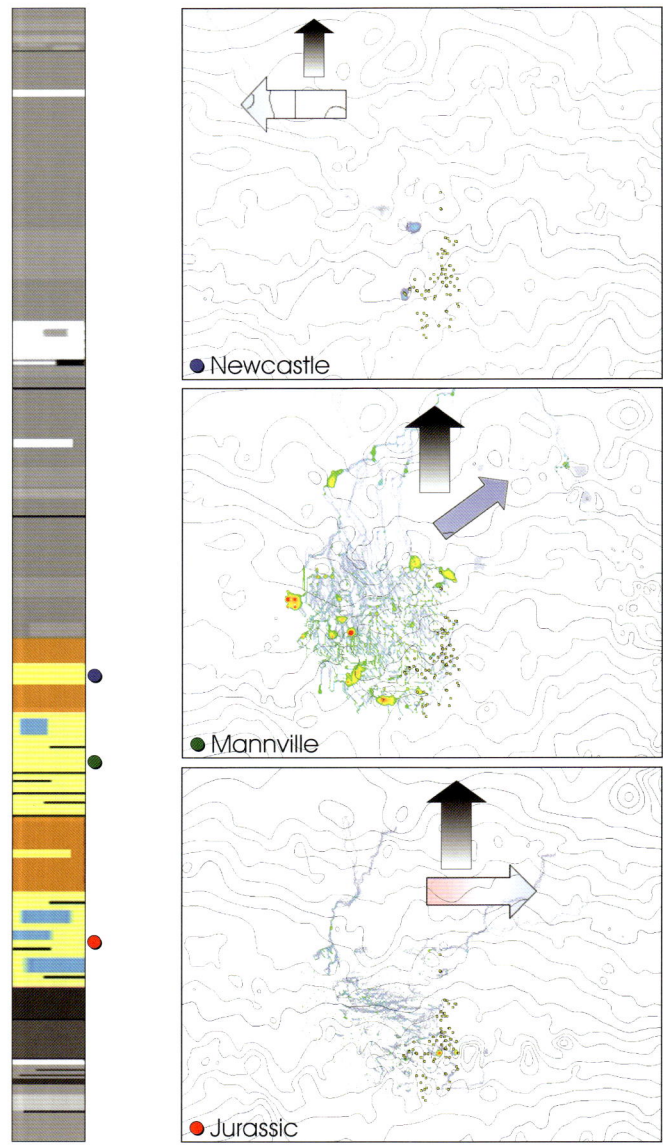

Figure 3-8. *Hydrodynamic results for the 8 micron scenario. Coloured arrows represent regional overpressure (decreases to the west for the Newcastle, to the northeast for the Mannville, and to the east for the Jurassic). Grey arrows represent the hydrostatic migration trend against regional dip. Pathways are light blue, pools are spectral (red-yellow-green).*

slightly to about 2.4 Mt, relative to the previous scenario. The overall secondary storage potential again increases to almost 3 Mt, with the Mannville playing the most significant role. However, the breaching of the Colorado aquitard indicates that the 8 micron scenario is the critical point for potential incursion of CO_2 into the Belly River, the shallowest aquifer within the geosphere.

Simulation results indicate that the hydrodynamic influence remains the same for the Jurassic and Mannville aquifers, although migration pathways away from the storage site area are much reduced due to deep pools breaching vertically, prior to spill-point. The hydrodynamic situation in the Newcastle aquifer results in a northwesterly migration pattern, as evident in the single pathway that spills out of the more northerly of the two pools above the storage site area. The cluster of yellow dots, east-southeast of the two pools, represents breach points for the Colorado aquitard.

SHALLOW BREACH MIGRATION: The 8 micron fracture network scenario, at the level of the Newcastle aquifer, gives an indication of where CO_2 would migrate under hydrodynamic conditions, if it were to reach the Newcastle aquifer via wellbores from below (Figure 3-9). Simulation results indicate that that CO_2 reaching the Newcastle aquifer is likely to be constrained to the eastern margin of the storage area, in contrast to the underlying Mannville and Jurassic aquifers. The simulation also indicates that, for a shallow breach, the Newcastle fails to retain the CO_2, with only two pools trapping the CO_2 at this level. The pool situated in the northeast quadrant of the storage area does not breach, but spills along a migration pathway to the northwest, with no further accumulations prior to exiting the modelled area. In contrast, the pool in the southeast quadrant breaches vertically. The remaining CO_2 bypasses the aquifer via 75 wells clustered on the eastern edge of the storage area.

This scenario indicates that the thick Colorado Group aquitard could breach for 8 micron fracture networks. Because of this, detection of shallow CO_2 might be expected to occur within the acreage of the Weyburn Field and to the east of the storage site.

SUMMARY: The hydrodynamic model significantly alters the model outcomes with respect to the hydrostatic case. The two most notable differences are that (1) non-hydrostatic pressure gradients within aquifers force the migration pathways from a simple topographically driven direction, and (2) the breaching locations change from a broad northwesterly corridor beyond the Weyburn Field to a much more constrained cluster of locations on the eastern margin of the storage site area within the Weyburn Field.

CO_2 Sourced Below the Watrous Formation

To better understand the effects of sourcing CO_2 in the injection zone below the Watrous aquitard, the hydrodynamic model was modified (downward) to include an 'attic space' (the combined Poplar and Ratcliffe aquifers) and the underlying seal (Midale Evaporite). The lowest layer of this model is an 18 m thick base layer, roughly equivalent to the Midale reservoir. The addition of the attic space allows exploration of the influence of the Poplar and Ratcliffe wedge geometry, as well as the subcrop termination of potential breach point locations above the Watrous.

The topography of the cap rock seal for the attic space was defined using data for the Base Altered zone from the geological mapping. Given the position of the subcrop beneath the Watrous aquitard in the project area, the attic space has a restricted subcrop expression that covers much of the southern half of the region. The attic space was populated with porosity, permeability, pressure, temperature, and salinity data. Global estimates of 0.06 porosity and 1 mD permeability from Phase 1 were used. A small amount of noise was added to the homogeneous permeability field, equivalent to a log-normal range of 0.5–5 mD (mean, 1 mD). This was required because a monotonic permeability results in a strong artifact in the pressure solution outcome. The initial CO_2 source term

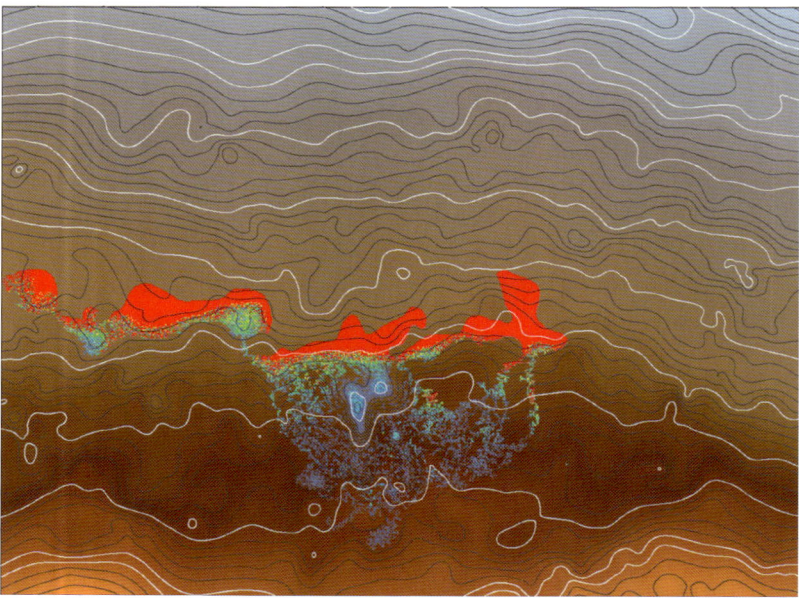

Figure 3-9. *Wells associated with potential breach points at the level of the Colorado aquitard. The yellow square is the location of the CO_2 storage area, and the source term of 1000 wells above the Watrous, relative to the outline of the field. Well locations (red points) are clustered along the eastern edge of the CO_2 storage area.*

for the attic space was defined to be all well penetrations of the Midale Evaporite within a designated 75-well pattern area, as noted previously.

FILL-SPILL ANALYSIS: A fill-spill analysis of CO_2 migration (not shown) identified six large pools along the subcrop edge, which fill with column heights of around 10–35 m. Many migration pathways braid and merge immediately above the designated source points. The general migration trend is northerly to the subcrop edge, and then a fill-spill cascade to the west along the subcrop edge. A stochastic analysis confirmed the stability of the pattern. Some smaller pools occur over the storage site area, with smaller columns in the range of 1–10 m. The CO_2 migration and accumulation pattern is confirmed by a high-resolution 3D mesh simulation of the attic space (Figure 3-10). Combined, the six large pools represent a potential attic space storage mass of about 1×10^6 t. Additional small accumulations bring the total up to around 1.4×10^6 t of potential storage. The relatively modest secondary capacity of the attic space reflects the poor porosity of the Ratcliffe and Poplar units, at around 6%, and the related low gas saturations within the pools.

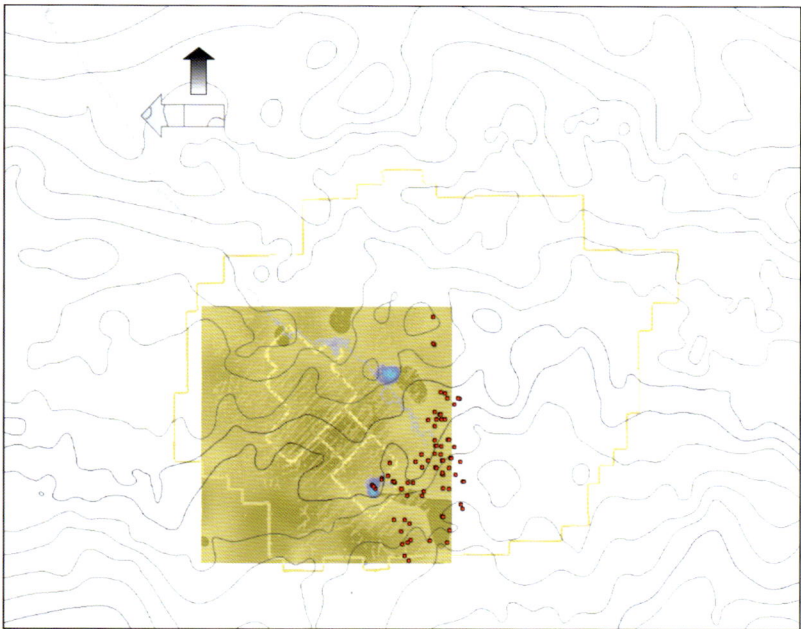

Figure 3-10. *3D mesh simulation of the attic space, showing migration (blue) and low saturations above the 75-well pattern area (green-yellow). Maximum saturations for pools (red) occur along the subcrop, related to higher columns of gas.*

HYDROSTATIC RUN: The breach criteria for the Watrous aquitard (attic space seal) are the same as for earlier scenarios. No breach occurred for lower well-path permeabilities in the Watrous aquitard. Instead, CO_2 spilled to the west (Figure 3-11). The first potential CO_2 breach occurred for a 1 micron scenario, but only at three well locations This represents a significant change in the model with respect to the source term. The previous set of hydrodynamic models assumed 1000 source points in the Jurassic aquifer, addressing the possibility of any well leaking above the storage site. While this approach allows for a general assessment able to identify all likely pooling locations (Jurassic, Mannville, and Newcastle), the influence of the attic space dramatically reduces the potential source points for a 1 micron scenario. As such, the earlier scenario does not address the behaviour of a small number of Watrous source points. However, given the sensitivity of the earlier simulations to hydrodynamics, the attic space scenario needs to include the aquifer pressure regime and heterogeneity.

HYDRODYNAMIC RUN: Under hydrodynamic conditions, the attic space results are effectively unchanged from the hydrostatic case. Low aquifer permeability (mean 1 mD, and assumed log-normal distribution) and a moderate overpressure regime (about 1.5 MPa over 50 km) results in a weak hydrodynamic system

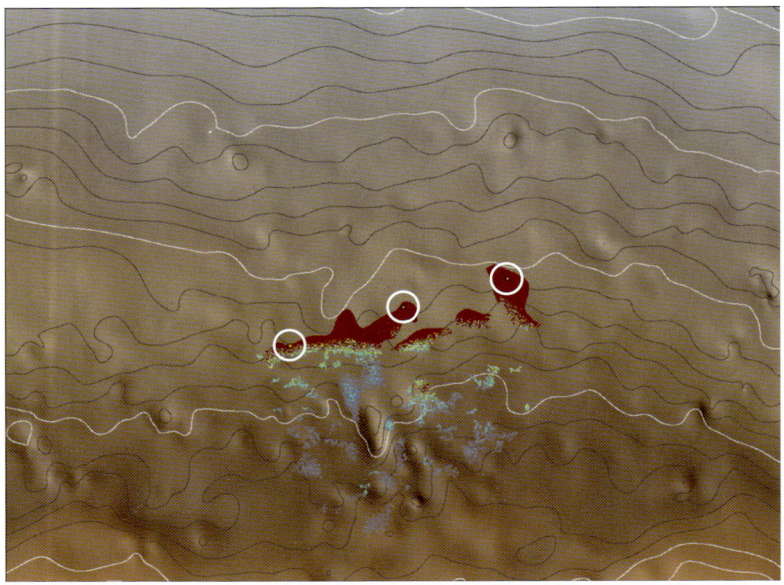

Figure 3-11. *Hydrostatic simulation for a 1 micron fracture scenario. Breaching occurs at three well locations (circled). The breaching prevents a westerly fill-spill migration, and reduces the potential source points above the Watrous Formation.*

and low formation water velocities (a few millimetres per year). Small changes in the migration path result in the loss of two small pools at the subcrop edge between the larger pools associated with leaking wells (not shown). However, this does not influence the outcome of a restriction in the number of potential breach locations close to the subcrop edge. As before, a tipping point is reached for the 8 micron scenario (155 wells leaking directly over the 75-well pattern area, associated with the area of densely configured wells).

Although the flow behaviour of the model is largely unaffected by the addition of the attic space (Poplar and Ratcliffe beds), the potential sources for leakage into the Jurassic aquifer and shallower beds are significantly restricted due to the influence of attic space trap geometry, resulting in a small number of large pools along the subcrop edge with significant column heights (Figure 3-12).

3.5 Storage Capacity and Mass Partitioning

3.5.1 Introduction

The intrinsic capacity of a CO_2 storage site is dependent on many key properties of the reservoir–seal system. Particularly important are the volume of the reservoir and the various factors that control partitioning of CO_2 among the

Figure 3-12. *Watrous aquitard breach points for attic space flow modelling. The 75-well pattern area (green) is the initial source term. The attic space breaches at 3 locations within the Weyburn Field (outlined) for a 1 micron scenario (blue), 5 locations for a 3 micron scenario (yellow), and 155 locations for an 8 micron scenario (red).*

Mass (% total)	Case I	Case II	Case III	GOR = 6000
Stored in gas	212 750 (56.3)	28 036 (15.7)	275 630 (52.9)	276 061 (56.5)
Stored in oil	82 286 (21.7)	78 813 (44.2)	141 950 (27.2)	95 328 (19.5)
Stored in water	83 080 (22.0)	71 524 (40.1)	103 940 (19.9)	117 024 (24.0)
Stored (% inj.)	378 116 (19.3)	178 373 (14.3)	521 520 (39.5)	488 413 (36.6)
Recycle (% inj.)	(80.7)	(85.7)	(60.5)	(63.4)

Table 3-3. *Summary of the partitioning of CO_2 mass (kt) within Pattern 1 at the end of EOR (2025).*

physical and chemical parameters of the geological setting. Within the context of these physical and chemical constraints, storage capacity evolves with changing pore volume and the partitioning of emplaced CO_2 among disparate physical (hydrodynamic, structural, residual) and chemical (hydrocarbon, aqueous, and mineral) trapping mechanisms. The pore volume changes are largely the result CO_2–fluid–rock transfer.

Two modelling studies will illustrate the methods for estimating local (single-pattern) CO_2 storage capacity and dynamic CO_2 mass partitioning. In the first, standard upscaling techniques were used, in the context of traditional compositional reservoir simulation (where only fluid-phase partitioning is considered), to provide an estimate of long-term cumulative storage capacity at Weyburn. In the second, reactive transport modelling was used to quantify storage capacity and dynamic mass partitioning on a local scale. This method considers fluid-phase sinks, mineral trapping, and reaction-dependent changes in porosity and permeability. The latter approach is not commonly applied in simulation efforts that support CO_2-EOR operations, but, as will be demonstrated, it provides important advantages in the context of modelling CO_2 storage activities.

3.5.2 Storage Capacity

The CO_2 storage capacity of Pattern 1 was simulated for three distinct EOR cases, each of which spanned the period 2000–2025. Case I was a single pattern simulation with no gas-oil ratio (GOR) control, Case II stopped CO_2 injection at 50% hydrocarbon pore volume, and Case III was the same as Case I but with separate GOR control for vertical and horizontal producers. The results of this simple case are given in Table 3-3.

3.5.3 Mass Partitioning

The dynamic mass partitioning of CO_2-trapping mechanisms strongly influences the long-term isolation performance of geological storage sites. Aspects considered include (1) a review of the types of trapping mechanisms, (2) an outline of fundamental system controls on dynamic mass partitioning, (3) a description of the distinct partitioning dynamics of end-member structurally open and closed systems, and (4) a discussion on these processes as they relate to isolation performance (Johnson, 2008). Both single- and full-field models were run, with emphasis on those elements that are most important to storage site planners and operators. Of necessity, only a brief overview is possible here.

Conceptual Overview

During CO_2-EOR and CO_2 geological storage in saline aquifers (that is, aquifers with saline formation water), the CO_2 mass is partitioned among a variety of physical and chemical trapping mechanisms. Physical mechanisms isolate CO_2 as a separate dense phase, capable of moving by virtue of its density (as in hydrodynamic trapping), or it may be immobile, as in structural (or residual) trapping. In both cases, there are pressure changes that may affect the reservoir seal. In contrast, chemical trapping mechanisms isolate CO_2 in dilute form as a phase component, which may be as solute concentrations within aqueous, hydrocarbon, and solid solutions (minerals). In solubility trapping, these concentrations are initially mobile, while in mineral trapping, they are immobile. Except for the case of hydrocarbon solubility, these downwardly mobile (aqueous solubility) or immobile (mineral precipitation) chemical sinks do not create a pressure perturbation on the reservoir seal.

In addition to the variables used in the models described previously (for example, hydrocarbon pore volume and GOR), other parameters that were considered include, but are not limited to, the following.

1. The effective volume ratio (C/I) of structural closure to total CO_2 injection –where C is the product of total closure volume, average porosity, and average CO_2 saturation, and I is the injection volume at reservoir conditions.
2. The vertical/horizontal (V/H) closure aspect ratio, and the fracture/matrix permeability ratio (F/M) – the mass ratio of physically- to chemically-trapped CO_2 increases with C/I, V/H, and F/M.
3. Reservoir composition – the mass ratio increases with formation water salinity, and decreases with aqueous solubility of CO_2 and lower hydrocarbon API gravity. The mass ratio also increases with decreasing bulk mineralogical concentrations of carbonate-forming cations (mainly Ca and Mg) which decrease the mineral trapping potential.

As will be appreciated, the long-term evolution of mass partitioning among distinct physical and chemical CO_2 sinks is decidedly not a one-size-fits-all case. For example, end-member structurally open and closed systems exhibit very different mass-partitioning changes, and hybrid settings span these extremes.

Superimposed are a variety of operational controls. Here, only one example is illustrated. Two scenarios are for an identical 10 000-year post-injection situation, as shown in Figure 3-13. The bottom diagram shows the end-member open case (C/I << 1, laterally extensive) and the top diagram the closed case (C/I >1, maximum realistic V/H closure aspect ratio).

For the closed case, structural trapping dominates, decreasing only slightly with time, through gravitational evolution of the vertical CO_2 saturation profile.

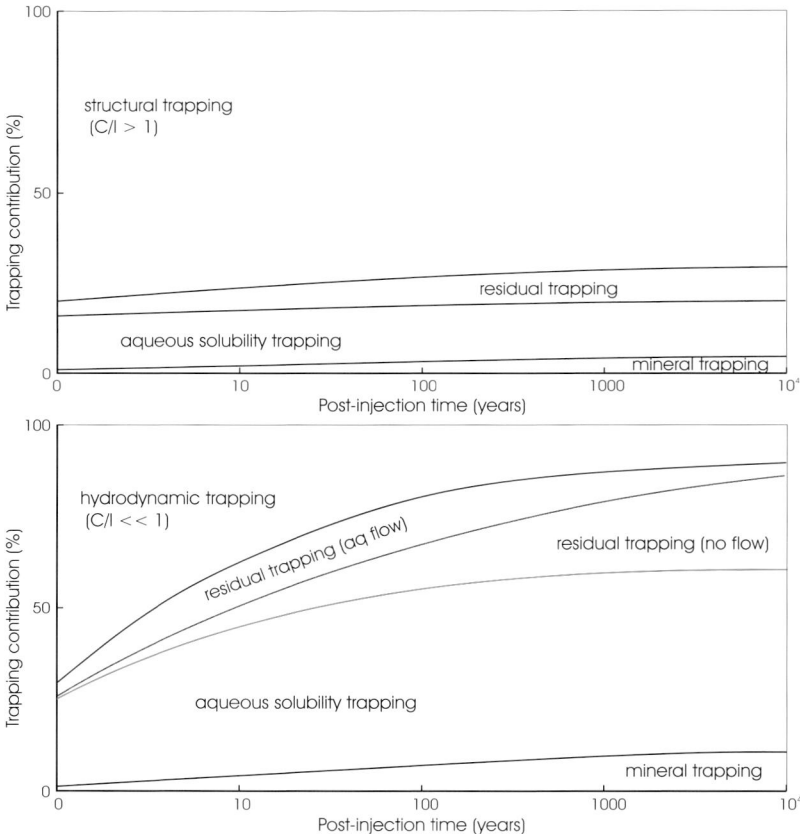

Figure 3-13. *Post-injection mass partitioning for selected end-members.*
Above. *Structurally closed (C/I >1) with maximum V/H closure aspect ratio.*
Below. *Structurally open (C/I << 1) with laterally extensive reservoir and seals. The upper blue and green lines delineate residual trapping for the case of insignificant regional aqueous flow, and the upper blue and purple lines delineate residual trapping for the case of vigorous regional aqueous flow.*

This leads to a slight increase in residual trapping below the CO_2–water contact. Aqueous solubility trapping is nearly constant, reflecting minimal direct interaction of the CO_2 column with underlying formation water. Mineral trapping is minor, but note that it may significantly improve the containment component of isolation performance by enhancing the seal integrity of typical shale cap rocks (Johnson et al., 2004, 2005).

For the structurally open end-member, the initial dominance of hydrodynamic trapping declines with unrestrained lateral up-dip plume migration. This expands the plume-water interface, and so increases solubility trapping, leaving behind an expanding wake of zones of residual trapping – whose long-term stability is strongly dependent upon vigour of the ambient flow field. As a result, the mass ratio of hydrodynamically to aqueous-solubility trapped CO_2 continuously decreases. In the case of weak regional flow, the initial increase in mass contribution of residual trapping will continue. However, with a strong flow field, residual trapping will eventually decline (potentially to elimination) through progressive consumption by solubility trapping. This important variation is illustrated by distinct 'no flow' and 'aq flow' projections in the figure. Mineral trapping is again minor in terms of mass contribution, although likely more significant here than in the closed systems in the expanded reservoir and plume–seal interface source regions of carbonate-forming cations.

Structurally Open versus Closed Systems

The preceding analysis provides a reference frame for addressing one of the fundamental questions associated with CO_2 geological storage – which generic setting provides optimal isolation performance: (1) a structurally open (laterally extensive) system, or (2) a structurally closed system? The pros and cons of each setting can be evaluated in terms of capacity, footprint, containment, and risk.

STRUCTURALLY OPEN SYSTEMS: *Advantages:* capacity (especially for large-scale injection, and extensive lateral dimensions), containment (due to relatively small pressure perturbation), wellbores (small potential impact), and relatively small ultimate mass ratio of physically-to-chemically trapped CO_2 (minimal mass fraction of concentrated, gravitationally upward mobile CO_2). *Disadvantages:* footprint (due to unconstrained lateral plume migration), risk (due to relatively large uncertainties in site characterization, especially in regard to the lateral extent of the primary seal), and minimum well densities (affecting dependent modelling forecasts).

STRUCTURALLY CLOSED SYSTEMS: *Advantages:* footprint (due to lateral confinement), and risk (as defined herein, especially for oil reservoirs because of relatively well constrained site characterization and dependent model forecasts using extensive well data). *Disadvantages:* capacity (especially for large-scale injections, because of their lateral compartmentalization), containment (due to the relatively large pressure perturbation), wellbores (large potential impact), and relatively large mass ratio of physically-to-chemically trapped CO_2 (maximum mass fraction of concentrated, gravitationally upward mobile CO_2).

There is a caveat to this conclusion, namely, that site characterization and dependent modelling forecasts must be continuously improved by sequential incorporation of time-lapse geophysical and geochemical monitoring results that will be obtained over the anticipated dynamic footprint. Structurally closed systems, especially active or depleted oil reservoirs, represent appropriate targets for smaller-scale injections, given their relatively well constrained site characterization and limited footprint, both of which permit more detailed calibration of integrated modelling and monitoring capabilities.

For projects that involve both initial CO_2-EOR and subsequent long-term CO_2 storage, it is important to understand the dependence of long-term CO_2 mass partitioning on the range of distinct operational strategies that will be deployed during initial EOR, and to take these dependencies into account during project design. Fine-tuning specific parameters of these strategies has a large impact on long-term CO_2 mass partitioning and storage performance, as it does on initial EOR performance, of course. Thus, both dependencies should be evaluated in tandem. For existing EOR operations, such analysis is still important, because the manner in which even the final stages of the EOR are implemented will still have a sizeable impact on ultimate storage performance.

A series of models were run to evaluate CO_2 mass partitioning, for both single-pattern and full-field domains at Weyburn. Many of the variables noted previously were part of the various scenarios examined.

In conclusion, the ultimate CO_2 storage capacity of a given reservoir and seal system is physically dependent upon its structural, compositional, transport, and hydrological characteristics. Most important among these are the structural closure volume, mineralogy, aqueous salinity, porosity, permeability, ambient flow fields, and pressure-temperature conditions.

Using the same reservoir simulation and reactive transport models employed to predict CO_2 migration, CO_2 storage capacity and dynamic mass partitioning at Weyburn have been forecast for both single-pattern and full-field domains. Interestingly, the relative utility of these models is somewhat reversed in this application, relative to the migration case and the design of monitoring programs. Specifically, although single-pattern simulations represent the primary method for tuning field-scale process models that control partitioning dynamics, it is full-field simulation that is of greatest utility to operators in forecasting the convergence of overall CO_2 mass partitioning, as an integral part of predicting storage capacity.

Results from Weyburn demonstrated the strong dependence of CO_2 storage capacity and mass partitioning on operational strategies during both

CO_2-EOR and post-EOR phases, spanning a diverse set of CO_2 and water injection schemes, hydrocarbon pore volume protocols, and GOR shut-in constraints. Equally significant dependencies on the geochemical model used and the values for key parameters – such as the mechanical dispersion coefficient – are also demonstrated.

3.6 Containment

3.6.1 Reservoir Integrity

Reservoirs where CO_2-EOR is practiced invariably have secure hydrodynamic (vertical) containment of the remaining hydrocarbon accumulation and secure lateral containment ensured by well-characterized structural controls, despite the presence of hundreds or thousands of production and injection wells. However, in the context of CO_2 storage, long-term containment is by no means assured, given what is known now about well integrity. Further, many aquifers in which CO_2 may be injected are laterally unconfined. Hence, long-term integrity of natural seals represents the most important constraint on isolation performance. Accurate long-term forecasting of reservoir integrity is also critical.

A variety of available geochemical models may be used to calculate the mass of material removed and deposited through water-mineral reactions. Detailed knowledge is required of both the chemistry of the formation water and the minerals of the rocks. Based on work reported in Chapter 4 and complementary work elsewhere (Johnson, 2008; Johnson et al., 2004, 2005), discussion here will be limited to presentation of Figure 3-14, which shows a three-dimensional view of calculated amounts of calcite dissolved from the Weyburn reservoir after 2 years of CO_2 injection. This will serve as an example of the type of information that becomes available through geochemical modelling.

The figure shows that after 2 years of CO_2-EOR, calcite dissolution has occurred in all gridblocks that contain and immediately surround CO_2 injectors, which are the only wells depicted. These gridblocks have lost of the order of 1 mol of calcite per cubic metre of rock, which is volumetrically insignificant. When extrapolated over 25 years of planned injection, and taking into account preferential dissolution along fracture surfaces, the resultant increase in porosity and (especially fracture) permeability may be significant.

Indirect evidence that this order-of-magnitude simulation result for calcite dissolution is reasonable can be obtained through the changes in formation water composition, predicted by the model, and observed by the monitoring program. The anticipated magnitude of this change can be estimated from that of the simulated calcite dissolution. For a rock having 10% porosity, each cubic metre contains 100 L of solution. Simulated calcite dissolution will result in about 40 g of Ca being added to each 100 L, or 400 mg/L. This magnitude is comparable to observed shifts in the calcium concentration seen at impacted wells (of the order of 500–1200 mg/L). The impact on the bicarbonate content will be about twice that for Ca (on a molar basis), or about 1200 mg/L, which is also in reasonable agreement with measured results.

Although the extent of carbonate dissolution during CO_2-EOR obtained by this modelling study is minor, its potential impact on matrix and especially fracture permeability is indirectly underscored by the composition of carbonate reservoirs. Specifically, CO_2-induced dissolution of the characteristically small amounts of non-carbonate minerals (silicates, etc) is unlikely to hydrolyze sufficient silica and alumina to facilitate significant (if any) precipitation of non-calcite-group carbonates (for example, dawsonite [$NaAl(OH)_2(CO_3)$]) or quartz in coupled dissolution–precipitation reactions. Thus, CO_2-induced mineral precipitation which might counterbalance (or even exceed) the effect of carbonate dissolution on the porosity and permeability is not likely to occur. However, it is certainly conceivable that this scenario could in fact play out where high silicate concentrations are localized, such as in the uppermost Three Fingers zone of the Marly unit. If that were to be the case, this would likely be an advantageous development, in terms of vertical CO_2 containment in the context of both CO_2-EOR and CO_2 storage at Weyburn.

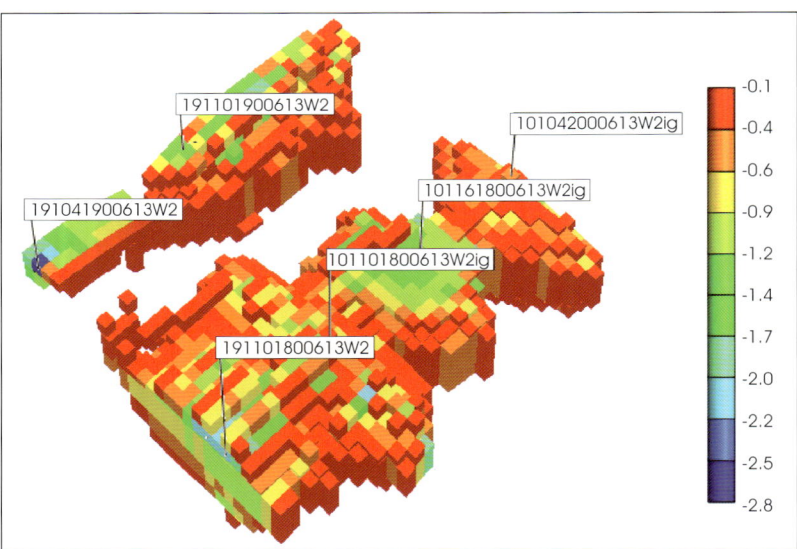

Figure 3-14. *Three-dimensional view of the calculated amounts of calcite dissolved per gridblock in Pattern 3 following 2 years of CO_2 injection. Gridblocks that are not visible have had <0.1 mol of calcite per cubic metre dissolved from them. Only the CO_2 injectors are shown.*

3.6.2 **Seal Integrity**

Predicting the long-term changes in permeability of reservoir seals during CO_2-EOR and CO_2 storage requires first identifying, then quantifying, its dependence on key parameters and processes. The most important factors influencing this change are, for convenience, subdivided into three groups: (1) intrinsic seal properties, (2) chemical conditions at the reservoir–seal interface, and (3) the pressure perturbation associated with CO_2 injection.

Relevant seal properties include geochemical characteristics, such as bulk mineral contents of carbonate-forming cations (principally Ca, Mg, Fe, Na, and Al), as well as geomechanical parameters, such as fracture normal stiffness. Cation contents represent the primary control on geochemical alteration processes. Chemical conditions at the reservoir–seal interface, which are dictated by the injection-stream (impure CO_2) and reservoir fluid-rock compositions, exert a secondary control. Magnitude and duration of the injection-induced pressure perturbation represents the fundamental control on geomechanical deformation processes. These also depend on the CO_2 influx and on reservoir parameters such as permeability, lateral continuity, compartment height (for structurally closed settings), depth, and thickness.

In the context of these dependencies, long-term enhancement or degradation of seal integrity hinges on the relative contributions of geochemical alteration (which may reduce or enhance microfracture apertures, depending upon seal composition), and geomechanical deformation (which incrementally widens and narrows microfractures during periods of pressure increase and decrease, respectively). As a result, long-term performance forecasting requires a predictive capability that quantifies this pivotal interplay of geochemical and geomechanical processes, which have highly disparate kinetics. With respect to Weyburn, we report only some aspects of the integrity of evaporite seals.

Evaporite Seals

Evaporite formations, principally anhydrites – such as the Midale Evaporite at Weyburn – serve as highly effective seals, isolating hydrocarbon accumulations in a variety of structural settings, often in association with carbonate reservoirs. Moreover, their effectiveness in isolating natural CO_2 accumulations for more than 10 000 years has been well documented (for example, Pearce et al., 1996; Baines and Worden, 2004; Rochelle et al., 2004), the best known example being the commercial Bravo Dome, which initially held about 225×10^6 m^3 of CO_2 (Roberts and Godfrey, 1994). These field observations are consistent with results from laboratory experimental investigations of CO_2–formation water–anhydrite interactions (for example, Wolterneek, 2010) and thermodynamic models of phase equilibria for the system H_2O–CO_2–$NaCl$–$CaCO_3$–$CaSO_4$ (Li and Duan, 2011).

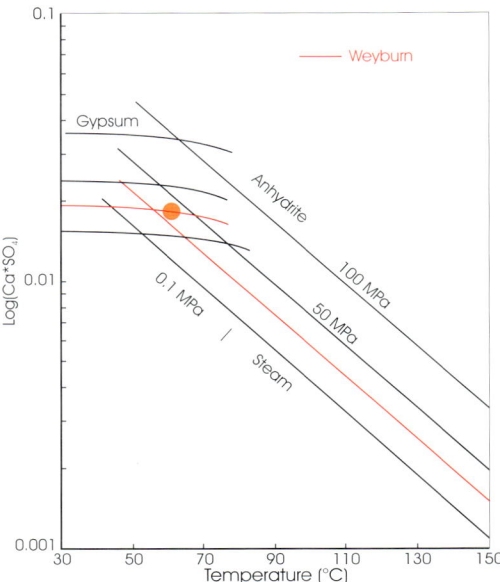

Figure 3-15. *Solubility of gypsum and anhydrite as a function of temperature for several isobars, including that for the Weyburn Field (in red). The solid orange circle denotes the measured temperature of the Weyburn reservoir, and average fluid composition obtained from a well sampled during the geochemical monitoring program.*

The solubility of anhydrite is effectively independent of CO_2 concentration in dilute aqueous solutions, compared to those equilibrated with supercritical CO_2 at typical reservoir pressure-temperature conditions. Because of the stability of anhydrite in the presence of CO_2-rich formation waters, concerns about anhydrite-to-calcite conversion is not an issue in the context of CO_2 storage (specifically, this relates to its deleterious effect on seal integrity as a consequence of the 20% volume reduction).

Although anhydrite-to-calcite conversion within the Midale Evaporite is highly unlikely, the transition from anhydrite to gypsum is quite plausible in view of pressure-temperature conditions in the reservoir. Specifically, the pressure-temperature dependence of gypsum and anhydrite solubility suggests that, although anhydrite is stable in the reservoir (slightly supersaturated), the local fluids are also in equilibrium with gypsum (Figure 3-15). The phase relations suggest that even a small temperature decrease (about 2–3°C) would destabilize anhydrite and effect its conversion to gypsum. Thus, even very minor cooling, associated with CO_2 or water injection, could potentially catalyze this conversion (for example, along seal microfractures). The conversion is characterized by a standard molar volume increase of about 60% and weakened mechanical stability, both of which can be considered advantageous with respect to seal integrity in the relevant context of constant confining stress.

In fracture flow experiments involving Midale Evaporite core and representative pressure, temperature, and confining stress conditions, pervasive anhydrite-to-gypsum conversion was observed both along fracture surfaces and throughout an extensive diffusion ('healing') zone within bounding matrix blocks. Further, this replacement process was indeed associated with mechanical deformation of the core, as anticipated from knowing the differing mechanical properties of anhydrite and gypsum.

For information on the integrity of shale seals, as affected by injected CO_2, the interested reader is referred to Pearce et al. (1996), Baines and Worden (2004), and Rochelle et al. (2004).

3.7 Recommendations

Models simulating reservoir composition and geochemical reactive transport are especially well-suited to predicting the long-term isolation at geological CO_2 storage sites. They may be used to ensure (1) that there is sufficient CO_2 injectivity, (2) that the ultimate required capacity exists within target-reservoir lease boundaries, (3) that in-situ chemical and stress perturbations imposed by CO_2 emplacement will not overly modify reservoir integrity (anticipated lateral CO_2 migration), or degrade vertical containment barriers (natural and engineered seals), and (4) that uncertainty bounds on these simulation results are of acceptable magnitude.

3.7.1 Migration of CO_2

The fundamental advance of reactive transport modelling over traditional compositional reservoir simulation is its explicit representation of transfer processes among fluids and the rock mass (both aquifer properties and minerals). These models can elucidate the impact of reaction-dependent changes in reservoir porosity and permeability, with respect to the integrity of the reservoir and seals. This impact is far greater during CO_2-EOR operations than other oil recovery techniques. In CO_2 storage operations it is possible to account for all CO_2–oil–formation water–rock/mineral interactions.

From a project design and risk assessment perspective, accurate forecasts of CO_2 migration rank high on the list of essential data. Reliable predictions of planned and potential movement of CO_2 are needed (1) to establish lease boundary requirements, (2) to construct effective reservoir, overburden, and near-surface monitoring programs, and (3) to demonstrate to stakeholders and the public that the long-term fate of injected CO_2 is understood within acceptable uncertainty.

3.7.2 Invasion Percolation Modelling

Invasion percolation models differ significantly in behaviour from traditional Darcy flow simulations, in that the simulation is dependent upon threshold and capillary pressures, rather than permeability and viscosity. The buoyancy of CO_2 promotes gravity segregation, causing the migrating plume to be sensitive to the topography of seals that overlie aquifers. This results in discrete fill-spill pathways of migrating and pooling gas, which can be identified using invasion percolating modelling. It has two important additional advantages over Darcy-type flow models. First, there is less loss of geological detail due to averaging or upscaling. Second, it is computationally faster because it assumes capillary-dominated flow, which allows for cellular modelling – the flow rates are typical of migration and provide a better approximation of non-continuum flow physics (capillarity, rather than viscosity flow), as well as very high resolution. Thus, where pooling of CO_2 occurs, invasion percolation modelling can determine both vertical and lateral breach points as a function of threshold pressure. The limiting condition for invasion percolation is a capillary force that exceeds the viscous force by a ratio 10 000 to 1.

An assumption that was not addressed in the Weyburn project is that the CO_2 is distinct from the formation water, that is, it is immiscible and insoluble in formation water. It has been shown in other studies that CO_2 is partially soluble in formation water over the range 1–5 wt. % with respect to formation water. Future research should probably address this assumption with respect to invasion percolation modelling.

3.7.3 Storage Capacity and Mass Partitioning

The storage capacity of a reservoir, with respect to CO_2, changes with changes in both pore volume and the mass partitioning of emplaced CO_2 among distinct physical (hydrodynamic, structural, residual) and chemical (hydrocarbon, aqueous, and mineral) trapping mechanisms. Full field-scale models are of greatest utility to site planners and operators in forecasting the convergence of overall CO_2 mass partitioning as an integral part of predicting storage capacity. Results of the Weyburn project show that it is essential to quantify the dependence of interdependent CO_2 storage capacity and mass partitioning on operational strategies during both CO_2-EOR and post-EOR phases. This allows the operator to evaluate the impact of various CO_2-water injection schemes, hydrocarbon pore volume protocols, and GOR shut-in constraints.

The ultimate CO_2 storage capacity of a given reservoir and seal system is physically dependent upon its structural, compositional, transport, and hydrological characteristics. Most important among these are the structural closure volume, mineralogy, aqueous salinity, porosity, permeability, ambient flow fields, and pressure-temperature conditions.

3.7.4 **Containment**
Long-term enhancement or degradation of seal integrity hinges on the relative contributions of geochemical alteration (which may reduce or enhance microfracture apertures, depending upon seal composition), and geomechanical deformation (which incrementally widens and narrows microfractures during periods of pressure increase and decrease, respectively). As a result, long-term performance forecasting requires a predictive capability that quantifies this pivotal interplay of geochemical and geomechanical processes, which have highly disparate kinetics.

3.7.5 **Program Integration**
Close integration of site characterization, modelling, and monitoring programs, through iterative inter-program feedback, is essential to achieving a defensible predictive capability. In particular, continuous minimization of discrepancies between predicted and measured results through refined modelling and monitoring (that is, successful history matching) is essential for long-term forecasting.

4

GEOCHEMICAL MONITORING

Summary

An essential component of both CO_2-EOR and CO_2 storage operations is the acquisition of a suite of geochemical monitoring parameters that can (1) document the continued secure isolation of injected CO_2 within the storage complex, (2) track CO_2 movement and phase partitioning within the target reservoir, and (3) calibrate the accuracy of modelling tools used to forecast long-term storage performance in terms of capacity and containment. This suite is conveniently divided into parameters within the overburden (near-surface soil gas and shallow groundwater) and those in the reservoir (formation water and oil properties). Determination of baseline conditions is essential, including both chemical and isotopic composition.

Soil gas sampling should take into account such features as topographic variations, floral heterogeneity, and geological lineaments. Soil gases can be monitored using a variety of methods, and spring and fall sampling is adequate.

Groundwater sampling may require a larger area than for soil sampling to take into account regional flow systems. Consideration should be given to developing a dedicated monitoring network if monitoring will be for long periods of time. Annual sampling is probably adequate.

Geochemical monitoring of reservoir fluids is principally concerned with changes in formation water composition. A sampling frequency of every six months is often sufficient. Data interpretation is usually straightforward in terms of CO_2–oil–formation water–rock interactions. Injection-induced signals are typically quite pronounced while variations due to natural processes are subtle, even when discernible.

Close integration of site characterization, modelling, and monitoring programs is essential for optimizing monitoring activities.

Recommendations arising out this research begin on page 152.

Authors: **J.W. Johnson** (Schlumberger Doll Research, Cambridge, MA, USA) and **B.J. Rostron** (University of Alberta, Edmonton, Alberta).

4.1 Introduction

The objective of a geochemical monitoring program is to provide appropriate geochemical information on initial baseline chemical conditions, and to support those involved with measuring possible migration of injected CO_2, including environmental considerations. There is thus a geochemical component to many parts of a CO_2-EOR or CO_2 storage project including, but not limited to, the rocks and fluids in and around the storage complex, and the overlying near-surface environment.

It is important to recognize that many of the reactions that take place in a reservoir – especially one that is subject to the injection of CO_2 – are kinetically slow. Therefore, about the only way to determine that they have taken place is to monitor the fluids and gases. Reactions involving the solution and precipitation of minerals are a good example. These can often be inferred by using many of the water–rock interaction programs available.

It should also be noted that the areal extent of any geochemical monitoring program is very dependent on the anticipated long-term migration of the CO_2 plume, as forecast by simulation studies – which, of course, require a geochemical component.

The design of a geochemical monitoring program therefore involves selecting appropriate parameters, sampling methods, and analytical techniques, to allow definition of the baseline conditions for the many materials present. The need for baseline conditions is both implicit in understanding future changes and as input to various numerical models. As injection of CO_2 proceeds, any changes in these parameters must be measured. There are no completed CO_2 storage projects, but geochemical monitoring must also be considered, possibly for several decades after injection ceases.

With respect to the Weyburn-Midale project, we report information on the design, deployment, and interpreted results of monitoring the shallow groundwater system, soil horizons, and both formation water and oil in the reservoir. The information from the research on shallow groundwater and soil gases effectively documents the potential influence of CO_2-EOR and CO_2 storage operations at depth on shallow water resources and near-surface soil horizons.

The geochemical database now available from work on the reservoir can be used to infer key fluid–fluid and water–rock reactions within the overall mass transfer process, and also as a resource for history matching for reactive transport modelling studies.

Attention was directed to those techniques not normally part of CO_2-EOR operations, but which should be part of CO_2 geological storage. The program was carried out in the context of existing and possible future regulatory requirements.

4.2 Groundwater Sampling
4.2.1 Introduction

A shallow groundwater water-well sampling program was initiated in 2000 to provide background groundwater chemistry for the project (Figure 4-1), with seven surveys conducted between 2000 and 2009.

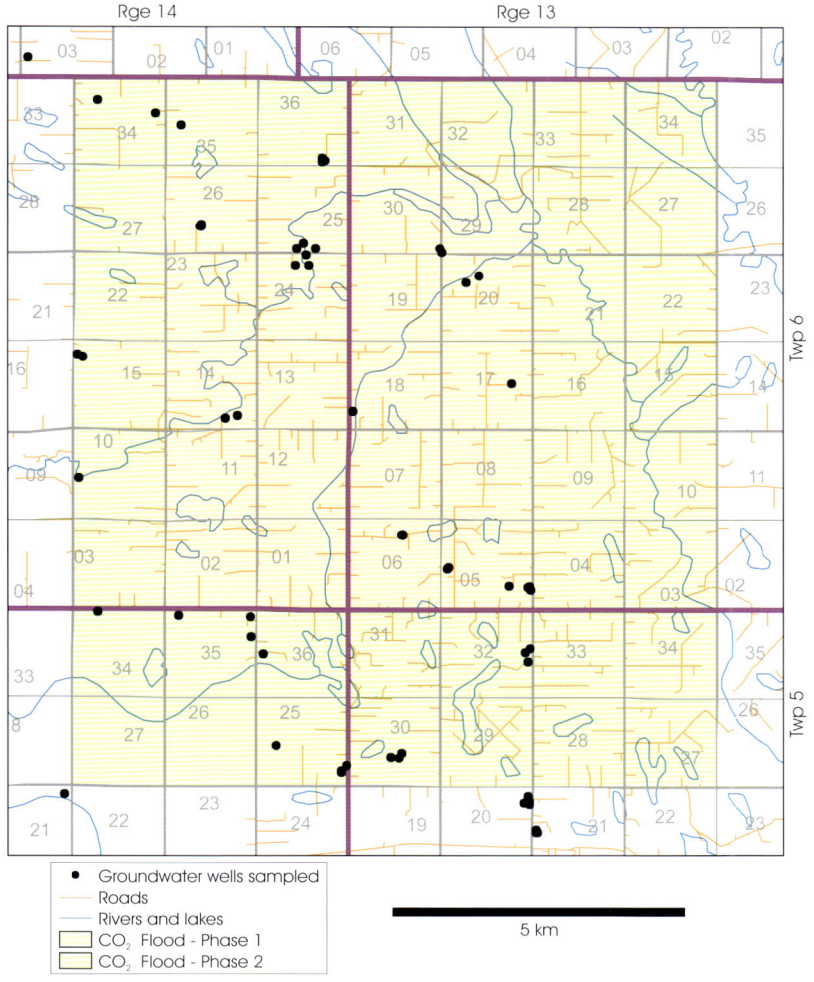

Figure 4-1. *Location of domestic and farm wells sampled during the project.*

4.2.2 Results

More that 60 shallow groundwater wells were sampled for water chemistry. The original intent was to start with a representative suite of wells to establish baseline water chemistry in the area, to be followed by regular sampling to monitor for any significant changes in chemical composition. However, this proved impracticable for several reasons, not the least of which was the switch by some groundwater users to water supplied by the Weyburn Utility Board to the city and nearby rural areas.

Groundwater composition in the area is highly variable, and is generally of the Ca–Mg–SO_4–HCO_3 type (Figure 4-2) with salinity generally in the range of 300–2000 mg/L, and typically elevated nitrate. A summary of the range in composition of individual ions is shown in Table 4-1, compared to data from elsewhere in Saskatchewan. Although several components exceed the Saskatchewan Drinking Water Standards (Saskatchewan Environment, 2006), the data are similar to those observed elsewhere in Saskatchewan (Ma and Morozov, 2010). There has been little change in the general quality of the groundwater over the sampling period, as illustrated in Figure 4-3. In particular, there was no significant long-term increase in CO_2 or HCO_3, both components that might reflect CO_2 breakthrough to the surface.

Parameter	Unit	Range	Standard/Objective	Exceedance No.	(%) *
Magnesium	mg/L	27–208	AO	1	4.1 (6.2)
Sodium	mg/L	9–710	AO	6	25 (30)
Sulphate	mg/L	59–1590	AO	18	75 (46)
Chloride	mg/L	7–340	AO	1	4.1 (8.5)
Nitrate (as NO_3)	mg/L	0.09–407	MAC	7	29 (21.6)
Iron	mg/L	0.005–10.9	AO	10	41 (64)
Manganese	mg/L	0.0012–1.9	AO	17	71 (70)
Arsenic **	µg/L	0.2–105	IMAC	3	12.5 (14.9)
Uranium**	µg/L	0.4–26	MAC	7	29 (14.5)
Zinc	mg/L	0.0036–1.41	AO	0	–
Salinity	mg/L	566–2980	AO	17	71 (51)
Total alkalinity	mg/L	252–668	AO	4	17 (26)
Total hardness	mg/L	303–1680	AO	12	50 (32)

Notes. * Per cent exceedance, as reported for various groundwater quality databases in Saskatchewan. ** Per cent exceedance for arsenic and uranium are based on well depths <30 m.

Table 4-1. *Range and exceedence of selected components of groundwater in Saskatchewan.*

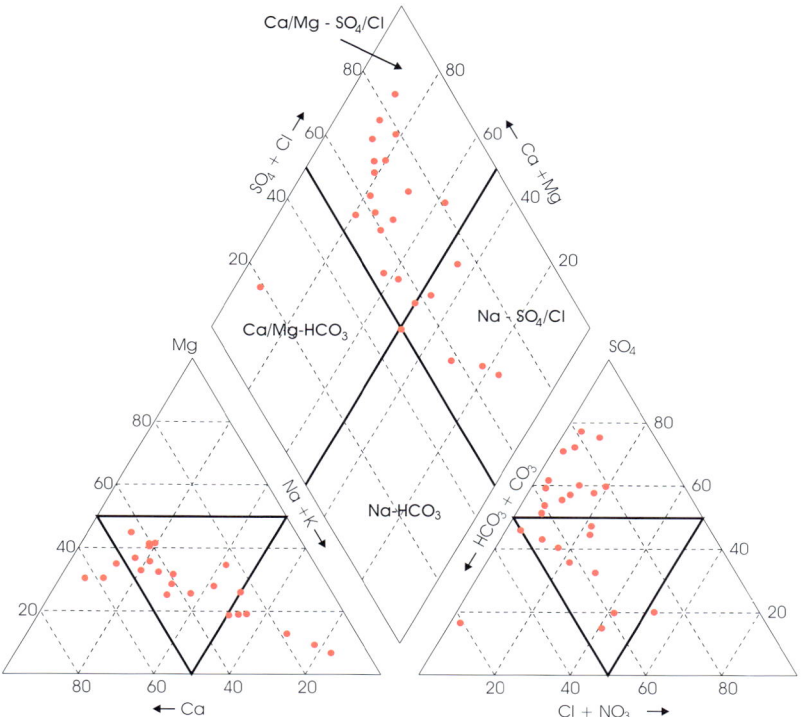

Figure 4-2. *Piper plot of water composition from domestic and farm wells in the project area.*

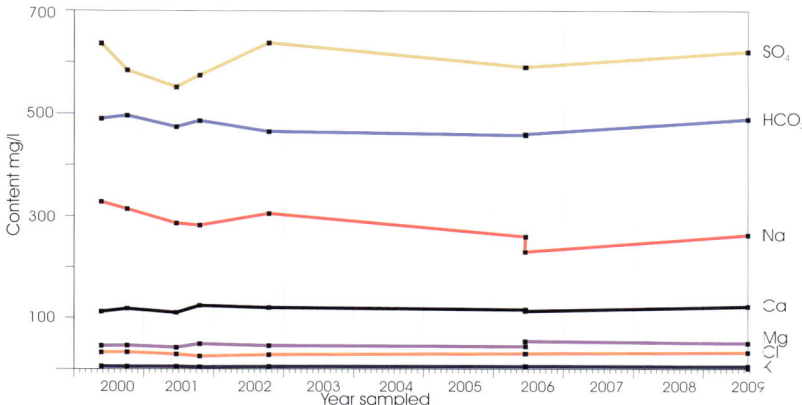

Figure 4-3. *Time-series for sampling results from a typical well in the project area.*

GEOCHEMICAL MONITORING 123

4.3 Soil Gas

4.3.1 Introduction

Soil gas monitoring provides an effective means of determining if any injected CO_2 has escaped the reservoir, migrated through overburden, and reached the near-surface biosphere. Clear demonstration that this unanticipated escape-and-migration scenario has in fact *not* resulted in a significant CO_2 release provides reassurance to both the site operator and general public. As a result, soil gas monitoring represents a likely requirement within future regulations governing CO_2-EOR and CO_2 storage operations.

In designing a soil gas monitoring program it is important to develop techniques that can cover large areas rapidly, often many square kilometres. Because possible CO_2 escape around the injection site may cover only a small area, the program should include a dense network of sample points near the injection site and a regional network for determination of the background values. Depending on the number of sample sites and the frequency of sampling, a program using both continuous and discontinuous measurements will be able to distinguish natural daily, seasonal, and annual baseline variations.

Continuous monitoring at specific locations may be necessary to track variations in soil gas concentrations, isotopic signatures, or surface fluxes that fingerprint CO_2. These fingerprints may be transient due to changes in pressure, temperature, soil moisture, precipitation, and wind, and might be missed by discontinuous measurements. Determination of carbon isotopes on gas components can often be used effectively to identify the source of CO_2.

During Phase 1 of the program, annual soil monitoring surveys were conducted from 2001–2005 and again in 2011, supplemented by a complementary study that deployed distinct measurement and modelling techniques.

Measurements included (1) soil gas content of CO_2, oxygen, argon, light hydrocarbons, helium, and radon, (2) stable and radiogenic carbon isotopes, (3) surface CO_2 fluxes, and (4) near-surface atmospheric CO_2 content. When interpreted in concert, these diverse data sets effectively document the presence or absence of injected CO_2.

During the program three techniques were employed: (1) discontinuous single-depth measurements, (2) discontinuous depth-profile measurements, and (3) continuous monitoring. Each is briefly described below, and the instrumentation and results are summarized in terms of CO_2 flux and carbon isotopes.

Discontinuous Single-depth Measurements

The standard procedure for collecting single-depth soil gas samples (Ciotoli et al., 1998; Annunziatellis et al., 2008) involves pounding a drive-point probe into the soil (see Figure 4-4 A, C). At Weyburn the depth was generally in the range of 60–90 cm. The gas content was determined using an infrared analyser. At some sites an additional subset of samples was taken for confirmatory laboratory analysis of CO_2, helium and radon.

All field instruments were calibrated prior to shipping to the Weyburn site. Sufficient sampling points were duplicated to assess the relative response of each instrument, and these results were used to standardize the measurements. These comparisons showed linear correlations with a difference in absolute concentrations of <5% (Figure 4-5).

Figure 4-4. *Soil gas sampling.* **A.** *Insertion of the probe.* **B.** *Infrared analysis of CO_2.* **C.** *Collection of gas sample in stainless steel canister for laboratory analysis.* **D.** *Buried continuous CO_2 monitoring probe at the end of the deployment period.*

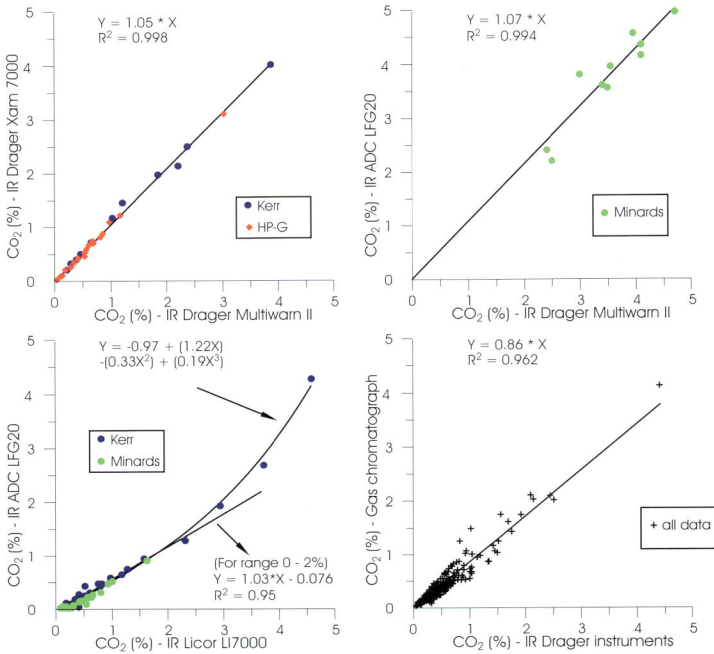

Figure 4-5. *Comparison of the CO_2 soil gas results obtained using various analytical instruments.*

Discontinuous Depth-profile Measurements

These measurements were made using horizontal wells similar to those described by Risk et al. (2002, 2008). The gas wells were installed in nested horizontal fashion from a pit large enough for an individual to sit in and operate a soil auger to dig the horizontal 3.75 cm diameter holes. For monthly samples, analytical repeatability was 1% or better.

Continuous Measurements

Continuous CO_2 monitoring measurements are ideal for documenting diurnal and seasonal variations in soil gas concentrations. At Weyburn, three such probes were buried in June 2011 and recovered at the end of October 2011, for an elapsed time of about 4.5 months. During that period measurements were made every 12 h (to evaluate seasonal, not diurnal, trends). The two different CO_2 sensors provided duplicate measurements, and two different ranges guaranteed good accuracy over a wide range of concentrations. Probes were placed at the bottom of the hole at a depth of 85 cm, and the hole backfilled with the removed soil. Holes were re-dug by hand to recover each unit (see Figure 4-4D).

4.3.2 Instrumentation

CO_2 Flux

The flux of soil gas CO_2 to the surface is a good measure of the productivity of the soil system. Unlike the content of soil gas CO_2, measured flux rates are not greatly affected by changes in gas transport because the conservation of mass dictates that produced gases need to come out, albeit with a small lag.

DISCONTINUOUS MEASUREMENTS: CO_2 flux at the soil–atmosphere interface was measured periodically using the closed-loop accumulation chamber technique (for example, Lewicki et al., 2005). This method involves placing an accumulation chamber on the ground surface (closed except for the bottom in contact with the soil) that is in a closed loop with an infrared analyser. A small pump circulates air from the chamber past the non-destructive analyser and back into the chamber to monitor the changes of CO_2 content over time. The slope of the change in content, together with the geometry of the accumulation chamber, is used to calculate the flux of CO_2 out of the soil and into the atmosphere. Measurement times ranged from 1–3 min., depending on the flux rate. Two types of flux meters were used at Weyburn (see Figure 4-6).

CONTINUOUS MONITORING: This method uses a forced diffusion chamber, which is a new soil CO_2 monitoring instrument that is conceptually similar to dynamic chambers. These chambers measure continuously (for example, at 30-min. intervals), and require only minimal downtime for maintenance and recalibration.

Figure 4-6. *LICOR (left) and URS (right) CO_2 flux meters used at Weyburn.*

Carbon Isotopes

The source of CO_2 in soil gas can be identified by determining the carbon isotopes of the gas, provided the end-member values for $\delta^{13}C$ and $\Delta^{14}C$ do not overlap with other possible sources. The $\delta^{13}C$ in CO_2 was measured on most soil gas samples. However, radiocarbon sampling was carried out only twice during the depth-profile soil gas study from two depths at each site, plus a background atmospheric sample. Because $^{14}CO_2$ is present only in trace quantities, larger sample volumes are needed.

4.3.3 Background and Regional Studies

The following sites were monitored during Phase 1: (1) a background site, located at Minard's Farm, 10 km northwest of the Weyburn Field, (2) a large regional grid within the Weyburn Field, (3) two detailed horizontal profiles (located within the regional grid), and (4) two detailed grids at abandoned well-sites (also within the regional grid). Three of these sites were monitored in the Final Phase: the background site, one location near one of the horizontal profiles, and one location near an abandoned well. Figure 4-7 shows the location of all of these sites. Also investigated, but not described here, was a site south of

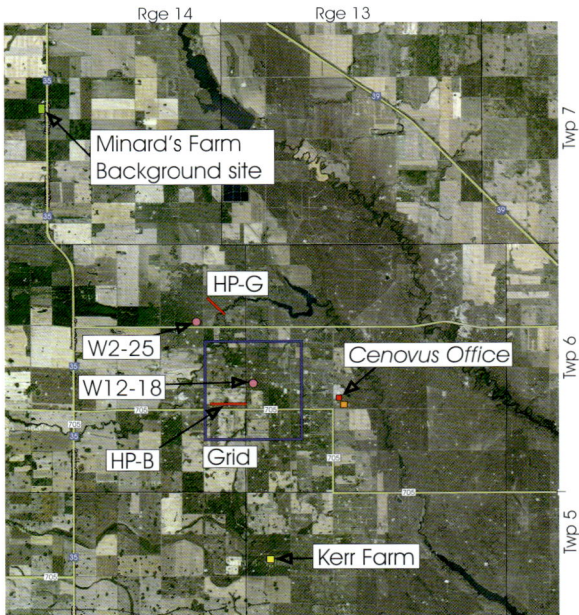

Figure 4-7. *Map showing the location of various wells and study areas in the project area.*

the Weyburn Field where landowners claimed there was a leak of injected CO_2 on their property.

Background Site (Minard's Farm)
This site was chosen because it is away from the main areas of oil production and has land-use, topography, and soil properties similar to those of the regional grid area. Sampling comprised 36 points on a regular grid with 100 m spacing.

Regional Grid
The grid was 3.6 × 4.0 km, with sample spacing of 200 m, comprising 360 sample points. The area has a maximum elevation difference of only 4 m, with mixed farming and significant areas of low-lying swampy ground or ephemeral standing water. The surficial geology comprises heterogeneous glacial till.

Abandoned Wells
Two abandoned wellsites were chosen for soil gas monitoring, each with 16 sampling points over a 400 × 400 m grid. The first site (Well 12-18) was completely abandoned prior to the first measurements and the borehole cemented shut, the surface infrastructure removed, and topsoil imported to return the site to its original agricultural use. This site is a grassy pasture that has been used for at least the past 10 years. In contrast, Well 2-25 suffered a casing failure and its operations were suspended prior to full abandonment. All infrastructures were still in place at the time of first sampling. Since then the well has been completely abandoned and all aboveground infrastructures removed.

Horizontal Profiles
Ten horizontal profiles were designed to investigate mapped lineaments (that may represent fault traces or other potential pathways for CO_2 migration). Some of the profiles were measured only once while others have been studied several times. Horizontal profile B (HP-B) has the most complete data set, having been sampled annually from 2001–2005 and again in 2011. It is located within the regional grid, runs E-W, and is 1500 m long, with 25 m sample spacing. It crosses cropped fields, low-lying swampy ground, and ephemeral standing water, and intersects a number of CO_2 and helium anomalies defined during the first sampling of the regional grid.

4.3.4 Results

Discontinuous Single-depth Measurements

Because soil gas CO_2 contents are impacted by physical factors, including atmospheric pressure, wind, and infiltrating rainwater, we first present graphs showing historical precipitation and temperature – proxies for soil moisture and soil temperature, respectively – for the Weyburn area during the sampling period (Figure 4-8).

Increased soil water content can influence soil gas content in two ways, first through improved conditions for metabolic processes, and second by reducing the exchange of soil gas with the atmosphere. Higher soil moisture has also been associated with higher radon contents, because water inhibits escape of this gas. Thus under wetter conditions one would expect the unsaturated soil to have

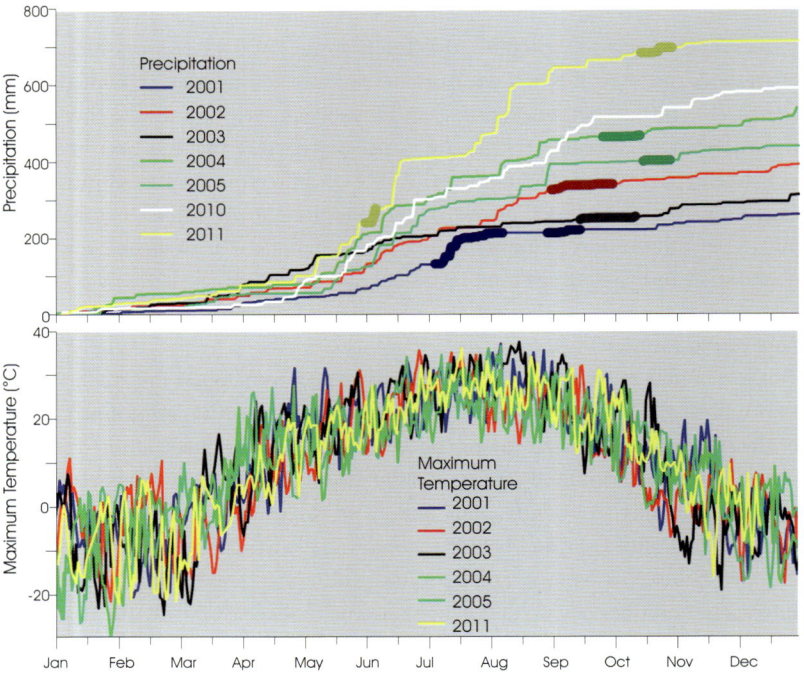

Figure 4-8. *Plots of precipitation (above) and maximum temperature (below) for the sampling period from a government weather station in the City of Weyburn, Saskatchewan. Note that the intervals in the precipitation data, marked with a thick line, denote the periods during which field work was conducted by the authors.*

higher CO_2 contents (increased production and accumulation), lower oxygen contents (greater consumption and slower atmospheric replenishment), and lower methane contents (decreased gas permeability results in slower methane replenishment).

The plot of the maximum air temperature shows that the conditions encountered during the sampling periods were all within the same general range of values. The primary influence of air temperature on soil gas production (or consumption) will be on microbial activity, all other variables being equal. Increased temperature will result in higher metabolic activity, thus increasing respiration (that is, consumption of oxygen and production of CO_2). Thus, the higher summer temperatures are expected to increase CO_2 production in the soil, which should decrease as temperatures drop during the fall.

The box and whisker plots in Figure 4-9 show the median value as a horizontal line within a box delimited by the lower and upper quartiles (25% and 75% of samples, respectively), and whiskers indicating non-outlier intervals. Trends over the two sampling periods are very similar, with excellent correspondence between the two methods described above.

Correlating statistical distributions of soil gas CO_2 (Figure 4-9A) with those for N_2 (Figure 4-9C) and O_2 + Ar (Figure 4-9D) yields insight into the origin of the observed CO_2 contents. Consider two end-member cases, one involving in-situ biogenic production of CO_2 as a byproduct of root and microbial respiration, and the other involving addition of deep injected CO_2 that is migrating into the soil environment. In the first case, production of CO_2 will occur together with the consumption of O_2 at a rate of approximately 1:1. However, because N_2 is not involved in this metabolic pathway, its concentrations will remain unchanged. In the second case, addition of external CO_2 will essentially dilute all gas species at a rate proportional to their original concentrations (for example, O_2 at 21:100 and N_2 at 78:100).

In Figure 4-10, N_2 and (O_2 + Ar) are plotted against CO_2 for all sampling campaigns. The solid lines indicate the trend expected for in-situ biogenic CO_2 production and the dashed lines show the trend expected for the addition of deep leaking CO_2 to the soil horizon. This plot clearly shows an excellent correspondence with the biogenic trend, and no correlation with the leakage trend.

Box and whisker plots for methane, ethane, propane, and ethylene (not shown) exhibit some seasonal variability. Several processes may have produced this trend, although a source from deep hydrocarbons should probably be ruled out in view of the results for CO_2 in soil gases, and of the fact that the ratios of these hydrocarbons are relatively uniform across the study area. They would presumably be different over an oil or gas field, at least according to some practitioners of surface geochemical prospecting for petroleum.

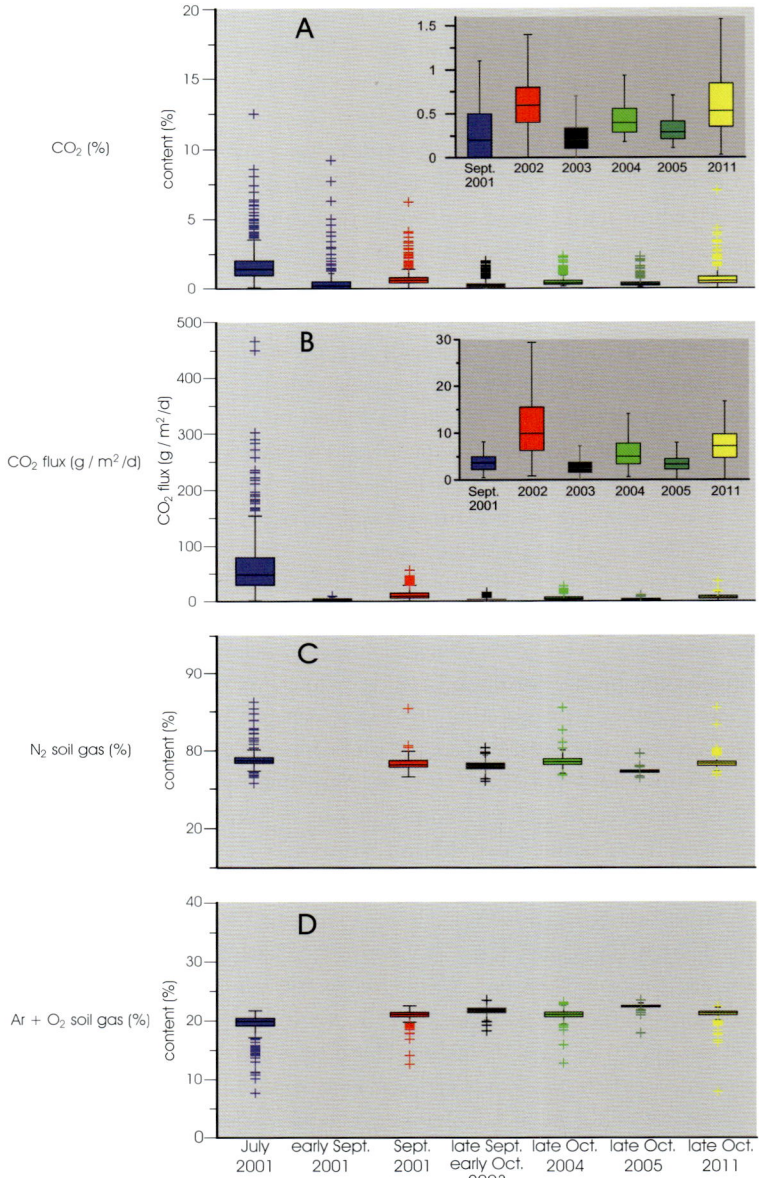

Figure 4-9. *Box and whisker plots for soil gas composition.* **A.** *CO_2 (%).* **B.** *CO_2 flux (g/m²/d).* **C.** *Nitrogen (%).* **D.** *Argon-plus-oxygen (%). The inserts in plots A and B show the same data without the July 2001 results and the outlier values, to better visualize the relative distributions.*

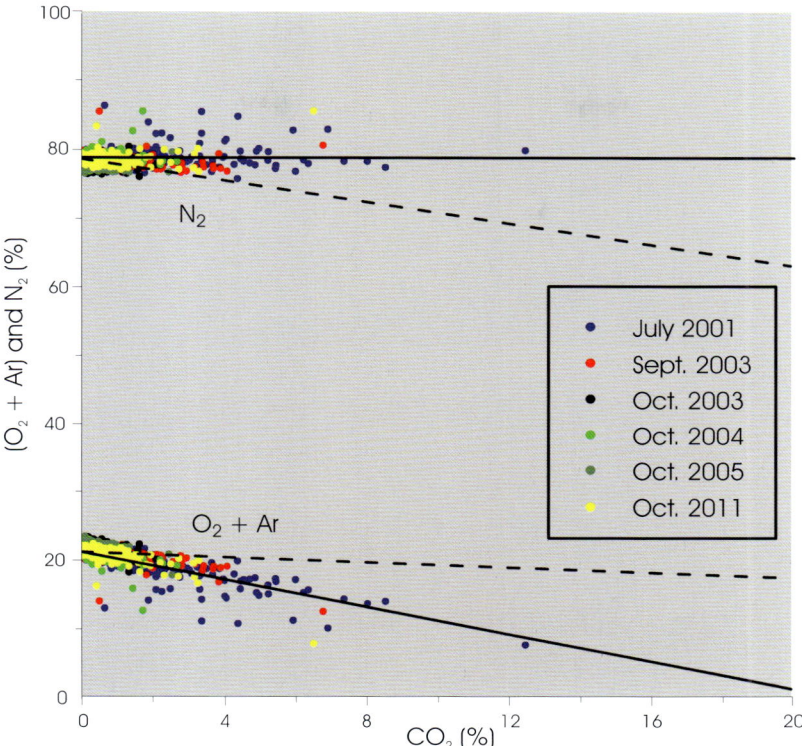

Figure 4-10. *Scatter plots of N_2 and $(O_2 + Ar)$ contents versus CO_2 content for soil gas samples collected during the project. The solid lines indicate the trend expected for biogenic CO_2 production, and the dashed lines the trend expected for the addition of deep, leaking CO_2.*

The statistical distribution of helium and radon for the sites sampled is shown in Figure 4-11. Both gas species are potential tracers of gas migration, helium because of its deep origin, and radon because of its short half-life (high radon anomalies may indicate rapid transport to the surface, within a carrier gas, prior to its decay). The low contents of both helium and radon fall within the range of natural variability and are not the result of deep leakage.

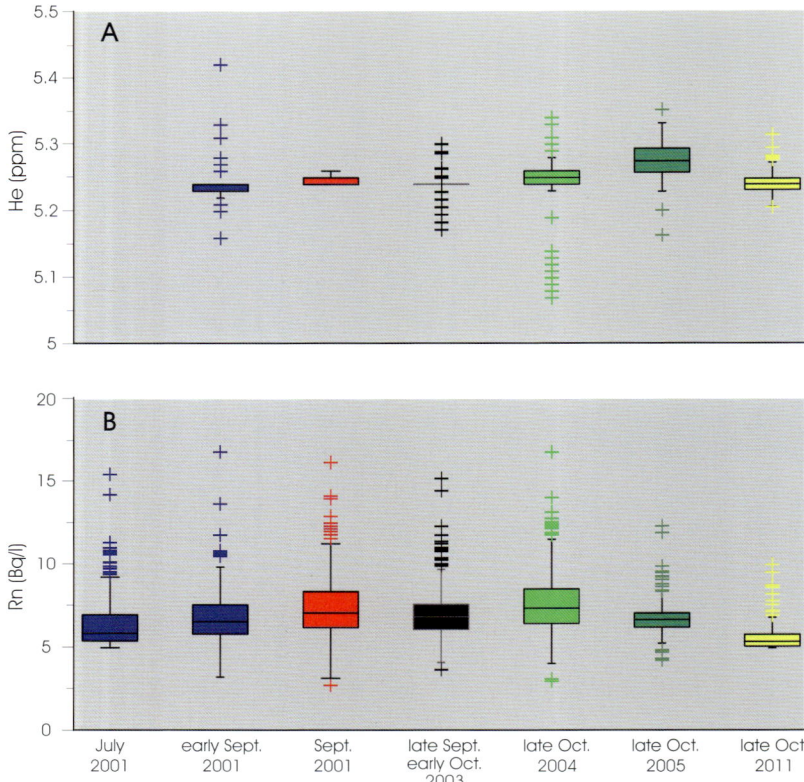

Figure 4-11. *Box and whisker plots of soil gas composition determined at the Weyburn site.* **A.** *Helium (ppm).* **B.** *Radon (Bq/L).*

Regional Distribution

Figure 4-12 shows box and whisker plots for the 2011 campaign for all sites, that is, the regional grid, individual wellsites, and the background site at Minard's Farm. Figure 4-13 shows plots of N_2 and (O_2 + Ar) against CO_2 for the same sites. Both these figures may be compared with similar data presented for the samples collected from 2001–2005 (Figure 4-9 and Figure 4-10, respectively).

The 2011 results are presented this way because it is assumed that anomalies observed both above and away from the reservoir are likely linked to near-surface processes, whereas those that occur only above the oilfield may be due to deeper processes. As a very broad generalization, the major distribution of data for the CO_2 flux and the contents of CO_2, oxygen and nitrogen in soil gas is similar, both from the different sites and between the samples collected over the entire period of the project (Figure 4-12). Further, the plots in Figure 4-13 confirm the source of the CO_2 as biogenic and not due to leakage from below.

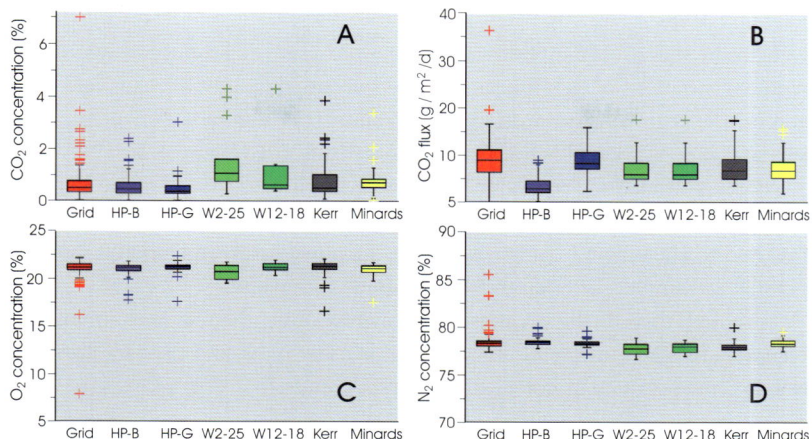

Figure 4-12. *Box and whisker plots for soil gas composition for 2011.* **A.** *CO_2 (%).* **B.** *CO_2 flux ($g/m^2/d$).* **C.** *Oxygen (%).* **D.** *Nitrogen (%).*

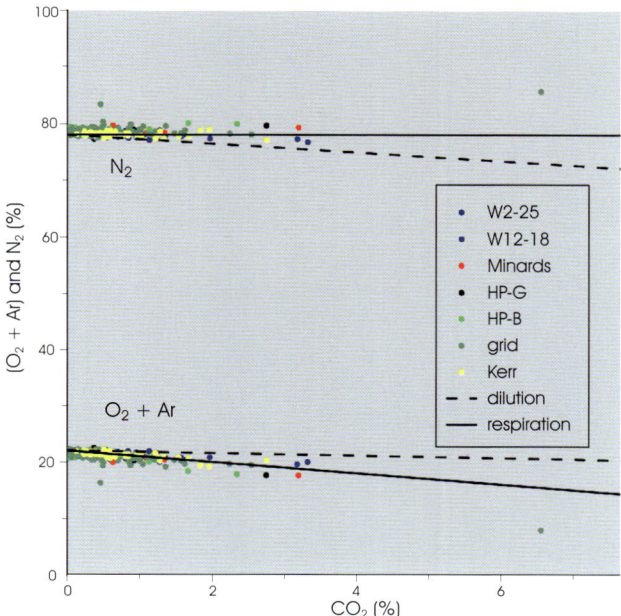

Figure 4-13. *Scatter plots of N_2 and (O_2 + Ar) contents versus CO_2 content for soil gas samples collected during 2011, distinguished by the type of site. The solid lines indicate the trend expected for biogenic CO_2 production, and the dashed lines the trend expected for the addition of deep, leaking CO_2.*

GEOCHEMICAL MONITORING   135

Regional Maps

Data from the 2011 sampling campaign were used to produce maps of the regional variations of CO_2 flux, and of the soil gas content of nitrogen and oxygen, in the regional grid around the injection site (Figure 4-14), at Minard's Farm (Figure 4-15), and at the abandoned wellsite 25-5 (Figure 4-16).

Water-saturated soil conditions restricted sampling to 209 out of the 360 grid locations in the regional grid. In all sets of maps, the oxygen scale has been reversed to allow better comparison with the other gases, and when it was not possible to sample a location the datum point is left white.

Although there is the occasional anomalous value, the relations among the components is as indicated in the box and whisker plots, and the data distribution is generally quite uniform without linear trends.

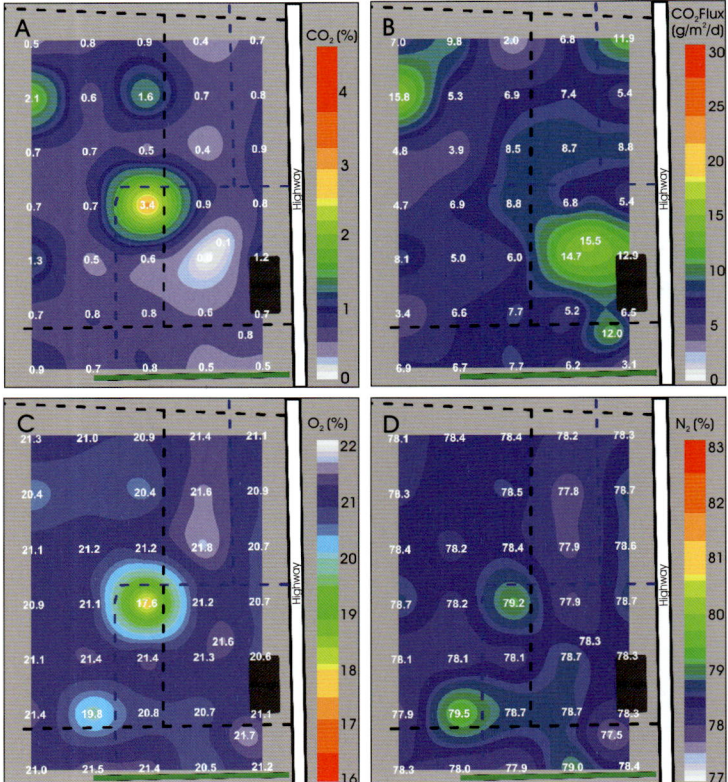

Figure 4-15. *Maps of the content of soil gas in 2011 at the background site at Minard's Farm. A. CO_2 (%). B. CO_2 flux (g/m²/d). C. Oxygen (%). D. Nitrogen (%). Dark grey box is a man-made dugout. Dashed blue lines are buried power cables, and dashed black lines are property boundaries.*

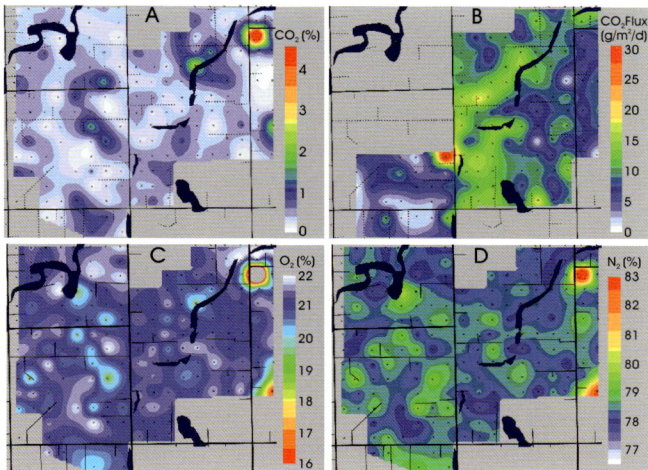

Figure 4-14. *Maps of the content of soil gas in 2011 in the regional grid.* **A.** CO_2 (%). **B.** CO_2 flux ($g/m^2/d$). **C.** *Oxygen (%).* **D.** *Nitrogen (%). Dark blue areas are surface water bodies.*

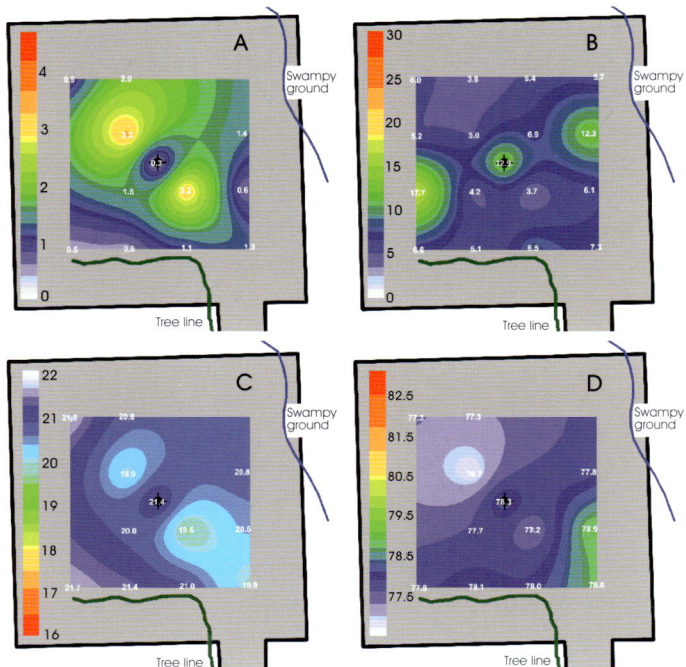

Figure 4-16. *Maps of the content of soil gas in 2011 at the abandoned wellsite 25-5.* **A.** CO_2 (%). **B.** CO_2 flux ($g/m^2/d$). **C.** *Oxygen (%).* **D.** *Nitrogen (%).*

Horizontal Profiles
Results for the horizontal profiles obtained during both sampling periods (2001–2005 and 2011) are shown in Figure 4-17. Although there is only about 2 m elevation difference across the profile, even this small difference has resulted in a correlation between local topographic depressions, CO_2 'peaks' and oxygen 'valleys'. This relation may be due to several reasons which cannot be resolved with the present data.

Discontinuous Depth-profile Measurements
Scatter plots of CO_2 content of soil gas against depth, as a function of various times during the growing season, show a broad linear trend of increasing CO_2 content with depth (not shown). This trend represents normal ranges of soil CO_2, and biological isotopic signatures that are characteristic of C3 photosynthetic pathway vegetation that is dominant at these sites.

Continuous Monitoring
Data for the content of CO_2 in soil gas from the three CO_2 monitoring probes that had been buried for about 4.5 months are shown in Figure 4-18. These locations were chosen to include a site near the highest recorded CO_2 value (Figure 4-18A), a site with high CO_2 and radon where a continuous radon monitor had been deployed for an extended period (Figure 4-18B), and a point near an abandoned well (Figure 4-18C).

All three probes show a fairly smooth increase in CO_2 over the summer months, with values that mimic the observed increase in atmospheric temperature. However, whereas temperatures peak around mid-July, CO_2 values tend to continue to increase until early September, after which they briefly stabilise, then begin to decrease. Values at the end of the samping period matched those from single soil gas samples taken near each probe.

CO_2 Flux
DISCONTINUOUS MEASUREMENTS: The CO_2 flux data from October 2011 are statistically slightly higher than those observed during the 2003–2005 sampling period, perhaps due to the effect of heavy rains in 2011. However, they fall within the normal range of autumn values observed during previous years. The overall range of values is relatively restricted, with most sites having values ranging between 5–15 $g/m^2/d$, with only one point reaching 36.5 $g/m^2/d$ (the highest measured at any site during the study).
CONTINUOUS MEASUREMENTS: Continuous surface CO_2 flux monitoring was conducted from June to November 2011 at three sites, two within the regional grid and one at the background site (Figure 4-19). For all three sites, soil CO_2 flux ranged from 0–10 $\mu mol/m^2/s$, and followed normal seasonal patterns in soil

temperature. The highest daily average rates were seen from July through September. During the final six weeks, the instruments were often under snow, but because reference sensors had been installed originally, accurate sub-snow data

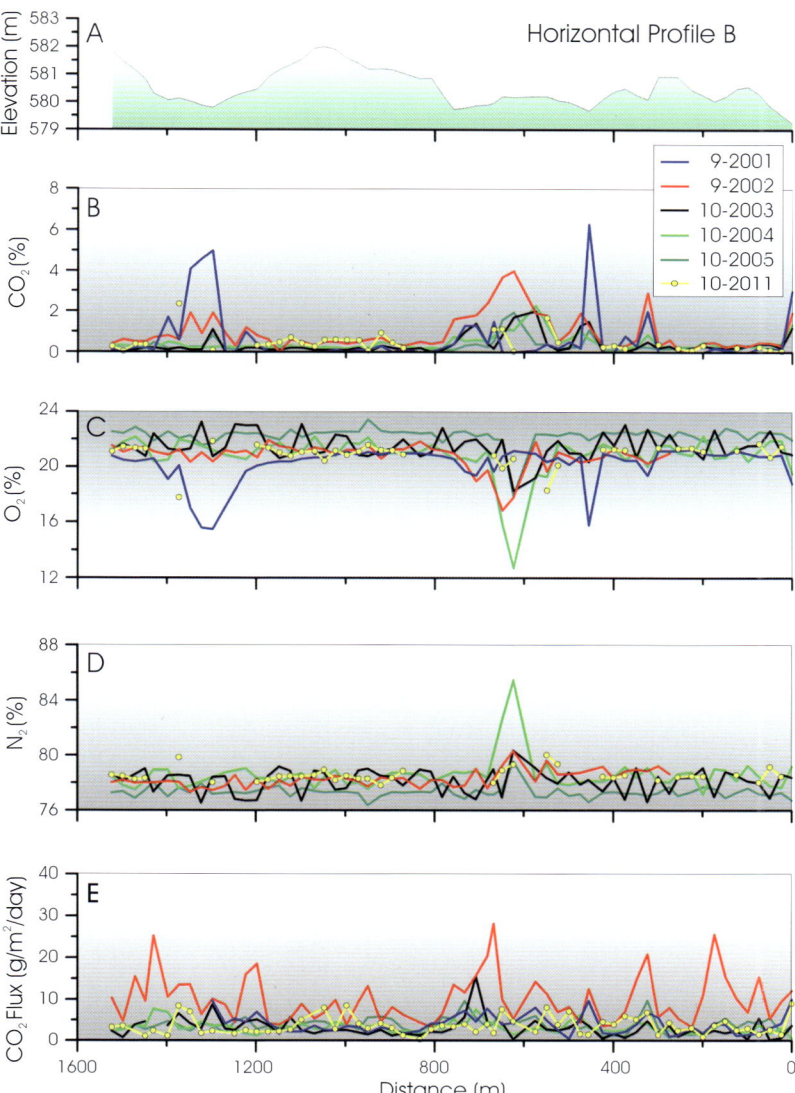

Figure 4-17. *Results for horizontal profile B.* **A.** *Topographic elevation (m).* **B.** CO_2 *content of soil gas.* **C.** *Oxygen content of soil gas.* **D.** *Nitrogen content of soil gas.* **E.** CO_2 *flux.*

were obtained. During early winter, fluxes continued to decline as CO_2 produced during autumn was vented, although it is also likely that some CO_2 production continued deeper in the profile where soils were not yet frozen. Soil CO_2 flux is highly responsive to major rain events, and these natural high flux anomalies are of critical importance for monitoring activities because this natural variation

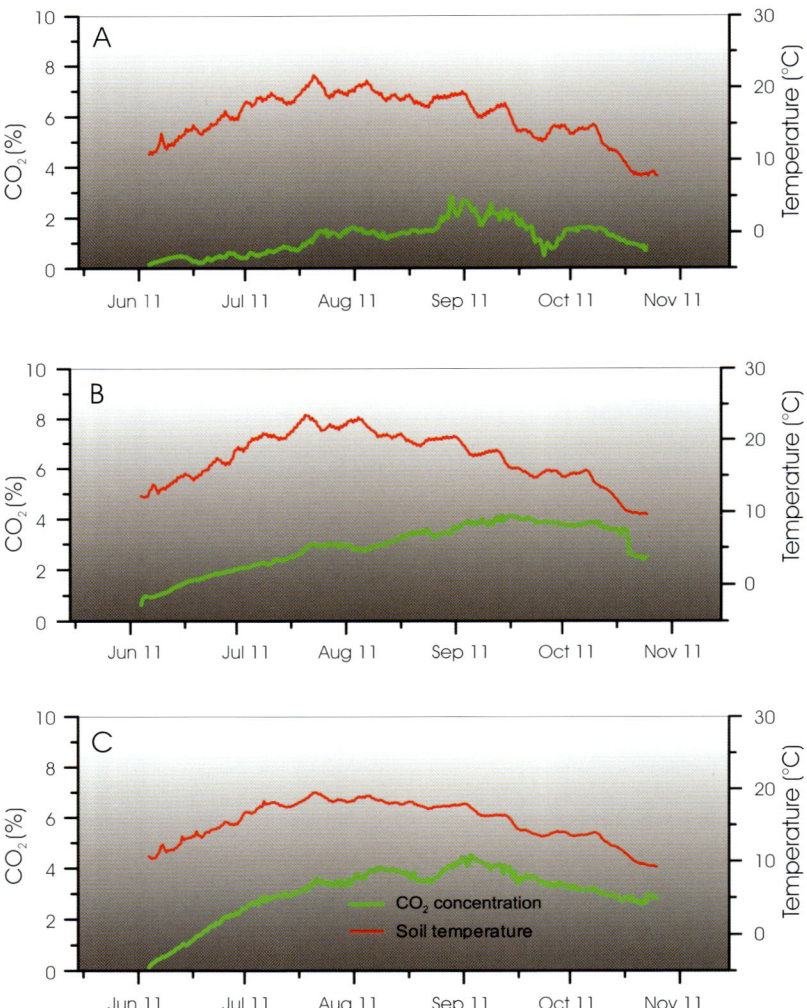

Figure 4-18. *Temporal variations of the content of soil gas CO_2 (green trend) and temperature (red trend) at the three buried CO_2 sensors for the period June–October, 2011.*

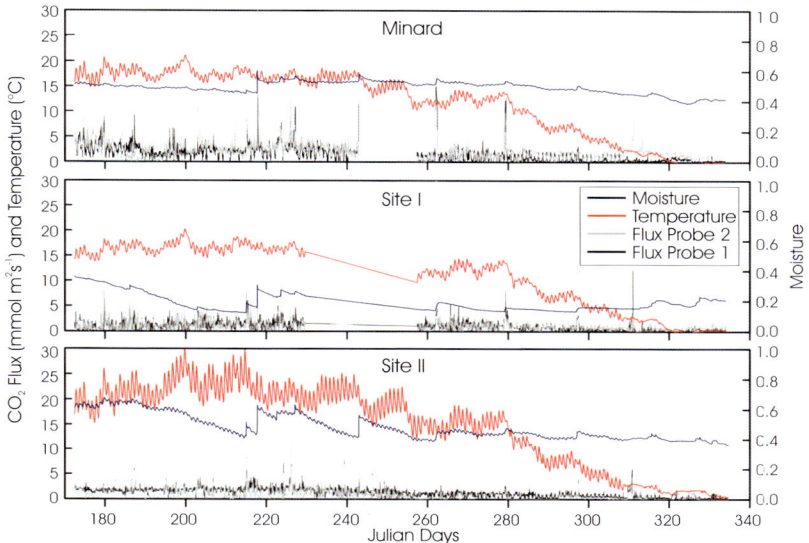

Figure 4-19. *Variation in the flux of CO_2 over the monitoring period (there are some gaps in the data).*

could easily be misinterpreted as leakage. Biophysical models provide a natural complement to high-frequency observational data, and can accurately reproduce short time-scale patterns. Agreement between observed and predicted evolution is quite good.

Carbon Isotopes

Limited data (Table 4-2) were obtained for ^{13}C in CO_2 at several sites in the project area. These results agree closely with three samples collected on the main grid, with $\delta^{13}C$–CO_2 values between –17.3‰ and –24.6‰, and some measured by Trium (2011) at the Kerr property, which averaged –23.4±1.36‰ and –22.6±0.91‰ over two field seasons.

To put these results in the proper context, the $\delta^{13}C$–CO_2 value of the injected CO_2 is –20.5‰, while different types of decaying vegetation produce values in the range of –35‰ to –21‰ for C3 type plants (such as wheat) and –21‰ to –9‰ for C4 type plants (such as corn). Thus, the measured values for soil gas CO_2 lie in the range of both natural near-surface biogenic variability and that of the injected gas, and thus the stable isotope data on their own are inconclusive. Combined with the measured soil gas CO_2 content, however, there is a clear isotopic depletion, with increasing CO_2 concentration similar to that observed by Trium (2011). These results further support the interpretation that their

Sample No.	CO_2 content (volume %)	$\delta^{13}C-CO_2$ (‰ VPDB)	Standard Deviation
A11	0.71	–19.1	0.095
B23	0.52	–21.7	0.026
G7	0.56	–21.5	0.180
B19	1.88	–23.1	0.068

Table 4-2. *Soil gas CO_2 content and $\delta^{13}C-CO_2$ values collected during the October 2011 field campaign.*

observed trend is caused by isotopic fractionation via biogenic decomposition of organic matter.

Radiogenic carbon isotopes were measured on 16 samples from several locations in the project area. Values for $\Delta^{14}C$ (the deviation in $^{12}C/^{14}C$ from the oxalic acid standard) range from about 9.86 to 50.33. This range reflects the complex source from biological degradation of (relatively) 'old' and 'young' soil organic matter, perhaps as expected.

4.4 Reservoir Fluids
4.4.1 Formation Water and Gases
Introduction

The primary objective of monitoring formation waters is to evaluate the way that CO_2 partitioning occurs among the hydrocarbon, aqueous, and mineral phases. This information is important for calibrating geochemical and reactive transport models to forecast and distinguish the several processes by which CO_2 can be stored in an aquifer (that is, by hydrodynamic, solubility, and mineral trapping).

During the project, a comprehensive geochemical database with more than 40 compositional and isotopic parameters was compiled, based on a baseline and 16 separate surveys. The most important evolutionary trends can be extracted, despite the complicated history of an active CO_2-EOR operation.

An important advantage at the Weyburn Field is the relatively simple, carbonate-dominated lithology of both the Marly (dolomite > calcite) and Vuggy (calcite > dolomite) subunits of the Midale reservoir. As a result, the most significant compositional and isotopic evolution can be framed in terms of three reactions related to CO_2-induced carbonate mineral dissolution. These three reactions represent two important processes: CO_2 aqueous solubility (reaction 1) and carbonate mineral dissolution (reactions 2 and 3), as follows.

$$CO_2 + H_2O \leftrightarrow H_2CO_3 \leftrightarrow H^+ + HCO_3^- \tag{1}$$
$$CaCO_3 \text{ (calcite)} + H^+ \leftrightarrow Ca^{2+} + HCO_3^- \tag{2}$$
$$CaMg(CO_3)_2 \text{ (dolomite)} + 2H^+ \leftrightarrow Ca^{2+} + Mg^{2+} + 2HCO_3^- \tag{3}$$

Solubility trapping (reaction 1) is a primary storage mechanism for CO_2 that effectively increases the dissolved inorganic carbon (and total alkalinity) and lowers pH. The lower pH causes carbonate dissolution (reactions 2 and 3), which also increases the dissolved inorganic carbon together with aqueous calcium and magnesium concentrations, while increasing pH. Because the kinetics of solubility trapping exceed by orders of magnitude those governing mineral dissolution, there is a net decrease in pH, although this drop is not as precipitous as in clastic reservoirs, which effectively lack the buffering capacity of carbonate minerals. In sum, then, the compositional evolution of Weyburn waters during CO_2-EOR will be dominated by increasing dissolved inorganic carbon, calcium, and magnesium contents, accompanied by a decrease in pH. Therefore, in terms of aqueous concentrations, H^+ (pH) and HCO_3^- (alkalinity) are always of paramount importance, because these two parameters are dramatically affected by CO_2–water interactions, independent of reservoir mineralogy.

With respect to isotopic ratios, $\delta^{13}C$ of dissolved inorganic carbon, like pH and alkalinity, should be measured because this parameter will typically have distinctive values for carbonate minerals, CO_2 gas, and bicarbonate in formation water in the reservoir, and injected CO_2. Figure 4-20 shows the general relation among these key markers at Weyburn.

In contrast, for typical quartz-dominated clastic reservoirs, key concentrations subject to significant CO_2-induced evolution include potassium, sodium, calcium, silica, and aluminum.

Sampling was conducted exclusively within the Weyburn Phase 1A area in the initial (northwest) sector of the field subjected to the CO_2 flood. The 17 sampling times included a 'baseline' data set taken in 2000 before CO_2 injection began.

Figure 4-21 shows the locations of the individual wells, together with the horizontal and vertical CO_2 injectors in Phase 1.

The availability of wells for sampling was complicated by oilfield operations. In addition, the drilling of horizontal wells *after* the initial baseline sampling was

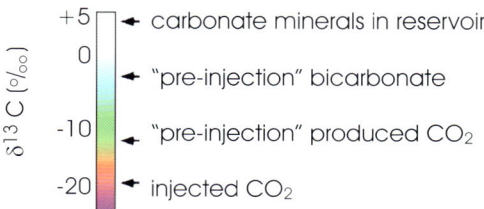

Figure 4-20. *General relations among values for $\delta^{13}C$ (‰) in reservoir carbonates, pre-injection bicarbonate and CO_2 in the reservoir, and injected CO_2.*

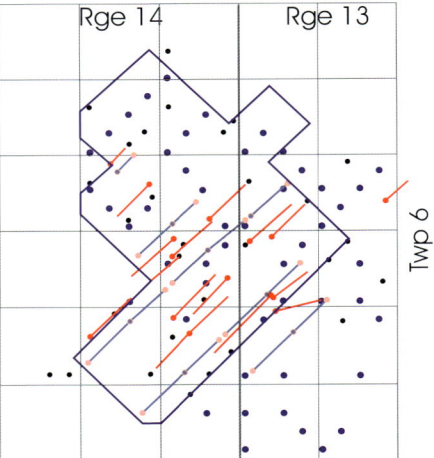

Figure 4-21. *Map showing the locations of horizontal (red lines) and vertical production wells (black dots). Horizontal CO_2 injection wells are shown as blue lines, and vertical CO_2 injection wells are shown as blue dots. Wells sampled for formation water and gas are shown as red lines.*

first deemed an impediment to evaluating the analyses. However, this proved not to be the case and sufficient data were acquired to allow time-lapse evaluation of the key parameters noted above.

Alkalinity

Figure 4-22 shows the changes in alkalinity from a median baseline value of 400 mg/L to 2200 mg/L after 12 years of injecting CO_2, or by more than a factor of five. Beyond 12 years, there is little change in the alkalinity. The magnitude of the increase is not uniform throughout the Phase 1A area. This is because of the significant impact of the parts of the field where CO_2 was injected, and also the different volumes at different locations. Specifically, the most marked increases are generally observed along the southeastern part of the area near the Phase 1A boundary, which corresponds to the region of largest volumes of injected CO_2. Discontinuous CO_2 injection in the north and northeast sectors have resulted in high variability in measured alkalinity over time.

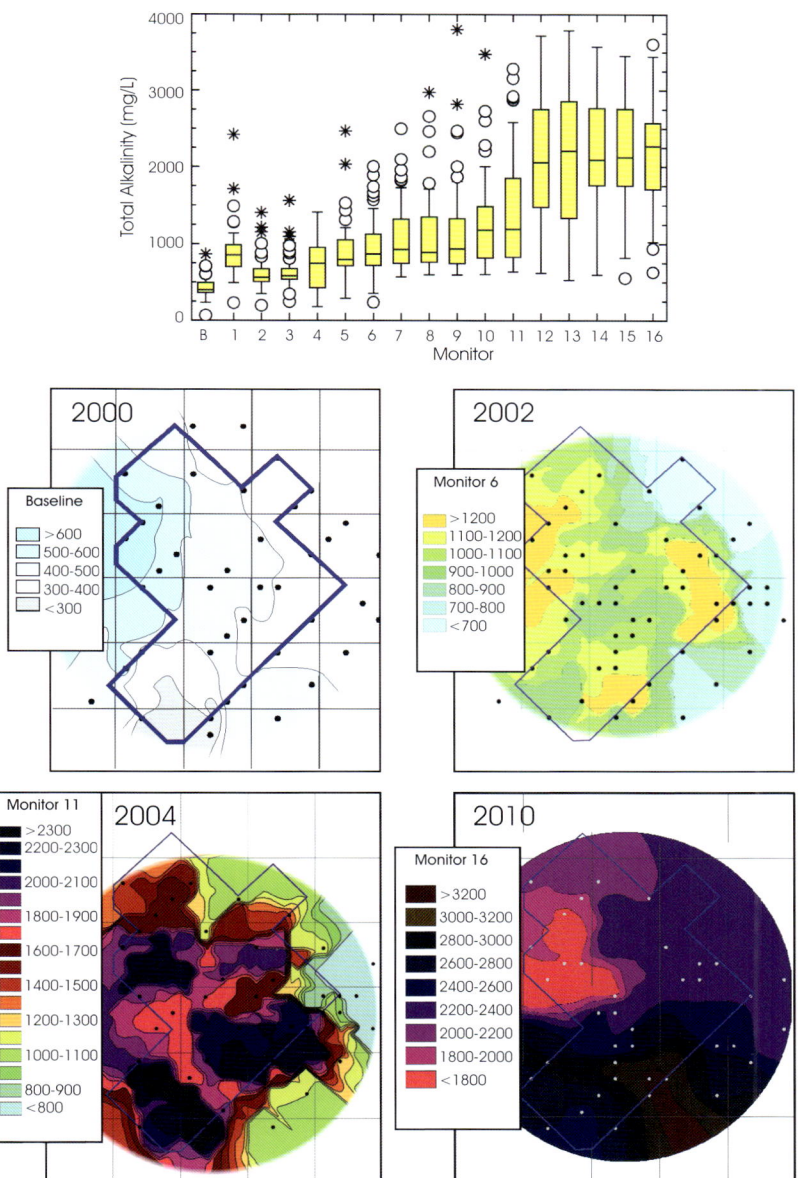

Figure 4-22. *Change in total alkalinity (mg/L) after a decade of CO_2 injection in the Weyburn Field.*

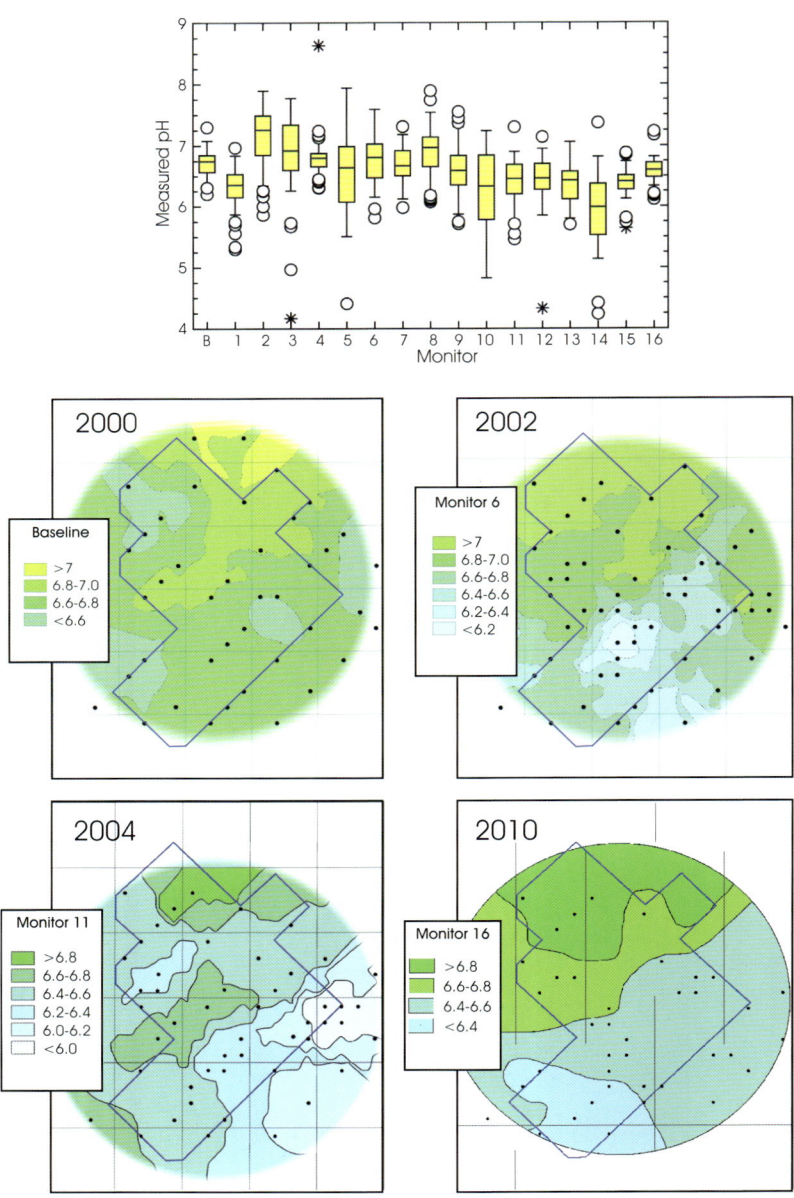

Figure 4-23. *Change in pH after a decade of CO_2 injection in the Weyburn Field.*

pH

In contrast to alkalinity, changes in pH (Figure 4-23) result from two opposing processes. There is a negative contribution from the aqueous solubility of CO_2, characterized by relatively large magnitude and fast kinetics, but a positive contribution from carbonate dissolution, characterized by relatively small magnitude and slow kinetics. Thus, the time-integrated change from baseline values to pH measured a decade later is very small, from 6.75 to 6.60. Despite this, there is considerable variability within and between each sampling event. A net decrease in pH indicates that the predominant contribution is from solubility trapping (reaction 1), while a net increase would, in principle, indicate a predominant contribution from carbonate dissolution (reactions 2 and 3). To determine the relative impact of these two processes requires additional information, such as the contents of calcium, magnesium, and bicarbonate. Wellhead pH must also be compared to calculated downhole pH.

As with alkalinity, the impact of variable CO_2 injection volumes and discontinuous CO_2 injection is manifested in the form of irregular patterns of pH values across the map area.

Downhole pH

When formation water is brought to the surface, it depressurizes and the CO_2 aqueous solubility is reversed, increasing the pH and potentially allowing dissolved minerals to precipitate. Mineral scales deposited in oilfield subsurface and

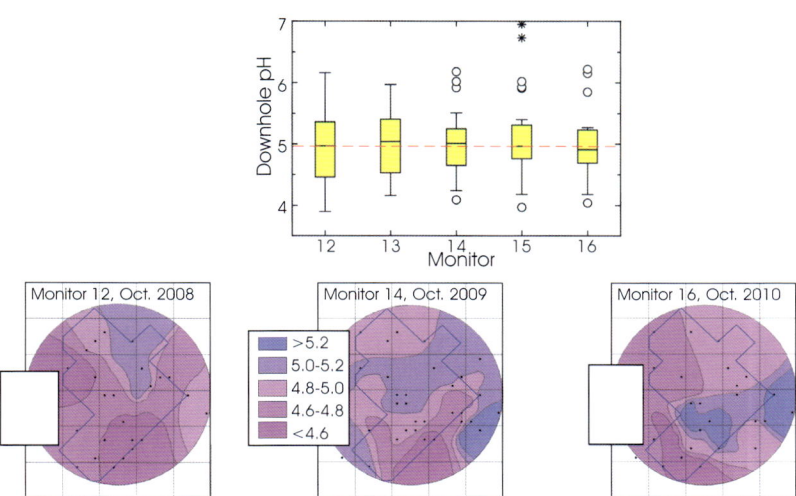

Figure 4-24. *Change in downhole pH over the period 2008–2010 in the Weyburn Field.*

surface facilities are most often the result of this process. Downhole pH can be determined using a computer code such as SOLMINEQ.88 (Kharaka et al., 1988).

Figure 4-24 shows that over the period 2008–2010, there was effectively no change in the downhole pH, with a median value of about 5.0. Differences across the field may relate to the number and distribution of horizontal and vertical CO_2 injection wells, and possibly also to water injection in some wells.

Calcium

The evolution of aqueous calcium content (Figure 4-25) is principally controlled by positive contributions from calcite and dolomite dissolution (reactions 2 and 3). These dissolution processes are definitely operative and result in a 50% increase in the calcium content of the formation waters (from 1400–2100 mg/L) over a decade of CO_2 injection. As was the case for alkalinity and pH, the maximum increase in calcium content is observed along the southeastern boundary of the area, where the number of CO_2 injectors is greatest. Somewhat lower calcium contents are seen in the northwestern part of the area where there are fewer CO_2 injectors.

Changes in the content of magnesium are very similar to those of calcium, increasing from a median baseline value of 375 mg/L to 525 mg/L a decade later, an increase of about 40%, reflecting progressive dolomite dissolution. As with calcium, the largest increases in magnesium content are found in the southeastern part of the area, in association with maximum volumes of injected CO_2.

Stable Isotopes ($\delta^{13}C$)

Changes in the value of $\delta^{13}C$ of dissolved inorganic carbon (Figure 4-26) result from a combination of two opposing processes. There is a negative contribution from CO_2 aqueous solubility, which is characterized by relatively large magnitude and fast kinetics, and a positive contribution from mineral dissolution, which is characterized by relatively small magnitude and slow kinetics. Over the decade during which monitoring took place, the median value of $\delta^{13}C$ remained constant around −2‰ for the first few years, but from then on there was a progressive decrease to −12.5‰. Thereafter, the median value remained relatively constant. At an overall field scale, the progressive rate of change indicates that the kinetically faster CO_2 aqueous solubility (reaction 1) exerts the predominant influence, with the slower effect of mineral dissolution also evident.

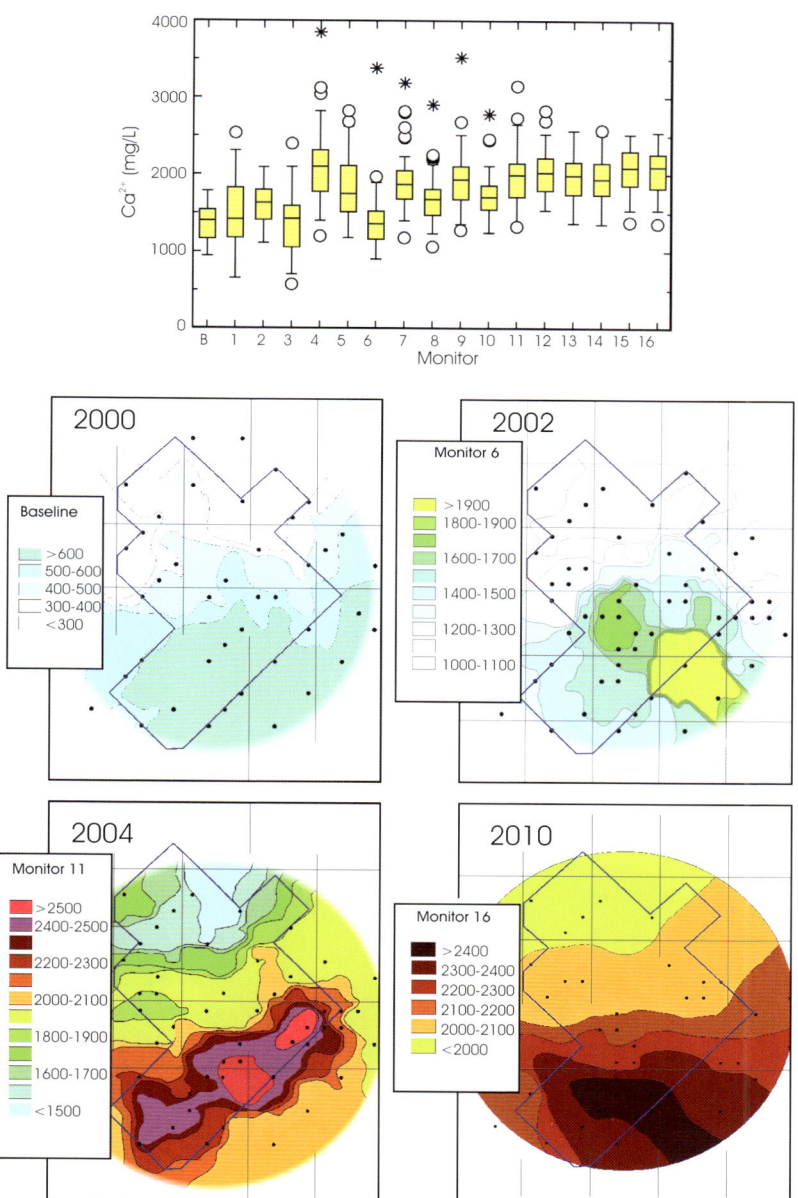

Figure 4-25. *Change in calcium content (mg/L) after a decade of CO_2 injection in the Weyburn Field.*

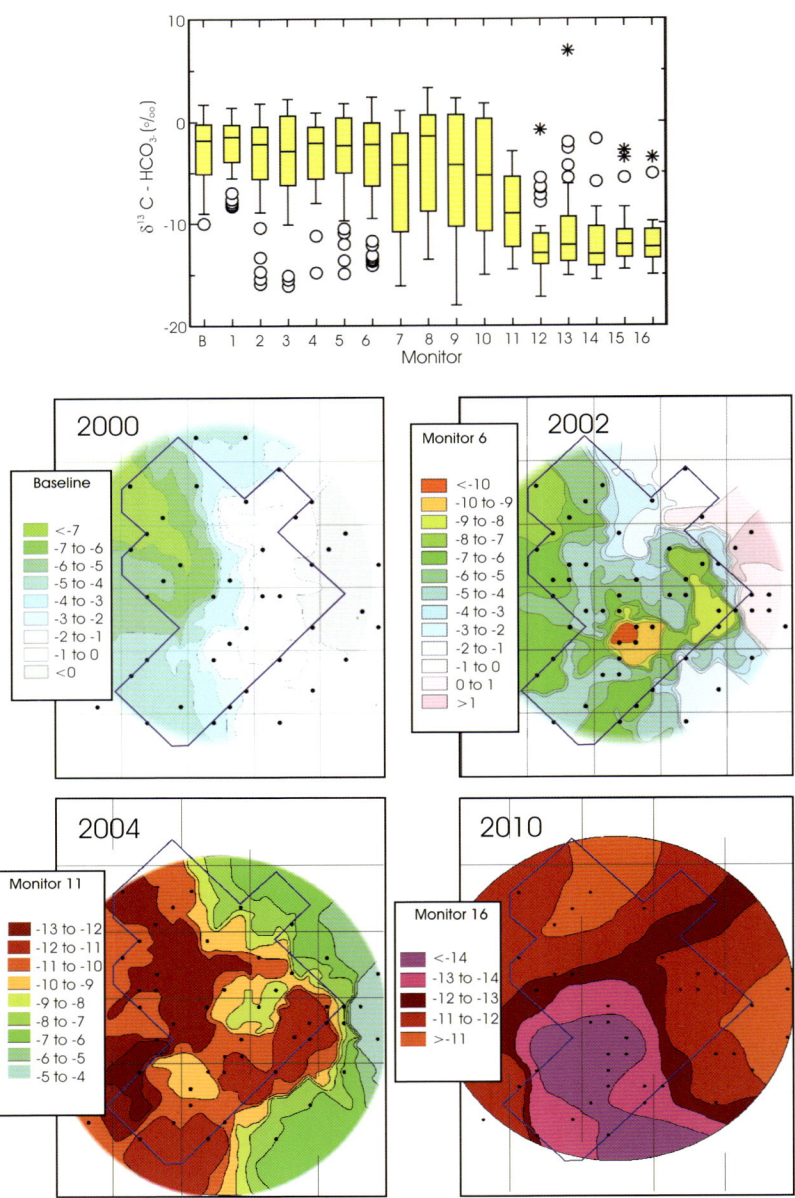

Figure 4-26. *Change in δ¹³C (‰) of dissolved inorganic carbon after a decade of CO_2 injection in the Weyburn Field.*

4.4.2 Produced Hydrocarbons

Normally, characterization of the produced hydrocarbons in a CO_2-EOR operation is under the purview of reservoir engineers, and so seldom forms part of geochemical monitoring. Indeed, in many formations selected for CO_2 storage, there may be no hydrocarbons. For these reasons the work carried out by the project on produced hydrocarbons will be summarized only briefly.

Hydrocarbon sampling was conducted exclusively within the Weyburn Phase 1A area in the initial (northwest) sector of the field subjected to the CO_2 flood. The wells sampled encompassed a range of the oil densities (921–858 kg/m^3; 22–33°API) that characterize the Weyburn Field (see Figure 4-27 for locations). The 14 sampling times included a 'baseline' data set taken in 2000 before CO_2 injection began. Table 4-3 shows some of the changes observed in the composition of separator gas. The composition of the produced oil also changes, getting lighter after CO_2 flooding in terms of density, viscosity, and molecular weight.

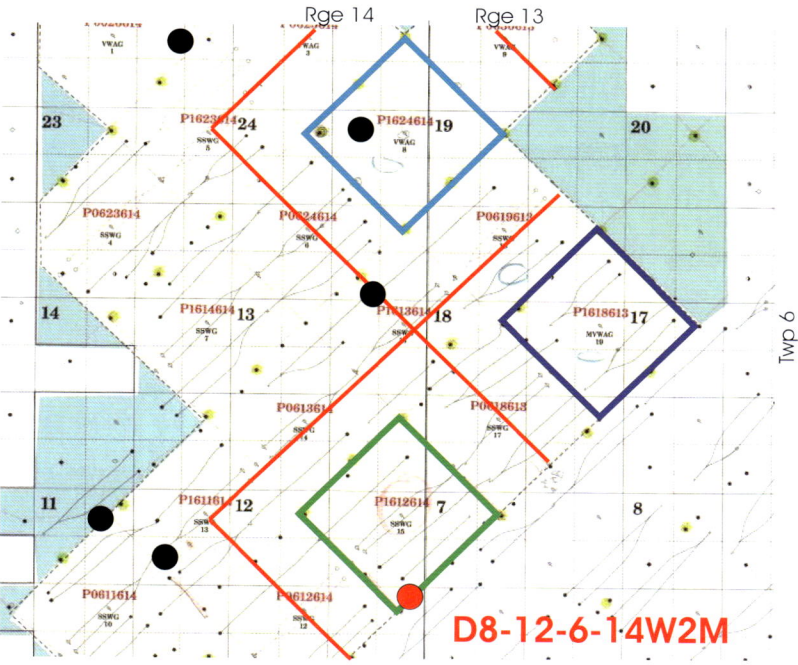

Figure 4-27. *Map of Phase-1A area wells sampled for hydrocarbons from 2000–2004 (black) and 2010 (red).*

Parameter	2000 (Baseline)	2010 (3700 days after injection started)
Field separator		
Pressure (kPa)	1100	2700
Temperature (°C)	17	27
GOR (m^3/m^3)	36	870
GOR (flashed oil) (m^3/m^3)	11.8	66.0
Gas composition (mol %)		
C_1	58.53	3.57
C_2	10.05	2.21
C_3	6.76	2.00
C_4	2.09	0.94
C_{5+}	0.57	0.27
N_2	18.14	1.02
CO_2	3.16	89.58
H_2S	0.70	0.41

Table 4-3. *Changes in gas composition after CO_2 injection for about 10 years.*

A 'live oil' had to be reconstructed by a non-standard method because when the well (D8-12-6-14 W2Mer) was sampled in April 2010 it was found that the separator oil and gas both contained very high amounts of CO_2. This was clear evidence that CO_2 breakthrough had occurred since the previous sampling of this well in April 2002. The synthetic gas composition used in each step of the reconstruction was determined from the sample bank of Weyburn hydrocarbons established previously and the gas added until the reservoir GOR was matched (about 36 m^3/m^3).

Tests conducted using the 'live oil' included (1) measurement of the fluid properties of the oil at pressures above the bubble point and at a constant temperature of 59°C, (2) determination of the equilibrium properties of the CO_2–live oil mixture, (3) differential liberation tests carried out on the CO_2-saturated reservoir fluid, from which formation volume factors and solution gas-oil ratios can be determined as a function of pressure, (4) determination of minimum miscibility pressure, (5) multi-contact tests to simulate CO_2–oil contact in the reservoir, and (6) fluid compositional, PVT and viscosity modelling.

4.5 Recommendations

Geochemical monitoring must be part of any CO_2-EOR or CO_2 storage program because these time-lapse data document the progressive CO_2-induced departure from baseline conditions. The key initial task of such monitoring is to determine the appropriate suite of geochemical parameters that will (1) document the continued secure isolation of injected CO_2 within the storage complex, (2) track CO_2 movement and phase partitioning within the target reservoir, and (3) calibrate the accuracy of modelling tools used to forecast long-term storage performance in terms of capacity and containment. This suite is conveniently divided into parameters within the overburden (near-surface soil gas and shallow groundwater) and those in the reservoir (formation water and oil properties).

For soil gas and groundwater, the injection-induced signal (even when discernible) is typically of small magnitude, and the identification and extraction of significant, temporal variations due to natural processes present a daunting challenge, one that often requires advanced simulation capabilities. In contrast, signals of CO_2 migration in and near the injection zone are easier to discern.

4.5.1 Soil Gas

The areal extent of a geochemical monitoring program is largely dependent on that of the anticipated long-term CO_2 migration as forecast by simulation studies. For overburden environments this is not a restriction, whereas reservoir monitoring is restricted to available production and CO_2 injection wells. The geochemical sampling of shallow overburden (for example, soil horizons) should take into account such features as topographic variations, floral heterogeneity, and the presence of unique anthropogenic and geological features, such as shallow wells and lineaments. Soil gases can be monitored using 2D- and 1D-grids, under either uniform or variable conditions, and with both static and dynamic sampling density, as well as long-term study. Further, both compositional and isotopic data are of value in interpreting possible surface leakage of migrating CO_2 (which was not found in the Weyburn study area).

For soil gas studies, and indeed all monitoring programs, a survey prior to CO_2 injection is of paramount importance, because these data provide the essential baseline for comparison against all post-injection monitoring results. Note that in order to capture the full range of natural background variations in near-surface environments, the 'baseline' survey is not a snapshot event, but rather involves a series of campaigns. Subject to regulatory requirements, and in the continued absence of adverse findings, seasonal (spring and fall) soil sampling campaigns are typically adequate.

4.5.2 Groundwater

The areal grid requirements for groundwater sampling are similar to those for soil gas, in addition to which regional flow systems must be taken into account. These may require a larger area than for a soil survey. Again, as with soil surveys, a baseline groundwater sampling program must be conducted prior to CO_2 injection.

Samples should be analysed for as complete a chemical spectrum as possible, including major and minor ions, trace components, and stable isotopes (specifically ^{18}O, ^{2}H and ^{13}C). Once the initial baseline conditions have been established, it is suggested that fluctuations in water levels and chemistry monitoring can be limited to a survey every 3 to 5 years.

In the Weyburn study, there was a problem with attempts to sample the same wells every time because some wells were either abandoned or were no longer in use; hence, obtaining a fresh sample without extensive pumping was not possible.

This situation suggests that consideration should be given to developing a dedicated monitoring network of small-diameter wells strategically located throughout the study area if the monitoring of the quality of groundwater above the CO_2 injection area will need to be continued for long periods of time. Subject to regulatory requirements, and in the continued absence of adverse findings, annual groundwater sampling is probably adequate.

4.5.3 Reservoir Fluids

Geochemical monitoring of reservoir fluids is restricted to available wells and is usually only to determine changes in formation water composition. Although gases and oils were sampled during the Weyburn project they are not discussed in detail in the chapter.

With respect to formation water, determination of baseline conditions prior to CO_2 injection is essential. Following initial plume arrival, changes in formation water composition are relatively slow because of the generally slow kinetics of water–mineral mass transfer processes. Therefore, a sampling frequency of every six months is sufficient in most cases.

In contrast to soil gas and groundwater monitoring, data interpretation and compliance with anticipated results is usually straightforward in the context of CO_2 migration and of CO_2–oil–formation water–rock interactions. Injection-induced signals are typically quite pronounced, relative to background values, and temporal variations due to natural processes are subtle, even when discernible. Analyses should include major, minor and trace components and a suite of isotopes including, but not restricted to, stable hydrogen, oxygen, and carbon isotopes.

Finally, it is important to recognize that during (and after) CO_2 injection, close integration of site characterization, modelling, and monitoring programs through iterative inter-program feedback is essential to facilitate continuous optimization of monitoring activities. Surface, overburden, and reservoir monitoring programs must continuously evaluate the anticipated and potential impacts of injected CO_2 on soil gas, shallow groundwater, potable aquifers, and produced fluids in the context of local regulatory requirements.

5

GEOPHYSICAL MONITORING

Summary

An effective geophysical monitoring plan can track the subsurface distribution of CO_2, the associated pressure plume, and injection-induced deformation of the storage container. The characteristics of the storage complex are of particular importance, and affect the type of geophysical method to be used, and whether this method should be surface-based or well-based.

Characterization of the local rock–fluid–stress system is essential to the design of an appropriate geophysical monitoring plan. Specific geophysical monitoring methods are sensitive to variations in the parameters of interest (primarily CO_2 saturation and pressure). These parameters are most readily measured on core samples, complemented by such in-situ measurements as time-lapse logging. The composition of the injected CO_2 and features such as vugs and fractures are important variables to be determined.

Several monitoring methods were considered for application in the project. In general, however, regional monitoring methods are best applied in conjunction with other higher-resolution monitoring methods (for example, seismic).

Downhole seismic techniques tested at Weyburn included (1) continuous downhole passive seismic monitoring to detect microseismicity, and (2) active-source downhole seismic methods to acquire high resolution subsurface images. However, surface-based 3D seismic methods provide the most effective means of monitoring the subsurface distribution of CO_2 over a large area. They can image and hence detect CO_2. In order to minimize differences in the results, acquisition geometry and near-surface conditions should be the same for all surveys. Three-component data should be considered, as should good quality shear wave data. Combinations of the AVA intercept (I) and gradient (G) can be used to help distinguish CO_2 saturation from pressure-related effects.

Recommendations arising out this research begin on page 210.

Author: D. White (Geological Survey of Canada, Ottawa, Ontario).

5.1 Introduction

This chapter describes field-based geophysical monitoring methods either utilized or considered for the project, with a main focus on monitoring of the deep portion of the storage complex. The physical properties of the rock–fluid system that make them amenable to geophysical monitoring are considered as a function of depth, both for a generic rock column and then specifically for the Weyburn Field reservoir rocks. Also considered are feasibility studies for utilizing various non-seismic monitoring techniques. The downhole geophysical methods employed in the project are described and their effectiveness assessed. The most extensive section focuses on 3D seismic methods, which constituted the primary geophysical monitoring method employed in the project. Included are descriptions of a variety of analytical methods that have been applied to isolate CO_2 saturation versus pressure changes within the reservoir. Finally, consideration is given to the implications on monitoring programs in the case where no pre-CO_2 injection baseline data were acquired.

5.1.1 Context

Integrated monitoring of the target reservoir and overburden is required to document and verify the conformance of the CO_2 injection and storage process, as forecast by computational modelling. In addition, monitoring results contribute directly to the ongoing refinement of the geological model used for predictive modelling. As noted in Chapter 4, various imaging and sampling techniques are required to adequately monitor the diverse environments that must be considered, including the storage complex and near-surface regimes. Monitoring is also required to assess potential changes in the mechanical and hydraulic integrity of geological seals, and to identify potential leakage pathways.

Geophysical measurements provide a means to monitor various aspects of the CO_2 injection process, and related effects on the subsurface rock–fluid system. In particular, they can be used to track the large-scale migration of CO_2 within the storage complex, or leakage of CO_2 to zones above. In addition, they are sensitive to induced changes in the subsurface stress field (for example, pore pressure and matrix strain), and can monitor associated microseismic activity. The sensitivity of geophysical methods to a range of subsurface phenomena makes them a powerful tool for monitoring purposes. However, this sensitivity makes it difficult to uniquely associate individual geophysical monitoring observations with specific changes in the subsurface. This non-uniqueness can be partially resolved by combining individual data sets, by comparison with fluid-flow simulation results, and by applying sophisticated analysis techniques such as seismic amplitude-versus-offset (AVO) analysis.

5.1.2 **Objectives**

The primary objectives of geophysical monitoring in the context of CO_2 storage performance include:
1. Tracking the CO_2 plume within the primary storage reservoir.
2. Detecting migration of CO_2 above the primary reservoir seal.
3. Providing constraints on perturbations in the subsurface pressure field.
4. Assessing the significance of injection-related microseismicity with respect to the integrity of the CO_2 storage container and potential effects on surface infrastructure.
5. Continuing calibration of the geological model used for predictive modelling.

5.1.3 **Scope and Layout**

Geophysical monitoring forms one part of an integrated monitoring plan. As with geochemical monitoring, the scope and layout of a geophysical monitoring plan must be considered in light of the overriding monitoring objectives, as determined through an initial risk assessment. In particular, the monitoring plan should take into account (1) the anticipated growth rate and size of the CO_2 plume, (2) the extent of the pressure plume, (3) the depth, and thickness of the primary reservoir and storage complex, (4) ongoing assessment of cap rock integrity, and (5) the sensitivity of various geophysical measurements to the parameters of interest. Further, the geophysical responses to changes in CO_2 saturation and pressure need to be calibrated.

The first step in designing a monitoring plan is to consider the boundaries of the area to be monitored. This is sometimes referred to as the area of influence. For the purposes of designing a geophysical monitoring plan, there are two primary aspects that determine the size of the area to be monitored: (1) the maximum anticipated areal extent of the CO_2 plume, and (2) the size of the associated pressure plume. Typically, the lateral extent of the pressure plume will be comparable to that of the CO_2 plume in the case of an EOR operation. However, post-EOR CO_2 injection or saline aquifer storage may produce a pressure plume whose extent exceeds that of the CO_2 plume by an order of magnitude, and thus will generally require the utilization of different primary tools for monitoring these aspects of the storage site. The plume dimensions are estimated by running fluid flow simulations, using a static 3D geological model developed through geological characterization of the potential storage site. The accuracy of these estimates is dependent on the level of geological knowledge for the prospective storage site. In the case of an existing EOR field, models used for simulation may be well-informed, whereas in the case of newly established saline aquifer sites, geological knowledge will likely be limited to what is available from a few existing or test wells. A strategy is required to deal with the inherent uncertainties involved in estimating the size of the CO_2 and pressure plume. A conservative approach may be adopted, whereby a monitoring area is designated which greatly exceeds the model-based plume size estimates. Alternatively, a smaller monitoring area may be chosen initially, and then be expanded if necessary, based on the results of the ongoing monitoring during CO_2 injection. This is consistent with regulatory requirements that dictate recurrent updating of monitoring plans over the lifetime of a CO_2 storage project.

Having ascertained the area to be monitored, a decision is then needed on the level of detail for the monitoring program. This will generally be dictated by local regulatory requirements. Typical questions to be answered might be: Is the detailed pressure field and CO_2 distribution and saturation within the reservoir required? Is the general outline of the CO_2 plume within the reservoir adequate? Do we only need to be able to detect CO_2 that has reached shallow depths above the storage complex? These requirements will fundamentally determine which geophysical monitoring techniques are suitable in a specific monitoring program. For the most stringent monitoring requirements, seismic reflection techniques are the only likely candidate, whereas for plume delineation and leakage detection, a variety of less-intense geophysical techniques (for example, electrical, electromagnetic, gravity) may be adequate. For pressure plume monitoring, interferometric synthetic aperture radar (InSAR) methods provide the only practical geophysical means of monitoring a large area. In storage situations where cap rock integrity is considered to be a significant risk (for example, a thin cap rock that could be subject to fracturing), microseismic monitoring could be an important part of the monitoring program.

Once the depth intervals of interest and the required levels of detail for monitoring have been established, the characteristics of the storage complex and overburden must be considered. These should be evaluated in conjunction with the expected CO_2 injection rates and injection lifetime, to determine which geophysical monitoring methods are capable of achieving the monitoring objectives for the site. For example, the depth, thickness, lithology (clastic versus carbonate), porosity, mechanical properties and ambient pore fluid properties will determine which geophysical monitoring methods are capable of resolving migration of the CO_2 plume within the reservoir, for different injection scenarios.

Similarly, these same characteristics of the overburden need to be considered in regard to detecting CO_2 leakage. In the overburden, CO_2 quantities would generally be expected to be smaller than in the reservoir, making detection more difficult. However, the difficulty of detecting smaller quantities of CO_2 will be offset by (1) the increased sensitivity of surface-based geophysical monitoring methods to the presence of CO_2 at shallower depths (due to the improved depth resolution), and (2) increased difference in pore fluid properties (CO_2 versus fresh water) at these depths.

In the case of time-lapse seismic imaging, elements of the overburden must be considered (such as karst, salt bodies, complex structures) that may diminish the ability to image CO_2 in the underlying storage reservoir. Again, most of this information will be based on core samples and logs from existing wells for the storage site. Surface infrastructure must also be considered in designing a monitoring program, as the various surface-based geophysical methods require varying degrees of land access, with their associated cultural and other problems.

A key element of any geophysical monitoring program will be calibration ('ground-truthing') of the various geophysical monitoring results. This entails in situ (that is, well-based) measurements, where the changes in pore-fluid saturation and pore pressure can be measured directly and compared to wellbore geophysical measurements. These measurements provide the means of linking surface-based monitoring results with actual changes in the subsurface.

Geophysical surveys are most commonly conducted using some regular spatial sampling interval, which determines the depth and areal resolution for the survey. The maximum depth and areal resolution achievable is also fundamentally limited by the distance from the target, regardless of the spatial sampling used in a given survey. Thus, geophysical monitoring surveys can be designed to achieve a required level of resolution with a minimum number of sampling points to reduce costs, subject to the maximum depth-based resolution restrictions. This is a standard process of geophysical survey design that is not unique to CO_2 storage monitoring.

The frequency of monitor surveys will be dictated by the rate at which the CO_2 distribution and pressure field show significant change. A minimum time between monitor surveys corresponds to the time required for there to be a change that is detectable by the adopted geophysical monitoring techniques. For example, in the project, the first time-lapse seismic monitor survey was conducted one year after the start of injection, at which time there was clear detection of the CO_2 distribution within the reservoir. During the initial phases of injection, monitoring surveys need to be conducted more often, (1) as a means of developing confidence in the monitoring results, (2) to verify that the initial predictions from the flow simulations are reasonable, and (3) to update the geological model where there are significant departures from the predicted behaviour. As the performance of the storage site demonstrates conformance, the intervals between monitoring surveys can be reduced. However, the frequency of monitoring may be increased at any time if there are significant changes in key indicators that suggest non-conformance or leakage from the storage complex. Following cessation of CO_2 injection, monitoring would likely continue until it can be demonstrated that the storage system is behaving predictably and is trending toward stability.

5.2 Geophysical Characterization of the Rock–Fluid System

The effectiveness of geophysical monitoring for CO_2 plume distribution is fundamentally controlled by the depth-dependence of the physical properties of CO_2, and their contrast with other constituent pore fluids (formation water and oil). To illustrate this, we first considered the depth-dependence of the physical properties of CO_2 and their contrast with the properties of formation water, which constitutes the most common rock pore constituent in sedimentary basins. Next, the physical effects of substituting CO_2 for formation water were considered for a simple homogeneous rock column (a high porosity sandstone), to demonstrate the sensitivity of the relevant physical parameters to the presence of CO_2. For the purposes of monitoring, this hypothetical scenario corresponds to the situation where CO_2 is injected into a saline aquifer for storage, or where CO_2 has migrated out of the storage reservoir. We focused on the properties relevant to gravity (density) and seismic monitoring, specifically P-velocity (Vp) and acoustic impedance (AI), because these are most relevant to the geophysical monitoring methods deployed in the project. The reservoir rock–fluid system was then characterized, as should be done for any fluid system in a CO_2 storage project. As described earlier, the Weyburn reservoir is a fractured carbonate and, of course, contains hydrocarbons. The presence of oil as a third pore fluid constituent complicates the characterization process significantly.

5.2.1 Fluid Properties versus Depth

Figure 5-1 (left) shows the relation of CO_2 and formation water density to Vp and acoustic impedance, assuming hydrostatic pore pressure. The depth dependency of properties for CO_2 changes across the phase transition from supercritical gas to vapour, which occurs at about 700 m depth in this case. The depth at which this transition occurs will depend on the pressure-temperature regime.

All of the properties of formation water have values that are an order of magnitude greater than for CO_2, except for density for supercritical-phase CO_2, which is 60–70% of the density of formation water. All of the properties of CO_2 show a general increase with depth, except for Vp in the vapour state, which has the opposite trend. Vp for CO_2 decreases with depth toward the phase transition, below which Vp increases with depth. Thus, seismic traveltime sensitivity to the presence of CO_2 will decrease with depth below the phase transition.

The contrast in acoustic impedance of CO_2 versus formation water decreases with depth, indicating that reflection amplitudes will be more sensitive to the presence of CO_2 at shallower depths.

The dramatic decrease in density of CO_2 across the supercritical gas-to-gas transition means that the volume of a fixed mass of CO_2 will increase dramatically at shallow depths. This will greatly increase the sensitivity of seismic and gravity methods for detecting CO_2 that migrates upward to shallow depths.

Figure 5-1 (right) shows the Vp-versus-depth curves calculated for various CO_2–formation water mixtures. Whereas the mean density (not shown) is directly proportional to CO_2-saturation, small CO_2 saturations result in disproportionately large decreases in Vp for the composite fluid. This effect is accentuated for depths near and shallower than the phase transition from supercritical gas to vapour. For example, at 500 m depth, 1% CO_2 saturation results in a Vp reduction of >50%, whereas 60% CO_2 saturation results in a Vp reduction of 90%. For comparison, 1% CO_2 saturation at 1500 m depth results in Vp decreasing by 10%, as compared to 75% for 60% CO_2 saturation.

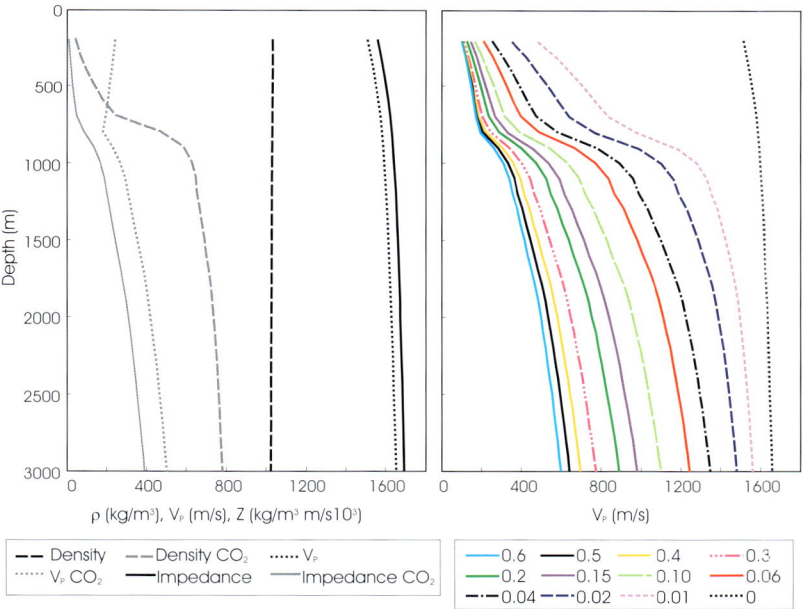

Figure 5-1. Left. *Density, Vp, and acoustic impedance (Z) versus depth for CO_2 and formation water. A temperature-versus-depth profile from Vigrass et al. (2007) was used, and pore pressures are assumed to be hydrostatic. The isothermal properties (density and Vp) of CO_2 were calculated using the NIST online tool (NIST, 2010). Equations 27, 28 and 29 of Batzle and Wang (1992) were used to determine the formation water properties for an assumed salinity of NaCl of 50 000 ppm.*
Right. *The change of Vp-versus-depth for various formation water–CO_2 mixtures. It is assumed that the CO_2 does not dissolve in the formation water. The acoustic velocity is given by the Reuss (isostress) average of the individual bulk moduli and the average density. The labeled line patterns refer to CO_2 saturation (that is, 0 represents pure formation water).*

5.2.2 Saturated Rock Properties versus Depth

The physical properties for sandstone (porosity fraction 0.2) were calculated using the Gassmann relation for fluid substitution and the fluid properties from the previous section. The characteristic properties for the sandstone are shown below.

Dry rock bulk modulus	12 GPa
Dry rock shear modulus	22 GPa
Mineral bulk/shear modulus	36 GPa, 70.2 GPa
Mineral bulk density	2650 kg/m^3
Fractional porosity	0.20

The variation of physical properties with depth was calculated for various degrees of CO_2 replacing formation water (Figure 5-2). Whereas the behaviour of density and acoustic impedance versus depth are similar to the case of the mixed fluids shown above, a more complex behaviour is observed for Vp.

Maximum observed decreases of approximately 5%, 4.5%, and 8%, are observed in density, Vp, and acoustic impedance, respectively. Both Vp and acoustic impedance show increased sensitivity to small values of CO_2 saturation as depth decreases, with an abrupt increase in sensitivity observed across the CO_2 phase transition from supercritical gas to vapour phase. Vp demonstrates complex behaviour where CO_2 is in the vapour phase: for CO_2 saturations <10%, Vp decreases with increasing CO_2 saturation; for higher CO_2 saturations (>10%) Vp actually starts to increase again. These results suggest that both seismic traveltimes and amplitudes will be increasingly sensitive to the presence of small saturations of CO_2 as depth decreases. As a result, the presence of CO_2 at shallow depths will be easier to detect, but the ability to quantify CO_2 saturation decreases. The sensitivity of all of the physical properties changes abruptly across the phase transition from supercritical gas to vapour phase CO_2. Thus, the knowledge of where this transition occurs is important for monitoring purposes.

5.3 Geophysical Characterization of the Weyburn Reservoir

Determination of rock and fluid properties at the specific stress conditions of the storage reservoir is an important first step in the successful implementation of CO_2 monitoring. Rock and fluid properties are required to accurately model and invert time-lapse geophysical measurements in terms of changes in the reservoir properties. For example, time-lapse seismic data can be inverted for changes in reservoir pressure and fluid saturation. Reservoir rock properties are commonly characterized by some combination of laboratory measurements made on core samples and of in-situ downhole logging. In the case of CO_2-EOR, the pore fluid system is complicated by the presence of hydrocarbons, in addition to formation water and injected CO_2.

This section illustrates how seismic properties were determined for the Weyburn reservoir. An analysis of the dry-frame properties of the reservoir rock is combined with determined fluid properties to obtain the saturated density, P-velocity, and S-velocity of the reservoir. The dry-frame properties are determined using data from ultrasonic core measurements, and the calculated fluid properties of oil–formation water–CO_2 mixtures. The complex properties of supercritical CO_2 and miscible oil systems are accounted for over a wide range of reservoir pressures and fluid saturations expected at Weyburn.

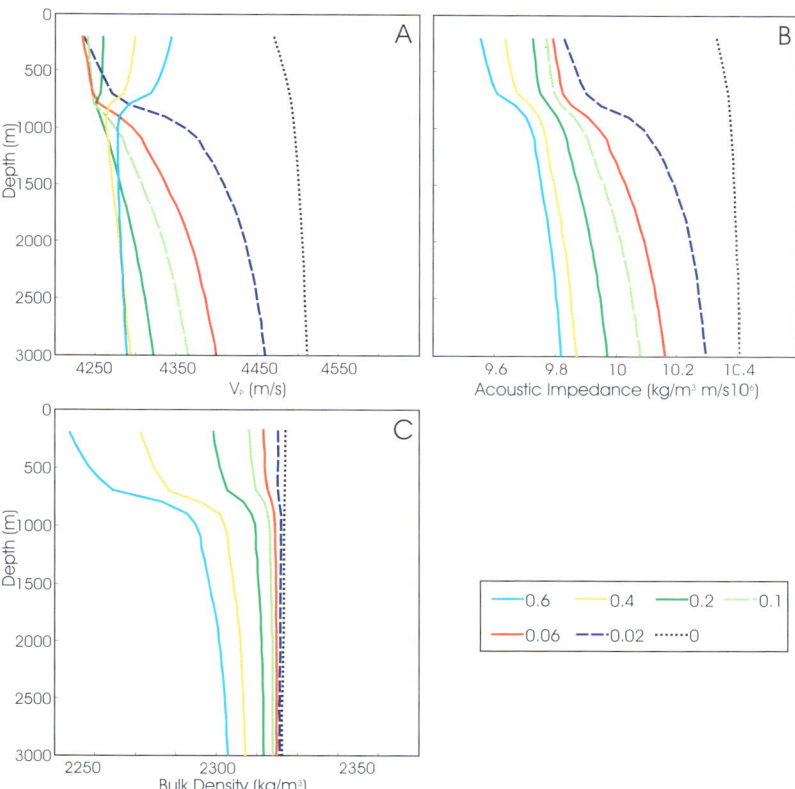

Figure 5-2. **A.** V_p, **B.** *acoustic impedance, and* **C.** *bulk density versus depth for a hypothetical sandstone with porosity fraction 0.2 and various CO_2 saturations (CO_2 replacing formation water). Pressure effects on the dry rock matrix were not considered. The labeled line patterns refer to CO_2 saturation (that is, 0 represents pure formation water).*

5.3.1 Reservoir Conditions

To determine the appropriate pressure range for analysis of reservoir properties, the expected confining and pore pressures within the reservoir must be determined. At Weyburn, the overburden pressure is approximately 34 MPa at reservoir depth, as determined from density logs. The minimum horizontal stress is estimated to be 18–22 MPa (McLellan et al., 1992), which is consistent with a horizontal stress of 22 MPa, which would result from the case of an elastic, isotropic reservoir with no lateral strain and with a Poisson's ratio of 0.33 (Brown, 2002). The average Poisson's ratio calculated from P- and S-wave sonic logs in this area is ~0.3, so a horizontal compressive stress of 22 MPa at Weyburn is reasonable.

It is important to consider how closely reservoir pore pressure approaches in-situ stress levels, because the dry rock frame is very sensitive to changes in effective pressure as the pore pressure approaches the minimum effective stress. At zero effective pressure, the rock approaches its fracture point, and all dry moduli values are drastically reduced. The pressure history in the Phase 1A area shows that through 2001, CO_2 injection pressures sometimes exceeded 25 MPa, with average reservoir pressures in the range of 17–22 MPa.

The determination of the rock and fluid physical properties comprises three main activities: (1) an analysis of the dry-frame properties of the reservoir rock, (2) analysis of properties of the formation water, oil, and injected CO_2, and (3) study of the saturated density, P-velocity, and S-velocity of the reservoir, obtained by combining the dry rock and fluid properties.

5.3.2 Dry-frame Properties

The bulk modulus, shear modulus, and density of the dry rock frame are required to calculate saturated rock properties by fluid substitution. It is assumed that only porosity and effective stress are relevant in determining how the dry-frame responds to varying reservoir conditions, with other variables (such as mineralogy, grain contact, pore structure, and temperature) remaining fixed at some average value within the reservoir.

Average dry-frame densities (ρ_{dry}) of 2.04 g/cm³ and 2.44 g/cm³ were determined for the Marly and Vuggy units, respectively, using grain densities based on cores (Brown, 2002), and average porosities of 0.29 and 0.1, measured from well logs. Average dry bulk (K_{dry}) and shear (G_{dry}) moduli values for the Marly and Vuggy units were determined from logs from the same well at an estimated effective pressure of 7 MPa. The dry moduli values were calculated by removing the fluid effect from the log values using Gassmann's equation (Gassmann, 1951). Pressure dependence of the dry elastic moduli was estimated from core measurements over effective pressures ranging from 2–35 MPa (from Brown, 2002). The resultant curves for the Marly unit are shown (for example) in Figure 5-3.

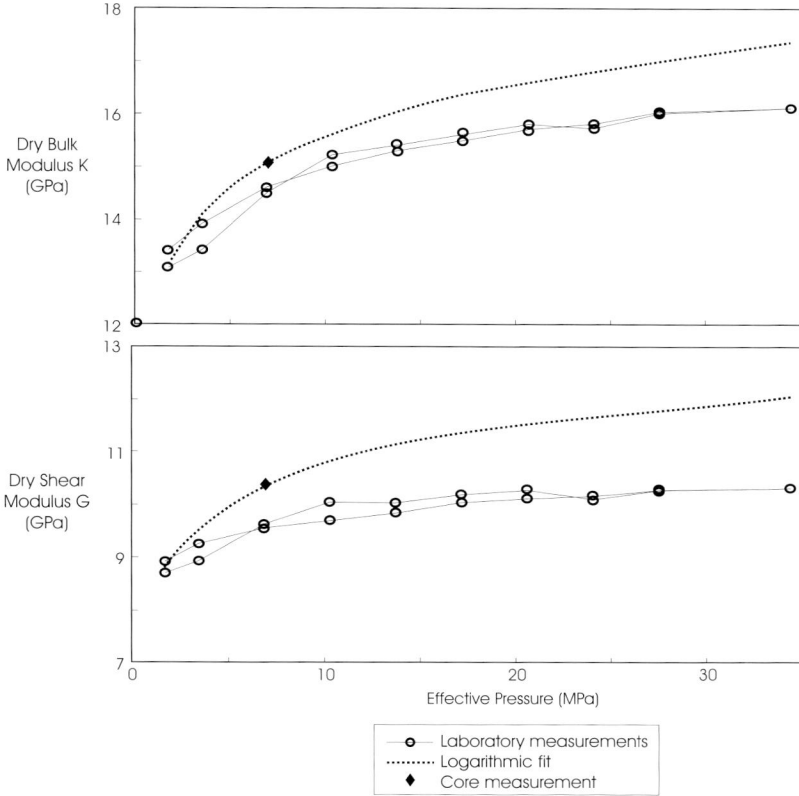

Figure 5-3. Top. *Dry bulk modulus (K) versus effective pressure for the Marly unit.* **Bottom.** *Dry shear modulus (G) versus effective pressure for the Marly unit. For both determinations, a logarithmic curve (dotted line) was least-squares fitted to the laboratory measurements (circles) of a core sample. The fitted curve was then adjusted so that it coincided with the well log measurements (diamond) at the ambient reservoir effective pressure. Each data set was fitted to a similar type of logarithmic curve (not shown).*

5.3.3 Fluid Properties

The density and bulk modulus of both pure, and mixtures of, formation water, oil, and injected CO_2 are required as functions of pressure and saturation, and in some cases temperature may be relevant. Thermodynamic properties of CO_2 can be complex for temperature and pressure regimes near the critical point (31.1°C, 7.38 MPa), which may commonly occur in CO_2-EOR and storage projects. Further complications arise when supercritical CO_2 is contaminated with even small amounts of impurities, which can change the properties significantly from those of its pure state (Lindeberg et al., 2000).

The behaviour of mixtures of oil and supercritical CO_2 is even more complex than that of pure CO_2. Weyburn oil readily absorbs up to 66 mol % of CO_2 for pressures exceeding the minimum miscibility pressure (14–17 MPa). This generally raises the bubble point pressure, the gas-oil ratio, and gas gravity of the oil–CO_2 mixture, and significantly lowers its bulk modulus. Higher concentrations of CO_2 can produce a second liquid phase consisting mostly of CO_2 mixed with light hydrocarbons, the properties of which are intermediate between pure CO_2 and the CO_2-saturated oil.

The density of a formation water–CO_2 mixture generally increases by no more than 1–2%, so for the relatively short time period (<10 a) involved in monitoring activities at Weyburn, the effects of CO_2 dissolution in formation water are negligible. The density and bulk modulus of formation water do not vary significantly with pressure, although formation water salinity can have a sizeable effect. For example, at 63°C (the temperature of the Weyburn reservoir) a salinity change from 20 000 to 230 000 ppm results in density and bulk modulus increases of 15% and 40%, respectively. A salinity of 85 000 ppm was used for modelling.

The density and bulk modulus of supercritical CO_2 are shown in Figure 5-4, for both pure CO_2 and for the CO_2 injected at Weyburn. Although the latter contains >96% CO_2, additional impurities are enough to lower both the density and bulk modulus by up to 5%. In this study, the computer program STRAPP was used to model all oil–CO_2 mixtures. Density and bulk modulus for Weyburn oil are shown in Figure 5-5. For oil–CO_2 mixtures, the density of Weyburn oil varies widely as a function of CO_2 saturation and pressure, as shown in Figure 5-6. For CO_2 saturations <66 mol %, a liquid phase exists in which density increases by a maximum of ~10% with increasing CO_2 saturation. A vapour phase also exists for pressures below the bubble point pressure. Above 66 mol % CO_2 the vapour phase becomes more continuous and more liquid-like at high pressures (around 15 MPa).

For pressures beyond the minimum miscibility pressure (~17 MPa), and for CO_2 saturation levels > 75%, the vapour phase becomes increasingly dominant until it becomes the only existing phase. The bulk modulus behaviour (Figure 5-6) is similar to that of the density, except that the bulk modulus values for the liquid phase have a wider range, and change more abruptly at the bubble point pressure.

Average bulk fluid properties of oil–CO_2 mixtures in the presence of two phases are obtained by averaging the properties of the individual phases according to their molar volumes (Figure 5-6). For both density and bulk modulus, the vapour phase dominates the bulk average at low pressures because of its larger volume. As a result, a sharp discontinuity exists at the bubble point pressure as the mixture transitions from a gas-like to a liquid-like state. At high pressures,

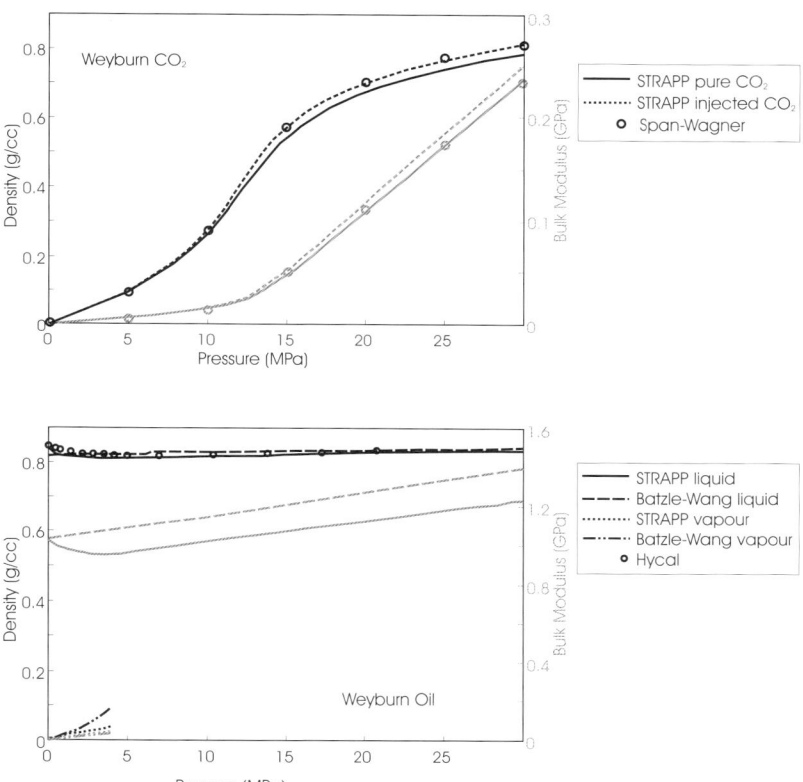

Figure 5-4. Top. *Weyburn CO_2 bulk density and bulk modulus as a function of pressure at reservoir temperature (63°C). The CO_2 (solid grey curve) has a lower bulk modulus than pure CO_2 (dashed grey curve) over the entire reservoir pressure range. However, the bulk modulus of pure CO_2 from the Span–Wagner equation of state (Span and Wagner, 1996) differs from that predicted by STRAPP by as much as 10%.* **Bottom.** *Weyburn oil density and bulk modulus as a function of pressure at reservoir temperature (63°C). The STRAPP oil density of the liquid phase is very close to the oil density measured by Hycal. The bubble point pressure from STRAPP lies within the range measured from the PVT analysis. Note that the oil density decreases up to the bubble point, then increases with increasing pressure. The Batzle–Wang relations (Batzle and Wang, 1992) predict a liquid phase with a density that is close to the Hycal curve, but they fail to capture the proper behaviour at low pressure. In addition, the Batzle–Wang vapour density deviates significantly from that predicted by STRAPP.*

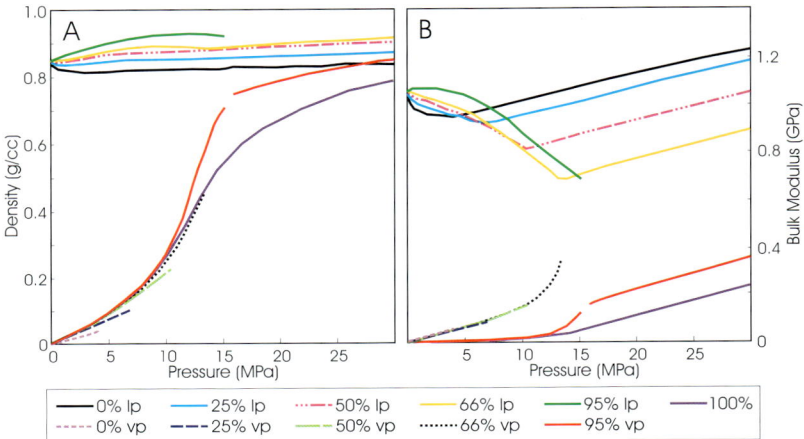

Figure 5-5. *Properties of Weyburn oil as a function of pressure and CO_2 saturation at reservoir temperature (63°C).* **A.** *Density.* **B.** *Bulk modulus. The labeled line patterns correspond to mol % CO_2 saturation (0% represents pure oil). For CO_2 saturations of up to 66 mol %, STRAPP predicts a liquid phase (lp) whose density varies within a fairly narrow range, and a vapour phase (vp) whose density and bubble point pressure both increase with increasing amounts of CO_2. For CO_2 saturations >66 mol %, the vapour phase becomes more continuous and more liquid-like as the pressure increases, while the liquid phase becomes unstable for pressures beyond the minimum miscibility pressure (about 15 MPa). The small gap in the 95% CO_2 vapour curve is a region of non-convergence in STRAPP. Such gaps are more prevalent for CO_2 saturations between 75% and 95%.*

the density actually reverses its increasing trend once the CO_2 saturation reaches 66% as the unabsorbed, residual CO_2 preferentially contributes more of its supercritical properties to the mixture.

5.3.4 Saturated Rock Properties

The saturated bulk modulus (K_{sat}) of the rock can be calculated for various pore fluid mixtures, using the dry bulk modulus (K_{dry}) and Gassmann's equation. Similarly, the saturated density can be derived from the dry density and the fluid densities calculated from the fluid properties analysis in the previous section. The dry and saturated (G_{sat}) shear moduli are assumed to be equal. It is then straightforward to derive the saturated Vp_{sat} and Vs_{sat} values. For illustration, the density and seismic properties calculated for the Marly reservoir unit are shown in Figure 5-7. For each scenario depicted, it is assumed that CO_2 displaces only

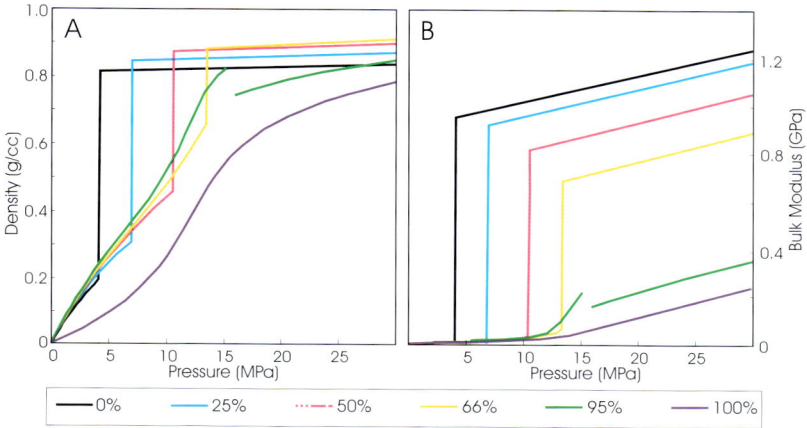

Figure 5-6. *Properties of Weyburn oil as a function of pressure and CO_2 saturation at reservoir temperature (63°C).* **A.** *Average density.* **B.** *Bulk modulus. The labeled line patterns correspond to mol % CO_2 saturation (0% represents pure oil). For low pressures, the vapour phase dominates because its larger volume more strongly influences the average of the two phases than does the liquid phase. Note the sharp discontinuity in density and bulk modulus at the bubble point pressure. In determining the average properties, an arithmetic mean was used for densities, whereas a harmonic mean was appropriate for the bulk modulus.*

the oil, and that the water saturation remains constant (at 0.47 in the Marly), except for the case when oil swells as it absorbs the CO_2. Weyburn oil can swell by as much as 30% (White et al., 2004) as it absorbs CO_2, which consequently displaces formation water from the pores. As can be seen, maximum saturation-related changes in Vp are ~5%, as compared to ~7% if the effective pressure decreases sufficiently, for example, down to 2 MPa, as might occur near an injector. Vs is much more sensitive to pressure changes than saturation changes.

The seismic impedance variations (which control seismic amplitude changes) show similar behaviour to the corresponding Vp and Vs curves. P-impedance values show maximum decreases of ~6% for either pressure- or saturation-related effects. The S-impedance is relatively insensitive to saturation changes, and decreases no more than 1.5%, but can decrease by as much as 6% if the effective pressure decreases from 9 MPa to 2 MPa.

Larger saturation effects on the seismic properties can be obtained if displacement of formation water by CO_2 is considered. For example, Brown (2002) estimated Vp changes of –6% for complete replacement of formation water by

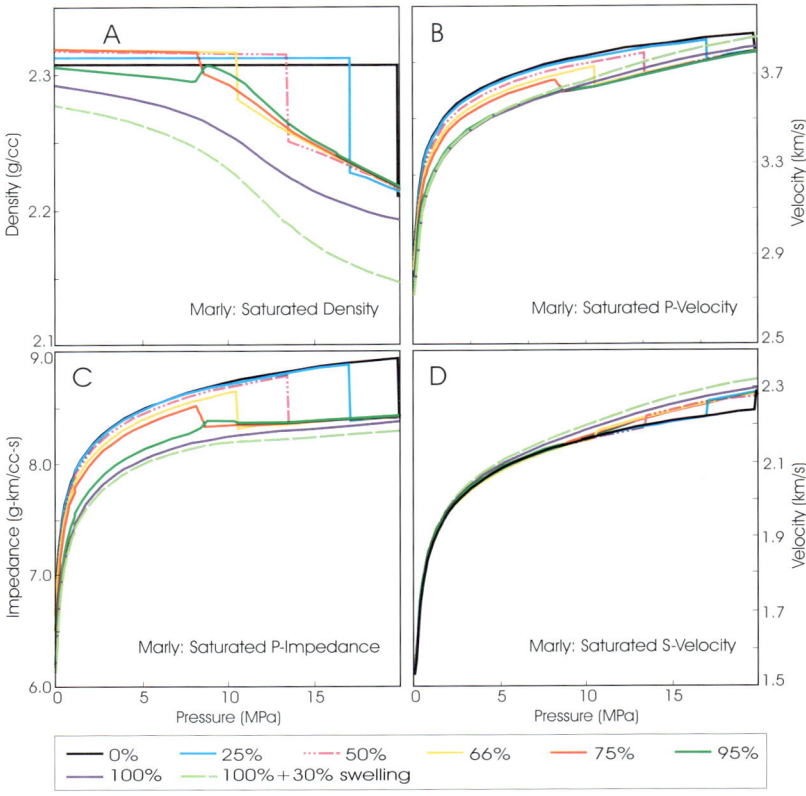

Figure 5-7. A. *Saturated density,* **B.** *Vp,* **C.** *Acoustic impedance, and* **D.** *Vs for the Marly dolostone as a function of effective pressure and CO_2 saturation at reservoir temperature (63°C) and average porosity 0.29. The labeled line patterns correspond to mol % CO_2 saturation (0% represents pure oil). Ambient reservoir effective pressure is assumed to be 9 MPa (for 15 MPa pore pressure and a conservative estimated value for minimum compressive stress of 24 MPa).*

CO_2 in the Marly unit, whereas White et al. (2004) estimated maximum saturation-related changes of up to 10% for 100% formation water replaced by 70% CO_2 and 30% formation water. The expected modes of fluid displacement in the reservoir are clearly important in determining which of these values is most representative.

5.3.5 **Conclusions**

The following conclusions derive from the general analysis of (1) the physical properties of constituent pore fluids versus depth, and (2) physical rock and fluid properties specific to the Weyburn reservoir.

The contrast in physical properties (density and bulk modulus) between formation water and CO_2 generally increases with decreasing depth. This implies that the sensitivity of surface-based geophysical methods will increase as the zone of fluid substitution decreases in depth. The contrast in fluid properties changes dramatically across the depth interval where the transition from CO_2 gas to supercritical gas occurs. Thus, knowledge of this depth is important for estimating geophysical responses.

Increase of pore pressure affects seismic rock properties significantly more than a decrease of pore pressure, because of the increased sensitivity of seismic properties near zero effective pressure. Thus in assessing the magnitude of potential changes in seismic properties in the injection zone, it is important to know, as accurately as possible, (1) the pore pressure, (2) the overburden pressure, and (3) the level of minimum compressive stress. Small pressure deviations in areas with low effective pressures can have large effects on seismic modelling and inversion results.

Rock physics analyses should include the complexities of CO_2 fluid physics, as well as the individual components of the live oil and injected CO_2, when calculating saturated rock properties. Supercritical CO_2 does not behave like a hydrocarbon gas, and miscible oil–CO_2 properties often deviate from those of simple oil–gas mixtures. For this purpose, the use of an accurate, multiphase compositional simulator is highly recommended.

Pressure effects on seismic properties are at least as large as CO_2 saturation effects for the scenarios examined. In the Marly unit, values for P- and S-impedance can change by 6% for a 7 MPa increase in pore pressure over ambient conditions. In the Vuggy unit, the effect is about half that amount. For large increases of CO_2 saturation, the P-impedance can change by 6% or more in the Marly, and 4% in the Vuggy – but the S-impedance change is significantly smaller. The scenarios that were examined included large pore pressure increases, and CO_2 saturation increases in rock partially saturated with oil. It should be noted, however, that such sensitivity studies are highly dependent on the range of pressures and saturations considered and the initial properties of the rock.

It is important to examine the effects of reservoir changes on seismic data modelled from realistic reservoir models. As noted above, the estimated change in seismic parameters is dependent on assumptions about the modes of pore fluid substitution, in the case where hydrocarbons are present in the injection zone.

5.4 Feasibility Studies
5.4.1 InSAR
Technology

Interferometric synthetic aperture radar (InSAR) is a satellite-based remote sensing technique that is capable of identifying and accurately measuring vertical ground displacements on a scale of millimetres per year, over areas ranging from 1 to 10 000 km². As a monitoring technique, InSAR has the advantage of being almost completely non-invasive. It has been successfully applied to measuring surface displacements that are caused by CO_2 injection (Rutqvist et al., 2010). The displacements mark the footprint of surface uplift, or subsidence, that occur due to subsurface pressure changes associated with activities such as CO_2 injection and oil production. The areal extent of the subsurface pressure

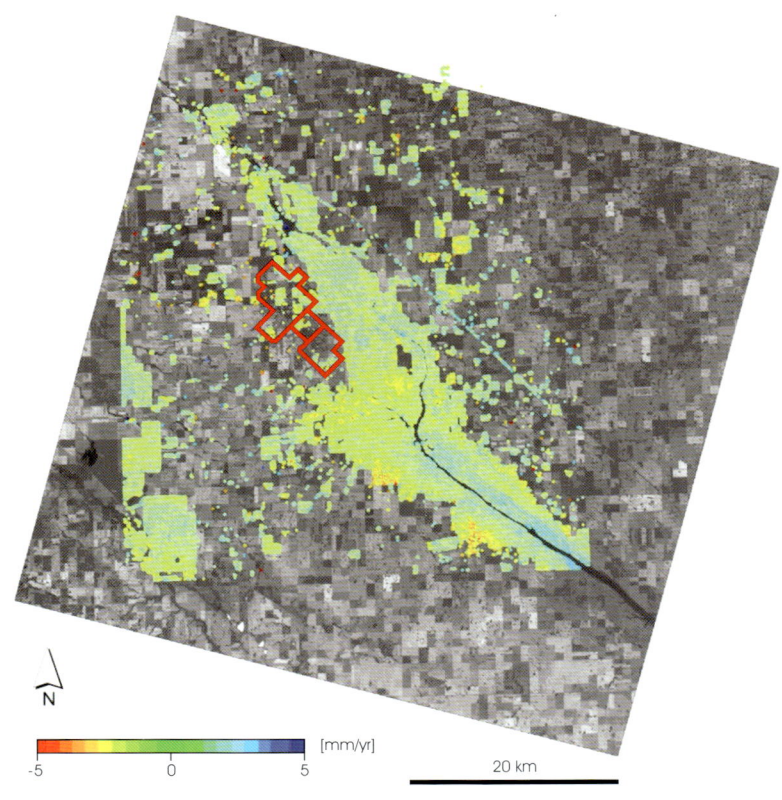

Figure 5-8. *Preliminary result of PSInSAR processing of ERS1/2 data in the vicinity of the Weyburn Field. Image provided by TRE Canada Inc.*

plume is typically much larger than the corresponding CO_2 plume, and thus will commonly define the 'area of influence' for a CO_2 storage project. Therefore, the monitoring of this pressure plume can be an important element in satisfying regulatory requirements. At the In Salah CO_2 injection site in North Africa, the application of InSAR led to the detection of reservoir non-conformance (Rutqvist et al., 2010).

Another advantage of this method is the potential it affords to perform historical analyses, because satellite radar data are generally archived and may therefore be available for prospective sites. This provides the opportunity to establish pre-injection baseline surface behaviour at CO_2 storage sites after the fact, and to apply feasibility analysis for monitoring specific sites under a variety of seasonal conditions, provided that historical satellite imaging coverage is good, and there are sufficient natural radar reflectors.

InSAR monitoring at the Weyburn Field was considered at the outset of the project, but due to the depth of the reservoir (~1450 m) and the estimated mechanical stiffness of the reservoir and overburden, it was thought that surface ground deformation would be difficult to measure. Measurement accuracy would also be hampered by seasonal variability in ground conditions, including ground cover (snow or vegetation), near-surface freezing in the winter, and wet-dry cycles in spring, summer, and fall. However, with the success of InSAR monitoring at the In Salah Field, where the reservoir is deeper than Weyburn (~1900 m depth), the possibility of InSAR monitoring was revisited. Toward this end, historical data were used to conduct a preliminary feasibility assessment. Figure 5-8 shows preliminary results of PSInSAR (permanent scatterers InSAR) processing of ERS1/2 satellite data. As can be seen, the coverage achieved is inadequate in the Weyburn Field monitoring area (outlined in red). These results indicate that installation of 'permanent scatterers' is required if InSAR monitoring is to be effective under Weyburn-like conditions.

Conclusions

InSAR provides an unobtrusive means of monitoring large areas for ground displacement associated with CO_2 injection operations. An understanding of what the ground displacement means in terms of the subsurface pressure regime and the spread of CO_2 can only be assessed in conjunction with a geomechanical model for the area (for example, Rutqvist et al., 2010; Morris et al., 2011).

In the absence of installed permanent scatterers, feasibility studies may be conducted with existing historical data to assess the likelihood of successful InSAR monitoring. In continental climates and landscapes like Weyburn, at least one full season of pre-injection monitoring is likely required to assess the seasonal variations in ground movements that are unrelated to CO_2 injection operations.

5.4.2 **Long-electrode Electrical Resistance Tomography**
Technology
Electrical tomography methods have been used successfully to monitor shallow subsurface migration of various fluids (for example, Ramirez et al., 1995; Binley et al., 1996; Slater et al., 1997; Kemna et al., 2000). These results suggest that electrical tomography may be technically feasible for deep CO_2 plume monitoring. The metal casing in boreholes can potentially serve as electrodes, which are used to pass an electrical current into the subsurface and to measure the resultant electrical field. Such measurements form the basis for electrical resistance imaging (long-electrode electrical resistance tomography or LEERT). Based on the higher resistivity of CO_2, relative to formation water, LEERT allows tracking of the CO_2 plume. The use of existing casings could potentially provide a non-invasive and cost-effective method for monitoring the spatial distribution of CO_2 within the reservoir.

To evaluate whether LEERT could be used successfully to monitor CO_2 distribution in the Weyburn reservoir, a numerical model was designed to assess the sensitivity and resolution of the method in a realistic field environment. Various conditions were tried to test (1) the effects of the distribution and number of electrode wells, (2) the depth of the reservoir, and (3) potential modifications to improve sensitivity at the reservoir level. Options tested included the use of 8 to 22 wells as electrodes, extension of the electrodes into the reservoir, and a shallower reservoir depth (820 m rather than 1450 m). Synthetic field data were calculated using a 3D finite-difference method. Sensitivity of the measurements was assessed against field experience that suggests that resistance measurements with magnitudes $>10^{-5}$ Ω and relative changes of $>2\%$ are likely required to achieve acceptable signal-to-noise ratios for tomographic inversion.

The modelling study indicates that none of the conditions which were considered produce data with adequate signal-to-noise ratio and sensitivity to image the resultant CO_2 plume, although in some cases the measurements were able to at least detect changes due to the presence of CO_2. Tomographic inversion of the synthetic data produced resistivity images that had little resemblance to the true resistivity calculated for the CO_2 models. The small magnitude and inhomogeneity of the induced current densities within the reservoir are the underlying causes of the low sensitivity of the modelled electrical responses and of the resultant poor tomographic images.

The lack of adequate current density within the reservoir for successful deployment of LEERT is likely due to several factors, including the long well casings (sensitivity is inversely proportional to casing length), the lower resistivity of the overburden relative to the reservoir, the large inter-electrode (well) distances, the presence of other well casings that distort the electric field, and the fact that the well casings terminated above the reservoir zone. Attempts to provide better coupling of electrodes with the reservoir through extension of the casing, or by placement of short electrodes within the reservoir, failed to provide any significant improvement.

Conclusions

LEERT is unsuited for direct CO_2 monitoring in the Weyburn Field, but may be more suitable in areas where the overburden is more resistive, the reservoir is significantly shallower, or a combination of both. However, the method may be better suited to monitoring the overburden for CO_2 leakage. In any case, pre-acquisition modelling is essential in assessing the feasibility of LEERT to specific geological and well geometry in an EOR field.

Interwell distance was a limitation in the case of the Weyburn Field, which has a very high density of wells. This does not bode well for the use of LEERT at other EOR sites which will likely have fewer well casings to use for electrodes. However, a reduction in the number of steel-cased wells that are not used as electrodes will reduce the associated distortion effects.

In light of the few field demonstrations of LEERT for deep CO_2 storage monitoring, and the non-commercial equipment, it currently falls into the category of research methods.

5.4.3 Time-lapse Gravity

Technology

Time-lapse gravity measurements have been utilized for monitoring subsurface fluids in the context of groundwater studies (for example, Liard et al., 2011), deep injection of water (Brady et al., 2006), and CO_2 injection at Otway (Sherlock et al., 2006) and Sleipner (Alnes et al., 2008). In the case of CO_2 injection, the density difference between CO_2 and formation water results in a local change in the gravitational field. This is used to infer the mass of CO_2, relative to formation water, at a particular location, and so indirectly assess the extent of the CO_2 plume. As shown above, the density difference between these two pore fluids increases sharply, moving upward across the zone where CO_2 changes from a supercritical gas to the gas phase. This increased density contrast improves the likelihood that time-lapse gravity monitoring is capable of detecting zones where CO_2 is present.

Surface gravity measurements were considered as a means of monitoring CO_2 injection at the Weyburn Field. However, the expected changes in gravity due to injection were deemed to be below the sensitivity threshold of the measurements, due to the depth (1450 m) and thin vertical extent (20–30 m) of the reservoir. Figure 5-9 shows the calculated maximum change in gravity due to CO_2 injection, and predicts gravity changes of <3 microgals (1 microgal = 10^{-8} m/s^2) for >1.3 Mt of CO_2 at 1400 m depth. This change is relatively small compared to the precision levels for absolute or relative gravimeters [approximately 2 microgals (Niebauer et al., 1995), and 5 microgals, respectively], or seasonal gravity variations of ±10 microgals attributed to changes in groundwater levels (Liard et al., 2011). Also, vertical gravity gradients of 3 microgals/cm require that injection-induced surface deformation must be accounted for. For example, at the In Salah Field, vertical strains of ~0.5 cm/a have been measured (Vasco et al., 2010). However, from Figure 5-9, it is also clear that at shallower depths, the CO_2-related gravity changes are an order of magnitude greater than the measurement uncertainties. This suggests that time-lapse gravity measurements may be best suited to monitoring for CO_2 leakage into the overburden.

Conclusions
Surface gravity monitoring may be useful in situations where the deep injection volumes are very large, or for monitoring leakage at depths shallower than about 1000 m, where there is increased sensitivity due to the dramatic reduction in the density of CO_2. In most practical cases, simultaneous monitoring of surface deformation and groundwater levels may be required to isolate the contribution of CO_2 to the gravity field.

A density model for the site will be required, in order to interpret what the observed gravity changes mean in terms of subsurface CO_2 distribution.

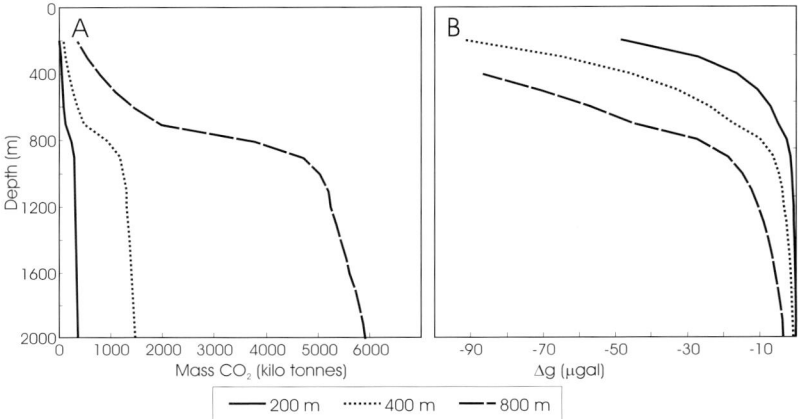

Figure 5-9. A. *Relation of mass of CO_2 versus depth. Each curve represents the mass of CO_2 associated with the same volume but at different depths.* **B.** *Maximum observed gravity change for a constant volume of CO_2 versus depth. The calculations are for a disc-shaped CO_2 plume of variable radius (200, 400 and 800 m) with a thickness of 20 m and a mean porosity of 0.2. It was assumed that the CO_2 completely replaced formation water as the pore fluid.*

Interpretation can be achieved through some form of gravity inversion (either formal inversion or forward modelling). Because gravity measurements provide an integrated measure of the subsurface density distribution, they will be most useful when applied in combination with other monitoring data sets, or a priori data, or constraints.

5.4.4 **Forward Seismic Modelling**

Technology

Well logs can be used as the basis for predicting the seismic response to various CO_2-injection scenarios, and so for assessing the likelihood of observing a measureable time-lapse signal. Synthetic Vp, Vs, and density logs can be generated for different pressure and saturation conditions using the process of fluid substitution described above. Once constructed, the logs in this study were used to generate synthetic seismic responses for a 1D layered medium corresponding to a range of source-receiver offsets. These modelling results can be compared with time-lapse seismic field data to gauge the accuracy of the rock physics model. As well, they can be used to generate a suite of modelled seismic attributes that can be used to invert field data for pressure and CO_2 saturation changes.

As a first step in calibrating the modelling results, the original Vp, Vs, and density logs are used to 'tie' the seismic data in the vicinity of a well to the known geology of that well. Figure 5-10 compares the log-based synthetic seismic traces to a portion of the pre-injection field data, in addition to comparing the log-based relative P-impedance with that obtained by inversion of the field data. The excellent match between the field and synthetic seismic data and impedances validates the modelling procedure.

Figure 5-11, Figure 5-12 and Figure 5-13 depict the offset-dependent time-lapse seismic responses corresponding to an increase in pore pressure of 7 MPa (5-11), an increase in CO_2 saturation to 95% (relative to oil) within the Marly unit (5-12), and the combined effect of these two changes in reservoir conditions (5-13).

In general, the modelling results support the conclusions from the analysis of saturated rock properties given in a previous section. Importantly, they demonstrate that detection of these changes within a relatively thin reservoir (specifically the Marly unit, which is generally <10 m thick) is possible using the resolution afforded by the field-based seismic wavelet. As can be seen, the magnitude of the time-lapse seismic response increases with offset, and the time-lapse amplitude change can be quite large when both pressure and saturation effects reinforce each other. The observed offset-dependence in the seismic response shown here forms the basis for the various amplitude-versus-offset analyses that have been applied toward pressure versus CO_2 saturation. These modelling results indicate that seismic anomalies near an injection well should be larger than those near a production well.

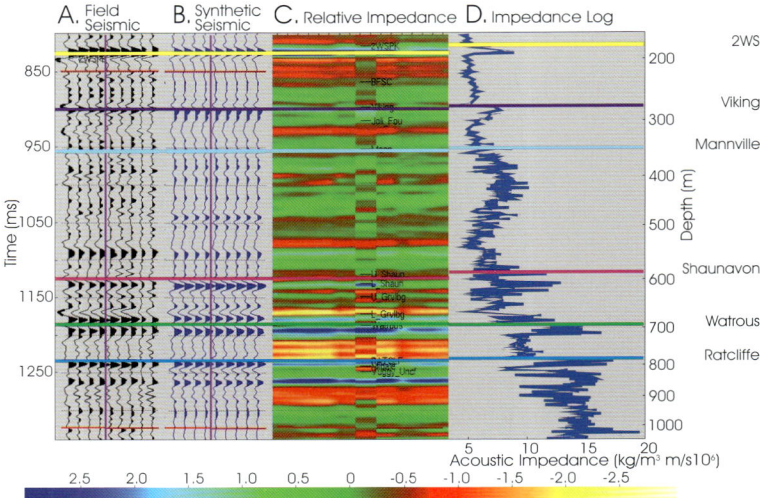

Figure 5-10. *Suite of logs tied to the migrated seismic traces extracted in the vicinity of a specific well.* **A.** *Seismic field traces.* **B.** *Synthetic traces generated from the P-impedances near the well.* **C.** *Relative P-impedance obtained by inversion of the field data from A (the P-impedance derived from well logs is plotted in the centre of panel C), and* **D.** *P-impedance calculated from the sonic and density logs.*

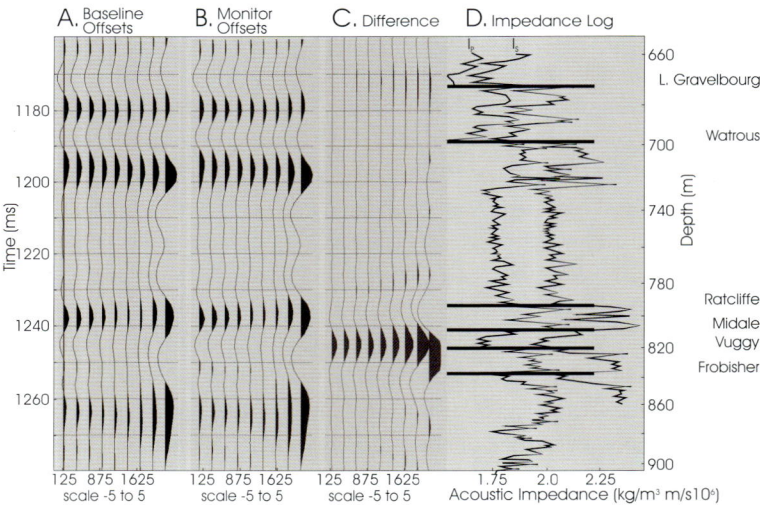

Figure 5-12. *Same as Figure 5-11, but for a saturation increase from 0 to 95 mol % CO_2 in the Marly oil phase, together with a 30% increase in oil saturation relative to formation water (due to swelling). The pore pressure remains constant at 15 MPa.*

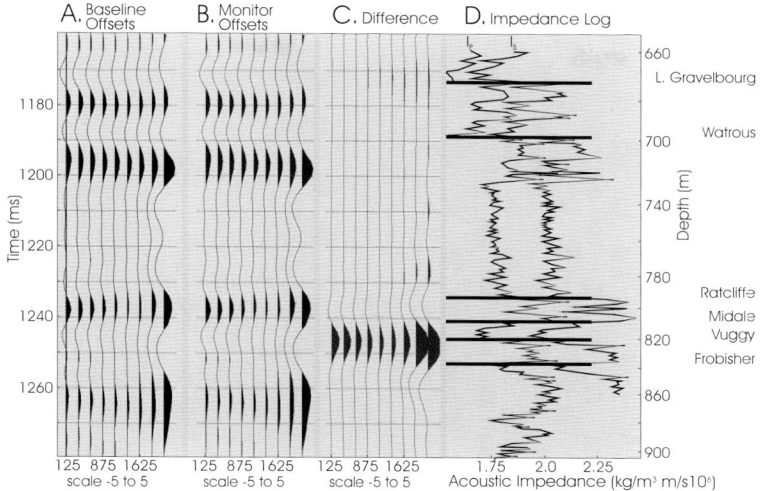

Figure 5-11. *Comparison of* **A.** *baseline and* **B.** *monitor synthetic seismic offsets using log data from the same well as Figure 5-10. The offset range corresponds to that of the Weyburn Field seismic data. Baseline traces were generated using a wavelet extracted from the 1999 field data and the baseline Vp, Vs, and density logs (that is, prior to CO_2 injection) after tying them to the 1999 migrated seismic field data, as shown in Figure 5-10. Monitor traces were generated the same way, but using a modified log suite corresponding to a pore pressure increase in the Marly and Vuggy units from 15 MPa (ambient reservoir pore pressure) to 22 MPa, as would occur near a CO_2 injector.* **C.** *Difference traces show a residual time-lapse signal only at the reservoir level, as expected, but with amplitudes about 10 times smaller than the baseline and monitor traces.* **D.** *P-impedance and S-impedance, determined from the baseline logs. Note the first breakthrough event in the difference traces is a result of the decrease in impedance caused by the pore pressure increase.*

Conclusions

Log-based fluid substitution modelling provides a means of testing whether time-lapse seismic observations are capable of detecting or imaging changes due to CO_2 injection in the subsurface. They take into account the resolution limitations imposed by an estimate of the field-based seismic wavelet.

In the case of the Weyburn Field, modelling indicates that changes within a relatively thin reservoir layer (the Marly unit generally <10 m thick) can be resolved.

The offset-dependence of the time-lapse seismic response can be modelled and used for the purposes of AVO inversion toward discriminating pressure

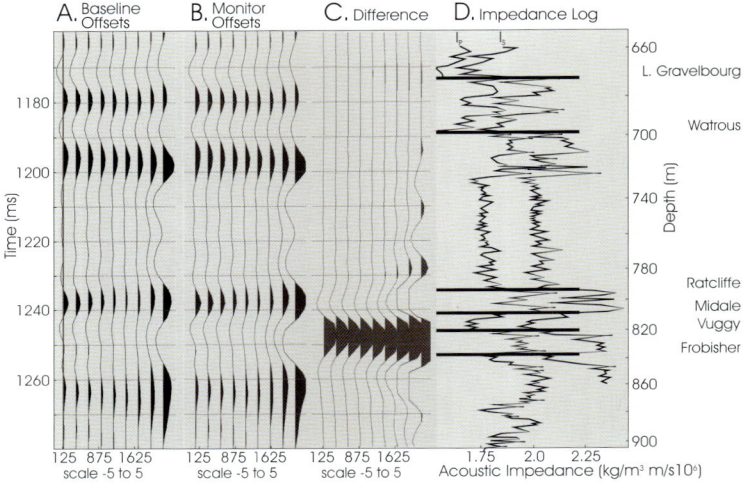

Figure 5-13. *Same as Figure 5-11, but with an additional saturation increase from 0 to 95 mol % CO_2 in the Marly and in the top half of the Vuggy. In addition, oil saturation has increased to 90% in the Marly and 36% in the Vuggy, which simulates 30% oil swelling in these units, with a further increase in oil saturation relative to formation water in the Marly unit, which might occur ahead of a water front during water injection. The time-lapse anomaly is now particularly large.*

versus CO_2 saturation. Seismic anomalies near a CO_2 injection well should generally be larger than those near a production well due to the superposition of effective pressure decrease and CO_2 saturation increase near the former. An exception to this may occur in the case of pore pressure decreasing below the bubble point near a production well.

5.5 Downhole Monitoring Methods
5.5.1 Passive Seismic
Technology

Injection of CO_2 and the production of hydrocarbons alter the effective stresses within and surrounding a reservoir. This can lead to the reactivation of pre-existing fractures, formation of new fractures by hydraulic fracturing or shear deformation, or potentially the reactivation of existing faults, any of which may impact the security of stored CO_2. Passive seismic monitoring is designed to detect seismic emissions (microearthquakes) generated by any of these processes (for example, Maxwell and Urbancic, 2001). Determination of associated event locations, magnitude, and in some cases focal mechanisms, can potentially identify active fracture zones or fault planes, or possibly deformation that indirectly maps the spread of CO_2. Seismic arrays designed for this purpose can potentially be deployed either at the surface or in wellbores, with the latter providing higher sensitivity due to their proximity to the injection zone. Event locations can be determined by matching the P-wave and S-wave arrival times across the array by ray tracing, and determining the propagation azimuth by hodogram analysis. Also, wave propagation effects, such as S-wave splitting, can be analysed to provide complementary information about stress-related anisotropy or anisotropy due to the presence of aligned fractures (for example, Verdon et al., 2010).

5.5.2 Microseismicity at the Weyburn Field

The project was the first large-scale CO_2 operation to include downhole microseismic recording as a monitoring tool. Details have been presented elsewhere (White et al., 2004; Verdon et al., 2010, 2011, and references therein) and are only summarized here. In 2003, a passive seismic array of eight geophones was cemented in an inactive vertical well that was located within 50 m of a vertical CO_2 injection well. The geophones were spaced at 25-m intervals from 1181–1356 m depth, about 200 m above the reservoir.

Water injection began in late 2003, and changed to CO_2 in late January, 2004. Passive monitoring commenced in August 2003 prior to the onset of water injection, and has been in near-continuous operation to the end of 2010 (Figure 5-14).

Approximately 200 microseismic events were located by passive seismic monitoring (Figure 5-15), indicating a low rate of low intensity microseismicity. The following key observations can be made.

1. The majority of events had moment magnitudes between –3.0 and –1.0, with a corresponding detection distance of <750 m for the larger events.
2. They occur mainly in episodic temporal clusters, and in most instances can be linked with specific operational activities in the field.
3. With the exception of events induced by drilling activities at the injection well, and later abandonment of that well, the majority of events are located near the production wells.
4. Event locations are not limited to the reservoir interval, but occur in rocks above and below.
5. There is a notable correlation of microseismic event locations with the time-lapse seismic amplitude anomaly (indicating the presence of CO_2 or an increase in pore pressure), extending to the west of the injection well. However, there does not appear to be a systematic correlation of seismic events with CO_2-related anomalies.
6. An anisotropic fabric consistent with vertical fracture sets, striking at 150° (SSW) and 42° (NE), respectively, was identified through S-wave

splitting analysis of the recorded microseisms. The 150° fracture set has the highest fracture density of the two sets. These orientations match those of two fracture sets known from core and electromagnetic interference logs from the reservoir (at 148° ± 11° and 40° ± 5°). However, the core and electromagnetic interference log results indicate that fracture densities are higher for the NE-striking fracture set.

5.5.3 Geomechanical Modelling

A simple coupled geomechanical–fluid flow model was constructed to assess the implications of the microseismicity and to determine the anisotropic pattern. The model simulated the geometry of the injection and production wells where microseismic monitoring has been deployed. Prior to CO_2 injection, production had reduced the pore pressures from 15 to 10 MPa. Following CO_2 injection (at 100 000 m³/d in each of three wells) for a period of one year, the pressure increased to about 18 MPa at the injection well but was still below 15 MPa at the producers. Two models were tested.

Model 1 was built using rock parameters determined from laboratory measurements and well logs.

Model 2 incorporated a reservoir layer that was mechanically 'softer' than for Model 1, thus allowing more stress to be transferred from the reservoir to the overburden (that is, it allowed 'stress-arching'). As a result, the changes in effective stress within the reservoir are reduced, while stress changes in the overburden are amplified.

The evolution of the potential for fracturing determined for these models is shown in Figure 5-16. Fracture potential provides an estimate of the likelihood

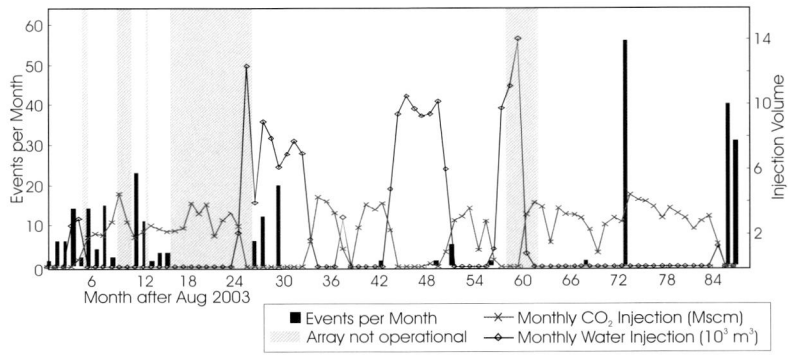

Figure 5-14. *Comparison of microseismic activity and injection history for water and CO_2.*

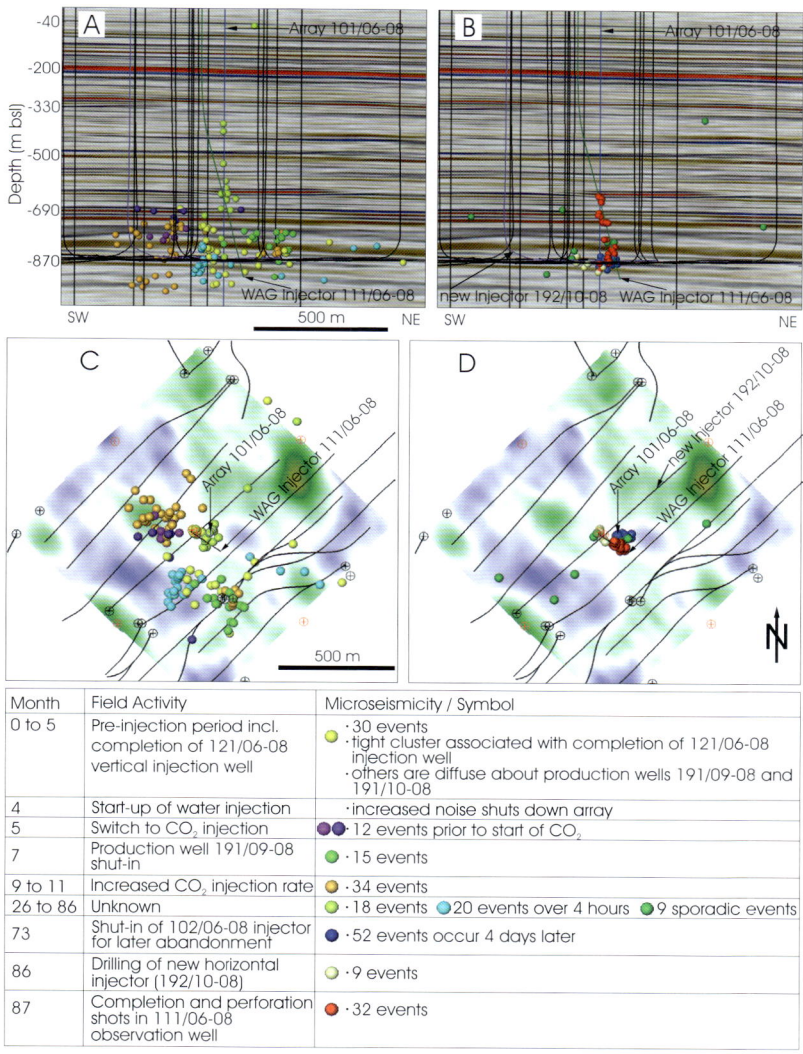

Figure 5-15. *Vertical section view of microseismic events recorded from:* **A.** *August, 2003 to January, 2006, and* **B.** *from January 2006 to December 2010.* **C.** *and* **D.** *Plan view of the same events. The 2004 minus 2000 time-lapse seismic amplitude difference map for the reservoir interval (from 3D surface seismic) forms the coloured backdrop. Green-to-orange and blue background colours represent negative and positive amplitude changes, respectively.*

of a material to experience brittle shear failure, and thus can be used as a proxy for microseismicity.

Figure 5-16 shows that the fracture potential for Model 2 increases in the overburden above the production wells after injection, in contrast to the results for Model 1. This is consistent with the microseismic observations, where events occur in the reservoir and overburden, predominantly near the production wells. The calculated shear wave splitting patterns (not shown) for the Model 2 overburden also show that fast directions are oriented perpendicular to the production well trajectories (that is, they strike NW) above the injection wells. This is consistent with the anisotropy determined from microseismic data. Therefore, Model 2 implies that the microseismicity observed in the overburden at Weyburn is caused by stress transfer through the rock frame, rather than by a pore fluid connection or CO_2 leakage. The fact that there are few events, most of which are of low magnitude, suggests that there are few large-scale fractures in the overburden. Further, there has been no seismicity detected >200 m above the reservoir, which is well within the detectability threshold of the geophone array, implying that if any fractures are being stimulated by CO_2 injection, they do not extend far above the reservoir.

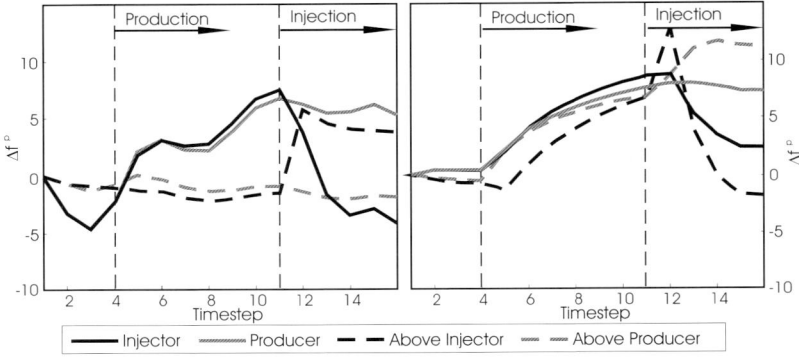

Figure 5-16. *Percentage change in fracture potential in the Weyburn reservoir and overburden through time.* **Left.** *Model 1 shows that the fracture potential does not increase anywhere after injection begins, except in the overburden near the injection wells. Therefore, this region should be the most prone to microseismic activity.* **Right.** *In Model 2, the fracture potential increases in the overburden above the production wells (grey dashed line). After a transient increase, the fracture potential above the injection well (black dashed line) decreases during injection. Figure modified from Verdon et al. (2011).*

Conclusions

Passive seismic monitoring provides a relatively inexpensive method of continuous long-term monitoring of storage security for CCS sites. It provides a means of defining the magnitude and spatial limits of injection-induced and production-induced seismicity. This information addresses general public assurance (such as: Is injection causing earthquakes?) and more technical details about deformation within the geological storage complex (for example: Is CO_2 injection damaging the cap rock?). Further information about time variability of the stress field can be obtained through associated shear-wave splitting analysis. Passive monitoring is not a tool that can be relied upon for directly tracking CO_2 movement, as indicated by the inconsistency of the spatial correlation of microseismicity and inferred CO_2 distribution in the reservoir, as observed at Weyburn.

Depth uncertainty in the location of microseismic events is a significant issue, especially when trying to ascertain that CO_2 is secure in the reservoir. Depth uncertainties can be reduced significantly by extending the vertical array of geophones and particularly by extending the array through the reservoir interval. Event locations would also be improved by deploying geophone arrays in more than a single well.

Geomechanical modelling is an essential requirement to adequately assess the significance of observed microseismicity and anisotropy. Results from the geomechanical modelling should be matched with field observations. In the Weyburn-Midale case, a geomechanical model constructed using mechanical properties based on core samples was incapable of matching the microseismic observations. This was likely due to increased elastic compliance caused by fractures and vugs that are not sampled by the cores. Subsequent analysis indicated that microseismicity located outside of the reservoir is not likely to be an indicator of either fluid migration or pore pressure changes out-of-zone, but is due to stress transfer ('stress arching') through the rock matrix.

Implementation of an effective passive seismic monitoring program would ideally entail some pre-injection steps. First, establishing pre-injection levels of microseismicity would provide a baseline against which subsequent CO_2-injection related seismicity could be compared. Second, early development of a geomechanical model of the reservoir would allow modelling to predict the seismicity associated with various injection scenarios. This model can be appropriately modified as microseismic monitoring results became available after the start of CO_2 injection.

5.5.4 Crosswell Geophysical Methods
Technology

Crosswell geophysical methods provide a means of acquiring detailed images of either seismic or electrical properties between adjacent wellbores. This is achieved by generating a field (for example, seismic or electromagnetic) using sources in one wellbore, that is then transmitted across the intervening region, and recorded with receivers deployed in another wellbore. Typically, the length of the zone in both the source wellbore and the receiver wellbore should be comparable to the distance between the wellbores. This leads to robust imaging results, with strong control on the resolution of the resultant image. Crosswell geophysical methods are capable of resolving details on the scale of metres to tens of metres. Because these surveys require temporary or permanent occupation of wellbores, they are considered high-demand or intrusive methods, and their use generally requires temporary suspension of production or injection operations. Alternatively, dedicated wells comprising either production wells destined for abandonment or purpose-drilled monitor wells can be used. More recently, operational wells have been deployed where the geophones or electrodes are placed outside the casing.

Crosswell tomographic methods have been applied at several CO_2 pilot sites, including the Nagaoka site in Japan (Spetzler et al., 2007; Saito et al., 2008; Onishi et al., 2009), the Frio site in Texas (Daley et al., 2007, in press), and the Ketzin site in Germany (Kiessling et al., 2010). Crosswell tomography is a method that is commonly used in the oil and gas industry. At Weyburn, crosswell seismic surveys were conducted using both vertical and horizontal crosswell arrangements. In the first case, a horizontal crosswell tomographic survey was carried out prior to the start of CO_2 injection (Li et al., 2001), with the intention of doing time-lapse monitor surveys. However, an attempt to re-enter the horizontal well with survey equipment at a later date was unsuccessful, and thus no time-lapse data were acquired. This instance emphasizes the increased operational risk associated with wellbore deployments in general. However, the one-time crosswell survey did provide a highly detailed image of seismic attenuation (Figure 5-17) that can be used to make inferences about patterns of reservoir permeability.

A vertical crosswell seismic survey was conducted in the Weyburn Field, using the geometry depicted in Figure 5-18. As can be seen, this geometry is generally suited to imaging the interwell zone, because the well interval over which the sensors and receivers were deployed (450 m) is comparable to the distance between the wells (600 m).

Because the reservoir interval is located below the deployment intervals, and thus can only be imaged using reflected ray paths, the lateral resolution that could be achieved was severely limited. Nevertheless, the reflection image

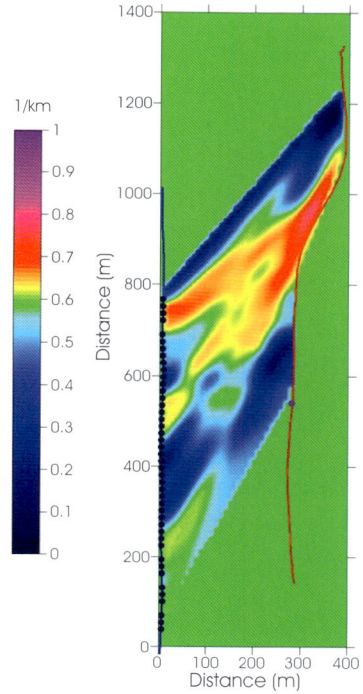

Figure 5-17. *Horizontal crosswell tomogram showing the seismic attenuation between wells (for 650 Hz energy). The image corresponds to a sub-horizontal slice through the Marly unit. The trajectories of the wells and the locations of sources and receivers are indicated (black dots on left and red line on right).*

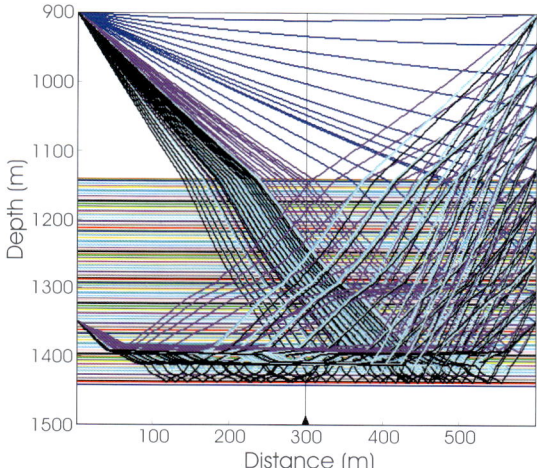

Figure 5-18. *Geometry of a crosswell seismic survey at the Weyburn Field. Sources and receivers were located over the interval from 900–1350 m depth. The reservoir interval is a ~30 m thick zone lying below this.*

does provide much better vertical resolution than surface seismic images (Figure 5-19). Although this crosswell geometry may be suitable for monitoring the zone above the reservoir, it predictably failed to provide adequately useful details about the reservoir interval. As a result, no follow-up monitor survey was acquired.

Conclusions

Under the right circumstances, and for moderate cost, crosswell methods have the potential to provide higher resolution images of the subsurface than surface seismic techniques. However, they are limited by (1) required access to boreholes, (2) the geometry of existing boreholes, and (3) the limited spatial range. In an EOR setting, borehole access requires interruption of production in active wells or access to abandoned wells. In many (if not most) cases, the wellbores do not extend through the reservoir, which limits the imaging aperture for transmission tomography at the reservoir level. Hence, for monitoring purposes, crosswell techniques are best suited for monitoring above the reservoir. For CO_2 storage outside a producing field, there may be few wells, and monitoring wells would have to be provided.

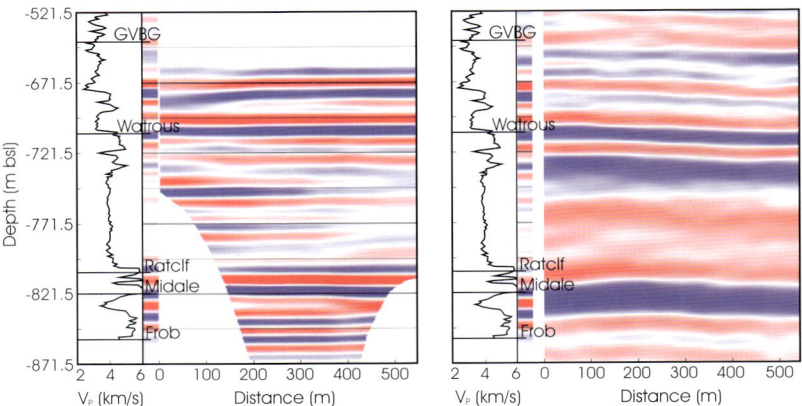

Figure 5-19. Left. *Comparison of a seismic reflection image obtained using a crosswell seismic method and the synthetic seismic reflections determined for the corresponding well logs.* **Right.** *A comparable seismic section based on the surface 3D seismic volume.*

5.5.5 Vertical Seismic Profiling
Technology

Vertical seismic profiling (VSP) is a monitoring technology appropriate for covering areas of up to several hundred metres around a wellbore. It is similar to crosswell seismic, in that a wellbore is used to deploy an array of seismic receivers, but differs because the seismic sources are located at or near the ground surface. Also, vertical seismic profiling most commonly utilizes reflections for the purpose of imaging, as opposed to transmitted seismic energy. The principal general advantage of vertical seismic profiles over surface seismic data is that the near-surface layer is avoided by the upcoming leg of the reflection travel paths. This reduces unwanted scattering and attenuation in the reflected wavefield. Also, the deployment of the seismic sensors at depth provides a low-noise recording environment, compared to surface recording. These factors lead to a higher frequency content, which is critical for deriving a high-resolution seismic image around the borehole.

Vertical seismic profiling data are usually processed for P-wave reflection images. As such, they are amenable to the same types of monitoring analysis as surface seismic methods, specifically (1) time-lapse traveltime and amplitude differencing, and (2) amplitude-versus-offset (AVO) or amplitude-versus-angle (AVA) analysis. Due to the comparatively narrow angular apertures provided by VSP surveys, AVO and AVA analysis has been relatively poorly explored.

5.5.6 3D VSP Surveys

Several VSP surveys were conducted in the Weyburn Field, although only one 3D VSP survey was repeated for the purpose of time-lapse analysis. This survey was run in 1999, prior to CO_2 injection, and was repeated in 2001, one year after the start of injection. Both the VSP survey and the surface 3D seismic survey used a downhole array of 80 three-component geophones.

Figure 5-20 shows the processed VSP data and a comparison of the VSP corridor stack from an adjacent well log. In order to directly compare and calibrate the normal-move-out (NMO)–corrected VSP to the surface seismic records, a VSP to common-midpoint (CMP) transformation was made by using the same CMP binning as in the surface seismic study. Any differences in timing and stacking velocities between the VSP and surface data sets were removed by applying depth-variant time shifts, similar to the conventional well-log 'stretching' during interpretation. As a result, the calibrated pre-stack VSP data set becomes directly comparable to the surface seismic in terms of both reflection-point locations and reflection times (Figure 5-21). After calibration, the VSP data show good quality and somewhat better resolution, particularly around the reservoir (red bar in Figure 5-21).

P-wave reflectivity at the reservoir level for the VSP and surface seismic results are compared in Figure 5-22. Although the results are similar at the gross scale, there are significant differences, likely due in part to the difference in vertical resolution of the two methods.

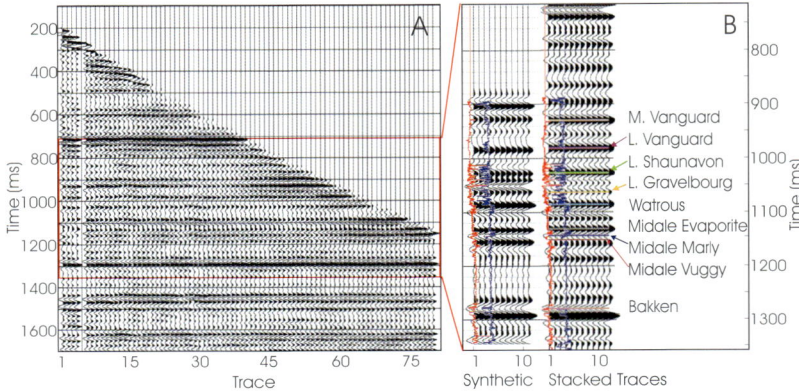

Figure 5-20. A. *Seismic reflection image obtained using crosswell seismic acquisition.* **B.** *On the right is the VSP corridor stack, and on the left is the synthetic seismogram determined from the corresponding well logs.*

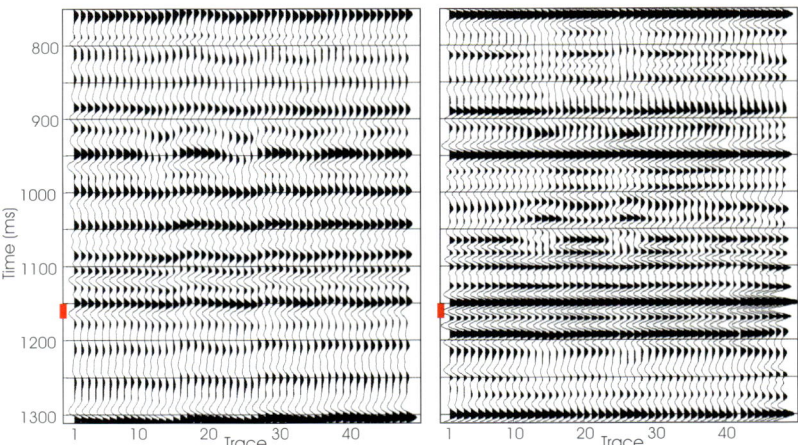

Figure 5-21. *Comparison of reflection stack sections.* **Left.** *1999 surface seismic.* **Right.** *1999 VSP data (after transformation into CMP form). The top of the reservoir interval is located at approximately 1150 ms.*

Traveltime and amplitude anomalies are well-developed in all of the dual-leg horizontal injection well patterns, where a significant volume of CO_2 has been injected (up to 14% hydrocarbon pore volume). In most cases, there is generally good agreement between the injection volumes and the areal extent and intensity of the anomaly. In contrast, seismic anomalies are not as prominent in the northern part of the area where vertical CO_2 injection wells are used, even though comparable large volumes of CO_2 have been injected. The absence of seismic anomalies in this area is interpreted as due to a combination of relatively low porosities in this region (particularly in the Marly unit), to most of the CO_2 injected there residing in the Vuggy unit, and to distortion of the outgoing plume shape that occurs with alternating CO_2–water injection cycles in a vertical injection well.

5.6.3 Overburden Monitoring

Time-lapse interval traveltimes provide a useful means of monitoring the overburden above the reservoir. Although they lack the resolving power of amplitude differences, they provide an integrated measure of change over the entire interval selected. Seismic amplitudes, though very sensitive to small changes in impedance, are also more prone to noise.

Time-lapse interval traveltimes were determined for the 2004 survey, relative to the 2000 baseline survey, for the depth intervals identified on the seismic cross-section in Figure 5-25 (in relation to the reservoir). Figure 5-26 shows the interval traveltime changes in map form for the two intervals identified in the seismic cross-section. Interval 1 is from the Upper Cretaceous, Second White

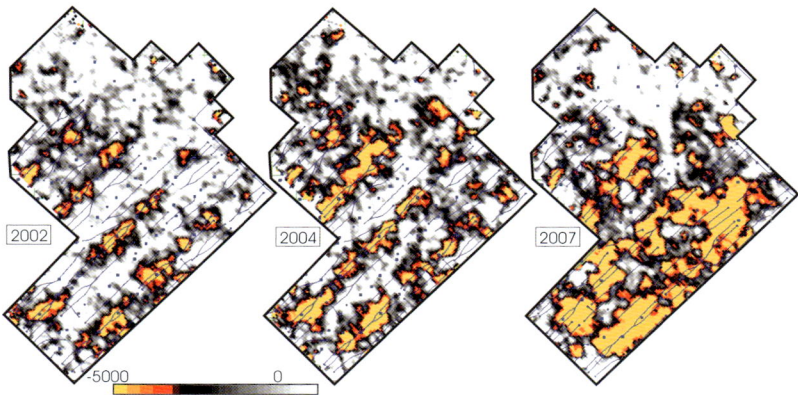

Figure 5-23. *Maps of seismic amplitude difference for the Marly unit (upper reservoir) horizon.*

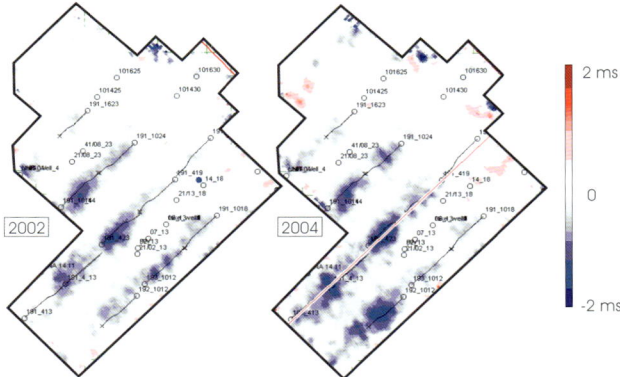

Figure 5-24. *Maps of seismic traveltime difference for the Bakken horizon (below the reservoir).*

Speckled Shale to the Jurassic, Lower Gravelbourg Formation. Interval 2 is from the cap rock (Jurassic, Watrous Formation) to the Bakken Formation (below the reservoir). Interval 2 has pronounced negative traveltime anomalies of up to about 2 ms, which are clustered about the trajectories of the horizontal CO_2 injection wells. In contrast, there are few (if any) significant traveltime anomalies in interval 1, which lies above the regional seal (Watrous Formation).

Using the cumulative traveltime delay and associated areal extent as a proxy for volume of CO_2 present, we can place an upper bound of 1.5% on the

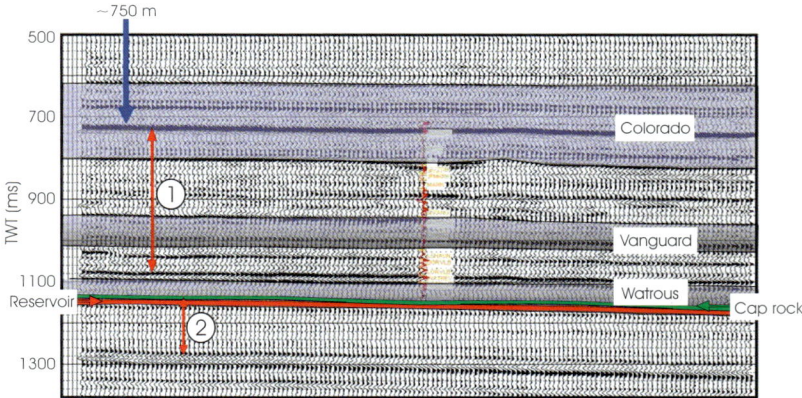

Figure 5-25. *Weyburn seismic cross-section. Purple intervals represent aquitards. Interval traveltime differences have been determined between interval 1 and interval 2. The reservoir cap rock is indicated in blue and the reservoir in red.*

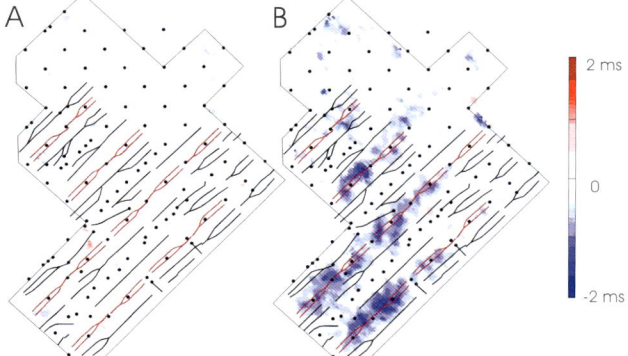

Figure 5-26. *Maps of interval traveltime differences for intervals 1 and 2, as indicated in Figure 5-25.* **A.** *Interval from the Second White Speckled Shale horizon to the top of the regional seal (Watrous Formation).* **B.** *Interval from the reservoir cap rock (sub-Mesozoic unconformity) to the Bakken horizon, including the reservoir interval. Horizontal CO_2 injection wells are indicated by red lines, horizontal production wells by black lines, and other wells (water injectors or production wells) by black dots.*

proportion of injected CO_2 that, hypothetically, may be allocated to this large geological interval above the regional seal. This crude calculation should only be considered as an illustration, because it certainly represents an overestimation of the upper bound for CO_2 that potentially resides above the regional seal. The calculation does not account for the increased seismic sensitivity either to CO_2 at shallower depths (see Figure 5-2) or the higher compliance of clastic rocks in the overburden. Further, changes in the interval traveltime for depth interval 1 may simply be noise, in which case there is no CO_2 above the regional seal. In any event, this exercise shows, in principle, how the overburden interval can be effectively monitored seismically for large volumes of CO_2 that may have migrated upward.

Conclusions

At Weyburn, 3D time-lapse seismic monitoring was an effective tool for mapping changes in the reservoir that are associated with the combined effects of CO_2 saturation and pressure changes. Seismic amplitude differences, and interval traveltime changes, provide complementary monitoring capabilities. Whereas amplitude differences provide the most detail at the reservoir level, interval traveltimes are very useful for monitoring the thick geological section above the reservoir. Interval traveltimes can be used to estimate CO_2 volumes within various levels of the storage container, as illustrated by a rather simple example.

More accurate apportionment of CO_2 volumes can be achieved by incorporating the depth dependence of pore fluid properties and seismic properties of the various stratigraphic intervals. Thus, 3D time-lapse monitoring provides the most comprehensive means of monitoring the subsurface, but is relatively expensive to deploy. Its specific role in monitoring will be site-specific, depending on the geological characteristics (such as porosity, thickness, and the presence of gas) and depth of the storage reservoir.

5.6.4 Discriminating Pressure versus CO_2 Saturation

Time-lapse seismic amplitude and traveltime difference maps clearly show injection-related changes in the reservoir. These changes have been attributed primarily to the replacement of original pore fluids with CO_2, but also to changes in reservoir pressure. The use of shear-wave information provides a potential means of separating pressure and CO_2 saturation effects, as seen in the time-lapse seismic response. Shear-waves are primarily sensitive to changes in the effective pressure, whereas P-waves are sensitive to both the pore fluid saturation and pressure. Thus, the pressure variations can potentially be isolated through shear-wave analysis.

Three different methods have been applied in attempting to isolate time-lapse CO_2-saturation and pressure changes directly from the seismic data.
1. Converted-wave (P-S) imaging.
2. Prestack amplitude-versus-angle (AVA) analysis and impedance inversion.
3. Prestack impedance inversion and pressure-saturation inversion.

Each of these methods and corresponding results obtained is described below.

5.6.5 Converted Wave (P-S) Imaging

In principle, converted-wave (P-S) processing is similar to the way P-wave images are obtained from time-lapse seismic data, except that the process is modified to utilize waves which have been converted from P-waves to S-waves when reflected at a subsurface boundary.

Figure 5-27 shows a comparison of P-S and P-wave sections for the 2001 data. The P-S data are characterized by generally lower frequencies, and thus lower resolution. This can be seen particularly at the reservoir level (Figure 5-28) where the resolution and signal-to-noise ratio is much lower on the P-S section. This difference is attributed to the fact that the P-S converted-waves arrive at later times (by about 30–40%) than the primary P-wave reflections, and thus are often buried deep in the background of later P-wave reflections. S-wave velocities are also lower, and often more variable than the P-wave velocities, particularly in the near-surface, which makes the S-wave statics problem particularly severe. The generally poor quality of the P-S data at the reservoir level precluded use of these data for pressure-saturation discrimination.

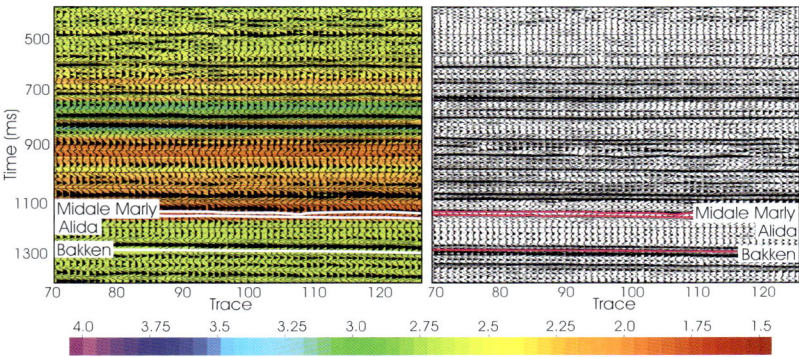

Figure 5-27. *P-S converted-wave section (left), compared to the P-wave section (right). Colour background on the left shows the log-based Vp/Vs ratio. The velocity conversion used to convert the P-S section to P-P time is based on log correlation.*

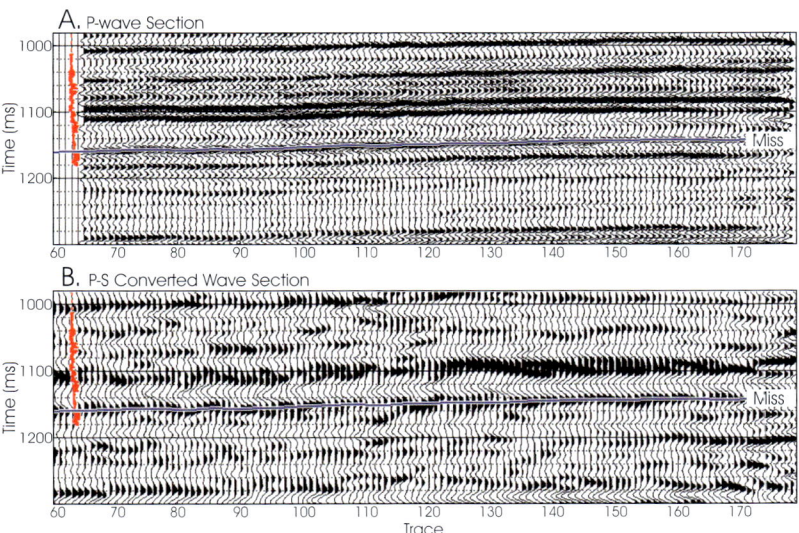

Figure 5-28. *P-S converted-wave section (bottom) compared to the P-wave section (top) in the vicinity of the Midale reservoir, which is located just below the Mississippian horizon.*

5.6.6 **AVA Analysis**

As an alternative to analysing S-waves directly (as considered in the previous section), S-wave properties can be determined indirectly by measuring the variation in response of the primary P reflections with incidence angle or, equivalently, source-receiver offset. This analysis is referred to as AVA (amplitude-versus-angle) or AVO and is applied to P-wave reflections before stacking.

The amplitude of the P-wave reflection (or P-wave reflectivity, R) can be expressed as $R(\Theta) = I + G \sin^2\Theta$, for small angles of incidence ($\Theta < 30°$) (Shuey, 1985). This means that a plot of R versus $\sin^2\Theta$ is a straight line, with intercept I and slope G. The intercept I represents the normal-incidence P-wave reflectivity, R_s. The normal-incidence S-wave reflectivity is $R_s = \frac{1}{2}(I-G)$ (Rutherford and Williams, 1989), if the 'background' velocity ratio can be described as $V_P/V_S = 2$.

Fluid substitution calculations (using the Gassmann equation) were applied to the logs from the same log set utilized for 1D modelling described earlier, but in this case the effect of shale content and effective porosities was incorporated, resulting in reduced sensitivity of the seismic parameters to injected CO_2. The associated seismic response was then calculated by ray-tracing over the 0°–30°

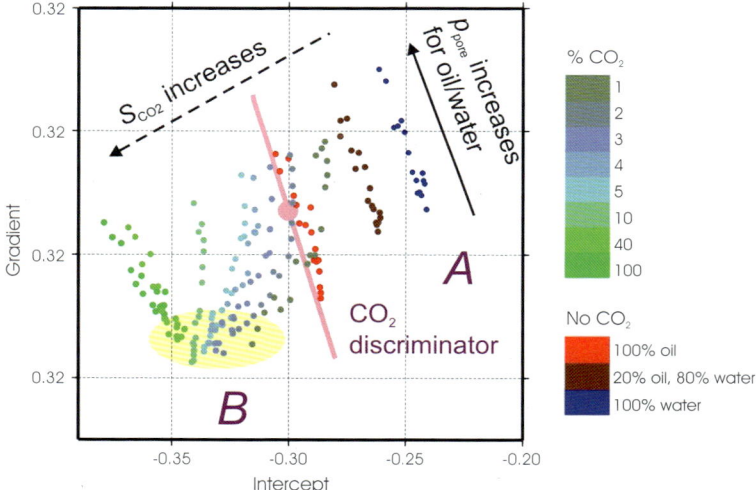

Figure 5-29. *Modelled AVA cross-plot from well-log based models. A 40-Hz Ricker wavelet was used. Solid and dashed arrows indicate the pore pressure increasing from 7–23 MPa, and CO_2 saturation increasing from 0–1, respectively. Yellow ellipse indicates the area of (I, G) values converging at low pore pressure. Pink line and large dot show the CO_2 discriminator. Note that a much higher confining pressure (32.5 MPa) was assumed, relative to results of the pressure-saturation modelling, resulting in a range of effective pressures from 9.5–25.5 MPa.*

range of incidence angles, for a range of fluid saturations and pressures, and the AVA intercept (I) and gradient (G) values were determined from the synthetic seismograms. The results are summarized in Figure 5-29. Because the variation trends for changes of pressure versus CO_2 are nearly perpendicular, it is suggested that I and G may be combined to produce a parameter that discriminates between CO_2 saturation and pressure effects.

A key part of preparing the data for prestack AVA analysis is the calibration of the amplitudes between various data vintages. In addition to the detailed prestack calibration, further equalization of amplitudes was required before they could be compared in detail. Scaling factors were determined by comparing the relative amplitudes of the monitor data to the baseline data at a reference horizon above the reservoir cap rock. Reflection amplitudes at the cap rock reflection for the 2001 and 2002 monitor surveys show maximum variations of ±10% (not shown), which is comparable to those determined at the Marly horizon (not shown). This suggests that the upper part of the reservoir is not resolved independently from the cap rock, or that the calibration against a reference horizon was not entirely adequate.

The AVA analysis focused on reflections from the reservoir cap rock and interpreted Marly horizon. Amplitudes were determined on a trace-to-trace basis for these horizons, and then a straight line was fitted to determine a local estimate of the AVA parameters, G and I. The Marly horizon becomes inconsistent and difficult to identify in the N-NW part of the study area. The small thickness of the reservoir, compared to the wavelengths used in this particular study (corresponding to 60–80 Hz), means that fluid variations within the Marly zone may in part be attributed to the cap rock zone.

Maps of P-impedance and S-impedance difference for the 2002 monitor survey, as determined from the P-wave and S-wave reflectivity from AVA analysis, are shown for the Marly horizon in Figure 5-30, left and right, respectively. As discussed earlier, S-wave impedance should be most sensitive to the pore pressure effects whereas P-wave impedance should be sensitive to both pore fluid substitution and pore pressure. Increases in the monitor–baseline ratios would be expected for both increased pressure and CO_2 saturation at the Marly horizon.

These results may be compared, for consistency, with the amplitude difference maps for the Marly zone shown in Figure 5-23. The amplitude differences in Figure 5-23 should generally correspond to regions of decreased impedance (the blue zones) in Figure 5-30. Although the impedance does decreases in Figure 5-30, and generally lies along the trend of the CO_2 injectors (except near the edges of the data coverage, where the results are less reliable), the expected correlation is limited. This suggests that the prestack trace-to-trace analysis required for the AVA analysis may be less robust than in the case of the poststack analysis results of Figure 5-23.

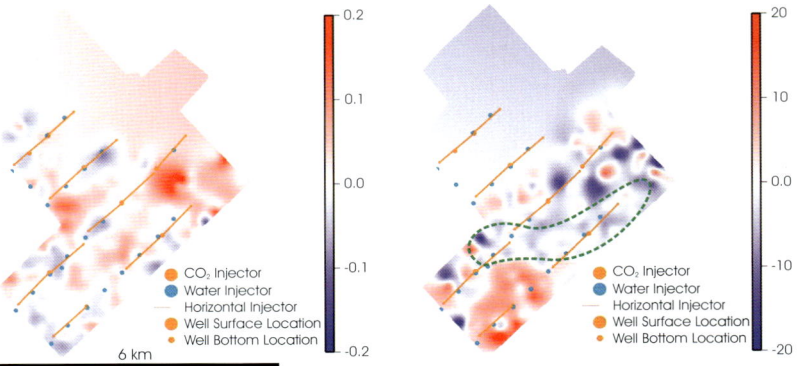

Figure 5-30. *Normalized P- wave (left) and S-wave (right) impedance variations at the Marly level for the 2002 monitor surveys, relative to the baseline survey. Approximate positions of the horizontal injection wells are indicated by the orange lines.*

5.6.7 CO₂ Pressure-saturation Inversion

As noted above, CO_2 pressure-saturation inversion is based on a petrophysical model which relates reservoir dynamic properties (pore fluid saturations and pressures) to seismic properties that can be extracted from the observed seismic data. The first step in this inversion is processing the customized prestack data of the different vintages, to produce migrated stack volumes that correspond to several offset ranges for each of the 3D data sets. Subsequent determination of AVO characteristics is made on the limited offset range stacks, as opposed to trace-by-trace measurement as employed in the previous section.

The next step is prestack P- and S-impedance inversion of each vintage of seismic data, which can then be inverted to obtain pressure and CO_2 saturation changes over time. The addition of the S-impedance is the key to separating pressure and saturation effects. However, the quality of the longest offset data from the Weyburn data set was too poor to use for prestack impedance inversion, limiting the quality of S-impedance inversion. Figure 5-31 shows an example of P- and S-impedance differences between the 2002 and 1999 data sets, at a depth corresponding to the top of the reservoir. Impedances decrease by as much as 10–12% and non-uniformly along each injector row. Within the reservoir, the S-impedance changes less than the P-impedance. The regions of decreasing P-impedance in Figure 5-31 compare more favourably with the amplitude differences in Figure 5-23, than did those in Figure 5-30, suggesting that the partial stacking process may stabilize the analysis. However, differences in the patterns may in part be due to the manner in which the amplitude differences are measured in Figure 5-23.

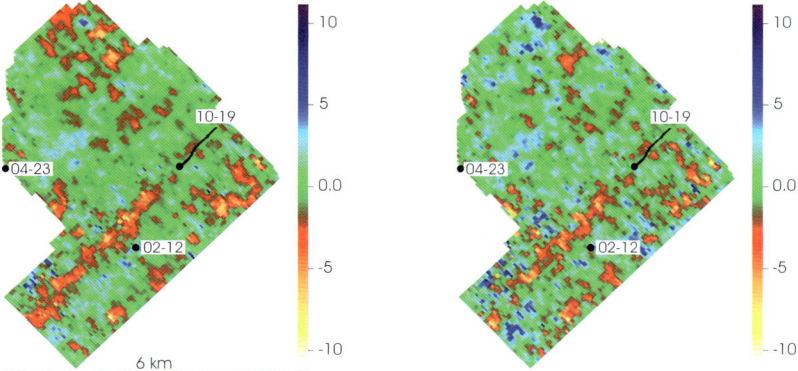

Figure 5-31. Left. *Time slices of the P-impedance difference.* **Right.** *Time slices of the S-impedance difference (both for 2002–1999 data), extracted along the Top Midale reservoir (1 ms below Top Midale), and showing distinct impedance decreases along injector rows in the Phase 1A area.*

The rock physics model needed to link time-lapse impedance anomalies to changes in pore pressure and CO_2 saturation was constructed for both the Marly and Vuggy units using the rock physics analysis described earlier. Figure 5-32 shows the model for the Marly unit, which depicts the dependence of the P- and S-impedance over a range of effective pressures and CO_2 saturations. Both P- and S-impedance are sensitive to pressure changes, but the S-impedance is insensitive to CO_2 saturation changes. The Marly P-impedance can decrease by 6% from its initial conditions for an effective pressure decrease of 7 MPa (pore pressure increase of 7 MPa). The Marly S-impedance is more sensitive than the P-impedance and can decrease by 9% over the same pressure range. An additional 6% decrease in P-impedance can occur, for an increase in CO_2 saturation from 0 to 1. Both impedances are less sensitive to pressure and saturation changes in the Vuggy (not shown) than in the Marly, primarily because the Vuggy limestone is a stiffer, less porous rock than the Marly dolostone.

Inversion for pressure and CO_2 saturation is accomplished by rearranging the order of the modelled data (as in Figure 5-32) to produce plots of effective pressure and CO_2 saturation change as functions of P- and S-impedance changes (Figure 5-33). The resulting inverse plots (Figure 5-33) are used to directly obtain the effective pressure and CO_2 saturation changes corresponding to the P- and S-impedance changes for each element in the 3D volume. The Vuggy inverse surfaces (not shown) are very similar to their Marly counterparts, but are more sensitive to impedance changes, because a given effective pressure or CO_2 saturation change corresponds to a smaller impedance change in the Vuggy than in the Marly.

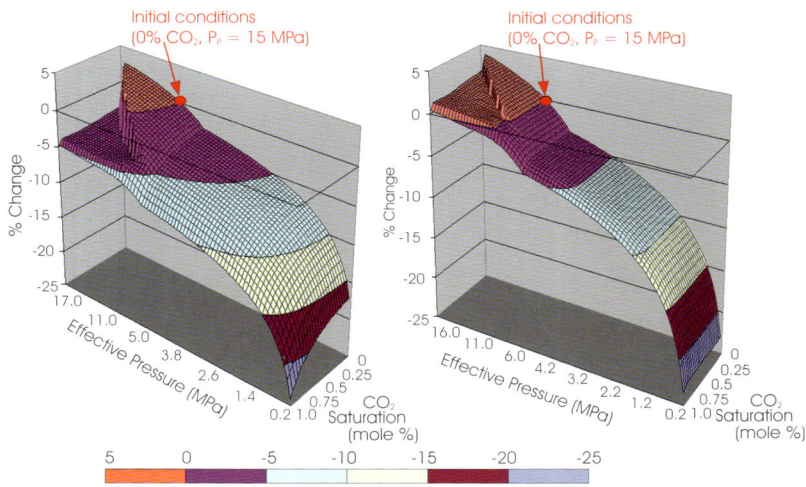

Figure 5-32. *Fractional change of P-impedance (Left) and S-impedance (Right) for the saturated Marly unit, plotted as a function of effective pressure and CO_2 saturation. The CO_2 saturation refers to the mol % of CO_2 that displaces in-situ oil within the unit, assuming an average porosity of 0.29 and an initial oil saturation of 53% (Edmonds and Moroney, 1998; Brown, 2002). The volume of CO_2-saturated oil was increased by 30% (partially displacing the formation water) to account for swelling, and the 30% swelling factor was linearly weighted between 0 and 66 mol % CO_2 (the saturation limit for Weyburn oil).*

Figure 5-34 shows the 2002–1999 changes in effective pressure and CO_2 saturation at the reservoir level, respectively, obtained by the inversion procedure. Pressure increases of up to 8–9 MPa are clearly evident, and generally tend to cluster around horizontal CO_2 injector rows. A maximum saturation increase of 100 mol % CO_2 occurs in the general vicinity of the pore pressure increases, but the clustering pattern is different and the noise level is higher, as anticipated. The largest effective pressure decreases near the reservoir level are usually associated with fairly uniform standard deviations of about 4.5 MPa or less. The largest CO_2 saturation changes, which generally occur near the largest effective pressure changes, display standard deviations of about 30–35%.

Conclusions
A variety of methods have been applied to analysing the 3D seismic data from Weyburn, in an attempt to isolate the contributions of pressure and CO_2 saturation effects to the 3D time-lapse images. Converted-wave (P-S) images produced poor results at the level of the reservoir, precluding further use of these data for

pressure-saturation discrimination. However, although not pursued, it is possible that these data could be used to measure converted-wave (P-S) interval traveltime changes for direct comparison with the corresponding compressional wave (P-P) values. Trace-based prestack AVA analysis and impedance inversion resulted in P- and S-impedance difference maps, as desired. However, comparison with time-lapse stack results and results from partial offset stacking and inversion showed a limited correlation in the resulting map patterns.

This suggests that the trace-based analysis may be less robust, although it should be noted that the manner in which the map comparisons were made was qualitative and not at exactly the same depth horizon. The third analysis method involved 4D processing to obtain migrated offsets stacks that were used in impedance inversion to obtain P- and S-impedance volumes. The resulting time-lapse impedance changes were inverted to generate volumes of pore pressure and CO_2 saturation changes over time. Maximum pore pressure increases

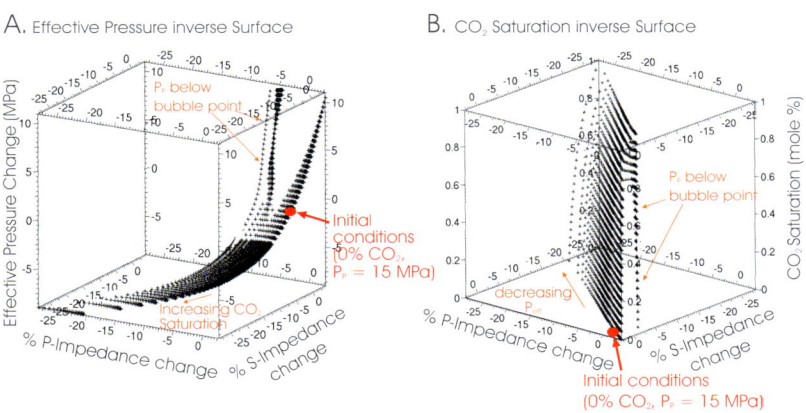

Figure 5-33. *Inverse surface for effective pressure changes and for CO_2 saturation changes derived from the Marly impedance plots in Figure 5-32.* **Left.** *Effective pressure inverse surface. The relatively flat-lying area on the left of the figure corresponds to a region of increasing CO_2 saturation and large decreases in P- and S-impedance from the initial conditions (red dot). This area is conducive to well-posed inverse solutions because small changes in P- and S-impedance result in relatively small changes in effective pressure. The steeply dipping area to the right is not a well-posed region because small changes in impedance result in large changes in effective pressure.* **Right.** *The subvertical surface is much more ill-posed than that for effective pressure changes, implying that it will be more difficult to invert for CO_2 saturation changes than for effective pressure changes in the Marly. The surface is less ill-posed for large effective pressure decreases and CO_2 saturation increases from initial conditions (red dot).*

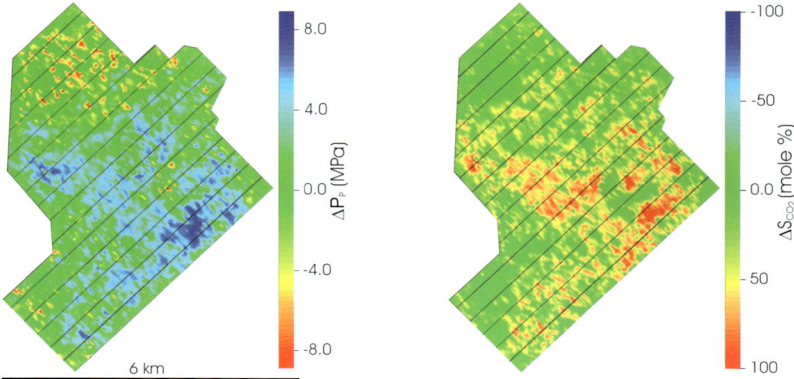

Figure 5-34. Left. *Time slice of pore pressure changes, and* **Right.** *Time slice of CO_2 saturation changes, both between 1999 and 2002 at the Midale reservoir level, and obtained by prestack inversion. Pore pressure increases of up to 8–9 MPa are clearly evident, and generally tend to cluster around the horizontal CO_2 injector locations. A maximum saturation increase of 100 mol % CO_2 occurs in the general vicinity of the pore pressure increases in the right map, but the clustering pattern is different and the noise level is higher, as predicted. A maximum saturation increase of 100 mol % CO_2 occurs in the general vicinity of the pore pressure increases in the left map, but the clustering pattern is different and the noise level is higher, again, as predicted.*

of around 8–9 MPa were found very close to horizontal injection wells, in contrast to reservoir simulation results which show generally steady pore pressures around injection wells since the start of CO_2 injection. This discrepancy could be due, in part, to the limited quality of the S-impedance inversion. Inversion results for CO_2 saturation changes were noisier than those for effective pressure changes, although the largest saturation changes (of about 90–100 mol %) occurred in the vicinity of the largest effective pressure changes. The apparent overestimation of pressure increases, and the observed coupling between the pressure and CO_2 saturation changes in the inversion, are likely due largely to the rock physics model underlying the inversion. Specifically, the model does not account for the enhanced sensitivity of the seismic response to fluid changes within low-aspect-ratio pores, which are an intrinsic property in a fractured carbonate reservoir.

An uncertainty analysis showed that in the area of the largest pressure and saturation changes, background standard deviation levels were about 4.5 MPa for effective pressure changes, and 30–35 mol % for CO_2 saturation changes.

5.6.8 **Assessment of Seismic Data Repeatability**

Several non-geological factors may contribute to artificial time-lapse seismic data differences. These factors include, but are not limited to, (1) the location of the source and receiver geophones, (2) the number of sources and receivers, (3) the type of geophones, (4) recording patches, (5) processing variability, (6) changes in noise conditions, and (7) changes in near-surface conditions. Surprisingly, relatively minor changes in acquisition geometry can have a significant effect.

5.6.9 **Comparison of Processed Stacks**

To assess the effects of relatively small differences in acquisition geometry, and to determine how these effects may be accentuated during similar but not identical data processing, two very similar 3D data sets were processed using the same processing steps. The effects of noise, or changes in subsurface geology, on the processed stacks are virtually eliminated because data in the smaller data set (99S) are identical to the corresponding portion of the larger data set (99C). Also, the shot and receiver locations are common to both of the data sets (corresponding to the majority of data traces).

Although the two data sets are very similar, there are small differences. For example, data set 99C has 132 additional distinct source points, relative to data set 99S (725 shots instead of 593), and the larger data set also included an additional 30 geophone stations (1006 versus 976). Different recording patches were also used for the two surveys. As a result, the larger data set has approximately 20 000 extra traces in the raw data volume (about 15% more), with resultant maximum fold values of 56 and 47, respectively (Figure 5-35). This figure demonstrates that the difference in the stacked traces increases with the difference in the number of contributing traces within a common depth point (CDP bin) between the two surveys.

When the medians of the maximum trace-to-trace correlation values for the two data sets were compared, it was somewhat surprising to find that the values were comparable to correlation values between the baseline survey and subsequent monitor surveys for the same area. This was true once trace-by-trace correlation shifts were applied, as is normally done during post-stack time-lapse calibration. Thus, the two sets of stack data that were acquired simultaneously have differences between them that are comparable to surveys acquired years apart.

Figure 5-36 shows a comparison between the amplitude differences, at the reservoir level, between the 2001 and 1999 baseline survey (left map) and the two post-stack calibrated data sets (middle map, processing; right map, acquisition). As can be seen, although generally much smaller, significant amplitude differences are observed relative to the magnitude of the time-lapse amplitudes that have been interpreted at Weyburn. Thus, differences exist even though common processing parameters were used (in particular, the velocity model and refraction statics model) for the two test data sets. Even when only the traces common to both test data sets are included in the final stack, there are still amplitude differences, although they are relatively small. This is due to the effects of 'multi-trace' calculations in the processing scheme (including surface-consistent deconvolution, surface-consistent residual statics, and trim statics) which will produce somewhat different results when there are differences in the input set of traces.

Conclusions
Processing of two almost identical data sets shows that even small differences in acquisition geometry can lead to differences that are unrelated to changes in the subsurface. In the example presented, part of the difference in the final data volumes is due to the traces that were not common to both data sets. These differences were accentuated by multi-trace processing steps that are based on the different data volumes. In particular, surface-consistent deconvolution, surface-consistent residual statics, and trim statics increased the data differences even when all other parameters were the same for the two processing flows. For legacy data sets, the non-repeatability of the data due to acquisition geometry differences may make robust time-lapse observations very difficult, unless the time-lapse signals being monitoring are large.

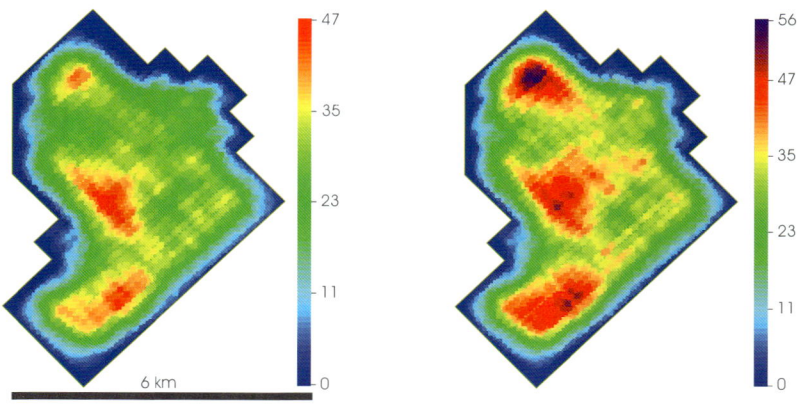

Figure 5-35. *Comparison of fold values for identically processed stacks.* **Left.** *Subset data volume (99S).* **Right.** *Complete data volume (99C).*

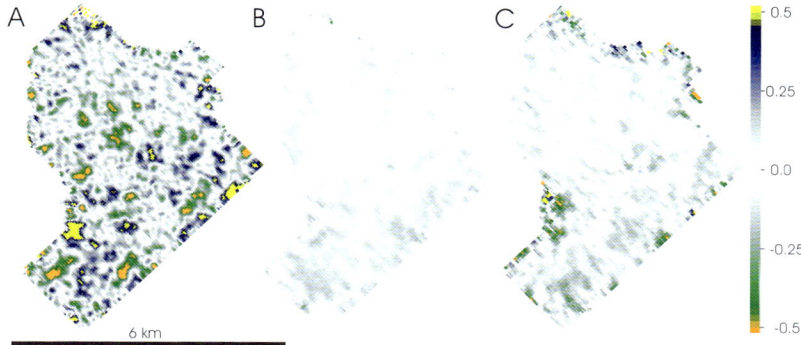

Figure 5-36. *Comparison of amplitude differences near the reservoir, after the application of time shifts to maximize the trace-to-trace correlations.* **A.** *Baseline data set.* **B.** *and* **C.** *Two test data sets (99C and 99S).*

5.7 3D Seismic Monitoring Without a Baseline

Most monitoring programs designed for geological storage of CO_2 incorporate pre-injection baseline surveys of various kinds, including 3D seismic. The underlying premise is that any CO_2-related changes will be observable by comparing subsequent monitoring surveys with the baseline. However, cases will likely arise where CO_2-EOR operations are to be converted to CO_2 storage many years after the onset of CO_2 injection. In these cases, it is possible that no consistent baseline surveys will be available. What are the consequences of not having a set of pre-injection baseline surveys in a monitoring program? Here, we consider the consequences, using 3D time-lapse seismic as an example.

The ability to track the CO_2 plume using seismic methods relies on the change in the seismic impedance that occurs when the existing pore fluids (formation water and oil) are partially replaced by CO_2. The spatial distribution of CO_2 over time can, in principle, be mapped seismically in two ways.

1. Through observation of time-lapse seismic changes, and the association of those changes with the presence of CO_2. The ability to look at differences allows isolation of the CO_2-related effects from variations in the seismic response due to geological heterogeneity. This method has been well demonstrated in the Weyburn project, and allows relatively subtle seismic changes to be detected.
2. By direct identification of CO_2-related amplitude effects in a single seismic survey using, for example, AVO analysis. This method has the advantage that CO_2 distribution can be inferred from a seismic survey taken at any time. However, it requires that the change in seismic response due to CO_2 replacement of formation water and oil is

significant, and that the change can be distinguished from that due to geological heterogeneity.

As described above, time-lapse seismic monitoring identifies changes in the seismic response that are caused by changes in the subsurface. When a pre-CO_2 injection baseline survey is the reference survey, it is simple to equate the observed changes with the effects of CO_2. However, when a post-injection survey is the reference baseline, interpretation becomes more ambiguous. This is illustrated in Figure 5-37, where amplitudes from the 2004 monitor survey are compared with amplitudes from the 2000 pre-injection baseline survey and the 2001 monitor survey (a post-injection baseline). Inspection of the 2004 to 2001 difference map shows areas of the reservoir that have changed since 2001. Focusing on the negative amplitude changes, the 2004 to 2001 difference map shows areas where the CO_2 saturation has increased. However, there are areas where the amplitude differences in the 2004 to 2001 map are negligible – but these are areas where the 2004 to 2000 difference map indicates that CO_2 is present. This could either indicate that CO_2 saturation has not increased in these areas, or that CO_2 saturation levels were high enough at the time of the 2001 survey that subsequent increases in CO_2 saturation produced a negligible change in the seismic properties from 2001 to 2004.

Thus, the absence of a true baseline appears to restrict the mapping to new areas where the CO_2 saturation has reached seismically detectable levels, and where existing areas of high CO_2 saturation remain blind to seismic monitoring. Of course, if the migration of CO_2 results in a large local decrease in CO_2 saturation levels, these areas should show time-lapse anomalies of opposite polarity to those that result when the original pore fluids are initially replaced by CO_2. All of these factors clearly complicate the interpretation for monitoring purposes, but some of the ambiguity can be removed if reservoir flow simulation results are used in parallel with the time-lapse seismic monitoring results. This is true even in the case where a true baseline is available. Above the reservoir, where CO_2 saturation levels are generally expected to be very low, time-lapse seismic monitoring should not be greatly compromised by the absence of a true baseline survey.

The other potential means of monitoring the distribution of CO_2 in the subsurface is through direct identification of CO_2-related amplitude effects in a single seismic survey. This has not been attempted using data from the Weyburn project, but some simple observations can be made in this regard, using some of the modelling results from previous sections.

For this approach to be effective, the CO_2-induced impedance changes have to exceed variations in the background impedance associated with geological heterogeneity. At the reservoir level in the Weyburn Field, pore fluid

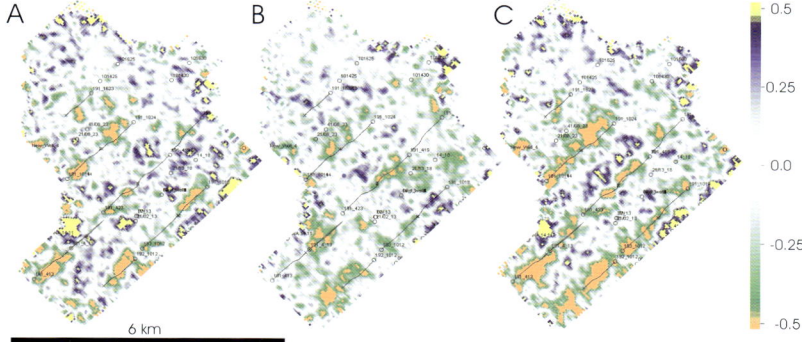

Figure 5-37. *Time-lapse seismic amplitude differences (5 ms mean) at the reservoir level.* **A.** *Map of differences relative to 2000 pre-injection baseline survey.* **B.** *Map of differences between 2004 and 2001 data (to simulate post-injection).* **C.** *Map of differences between 2004 and 2000 pre-injection baseline survey.*

substitution modelling suggests that the maximum changes due to combined pressure–CO_2 saturation impedance are about 12%. This change is comparable to the pre-injection variation in seismic impedance within the reservoir, and thus is unlikely to be easily distinguished. The spatial pattern of impedance variations could provide some clue in this regard.

At shallower levels, maximum CO_2-related impedance changes are comparable to those at the reservoir depth (at least for the hypothetical sandstone used for modelling). Thus, only in relatively homogeneous porous formations would we expect to see clear indication of CO_2-related amplitude anomalies.

5.7.1 Conclusions

In the absence of a pre-CO_2 injection baseline survey, 3D time-lapse seismic methods can still be used to track the spread of CO_2 in the reservoir and overburden. However, areas of the subsurface that had high levels of CO_2 saturation prior to the post-injection baseline survey will likely remain unchanged in the subsequent monitor seismic surveys. This is a result of the non-linear sensitivity of the seismic parameters to CO_2 saturation levels. Thus, time-lapse monitoring should be effective in tracking the CO_2 front over time. The direct identification of CO_2-related amplitude effects from non-time-lapse seismic data may be possible for porous formations that are very homogeneous (that is, for those formations with inherent impedance variability of only a few per cent). AVO methods applied to non-time-lapse seismic data for the direct detection of CO_2 were not considered in the Weyburn project.

5.8 Recommendations

Although the design of an effective geophysical monitoring plan is site specific, and will likely be driven by a risk assessment for the storage site, it will most often entail some level of monitoring the subsurface distribution of CO_2, the associated pressure plume, and any possible injection-induced damage of the storage container. The characteristics of the storage complex are of particular importance with respect to geophysical monitoring. Factors to be considered include, but are not limited to, rock type (clastic or carbonate), porosity, depth, thickness, and resident pore fluids, as well as the quantity of CO_2 to be injected, the expected areal extent of the CO_2 plume, the surface infrastructure or physiography, and the number of wells in the field. All these variables will affect the type of geophysical method used and whether it is surface-based or well-based. Knowledge gathered from the application of geophysical methods in the Weyburn project provides a basis for recommendations for their application in the context of validating CO_2 storage.

5.8.1 Geophysical Characterization of the Rock–Fluid System

Characterization of the local rock–fluid–stress system is essential to the design of an appropriate geophysical monitoring plan and for the subsequent understanding of the results. These measurements provide the means of assessing the sensitivity of specific geophysical monitoring methods to variations in the parameters of interest (primarily CO_2 saturation and pressure). The measurements are most readily done on core samples under controlled conditions in a laboratory, and may be complemented by in situ measurements, including static measurements or time-lapse logging during changes of in-situ pressure and fluid saturation. The latter measurements have two advantages: (1) they are acquired at the specific subsurface conditions, and (2) they are at a larger scale than the core measurements.

Variations in the composition of the injected CO_2 can have significant effects on the properties being monitored, and thus the composition of the injectant should be well known. Larger-scale features (for example, vugs and fractures) are not well sampled at the core-scale, and can change the physical properties of the reservoir. They need to be considered in regard to geomechanical and monitoring parameters for the reservoir. Estimating these effects can be difficult, and in the case of the Weyburn project their effect was inferred during modelling of the microseismic-based stress observations and in modelling the baseline seismic data.

5.8.2 Feasibility Studies

Several monitoring methods were considered for application in the project. Based on feasibility studies, the following recommendations were derived.

The InSAR (interferometric synthetic aperture radar) technique is a relatively inexpensive means of monitoring injection-related surface deformation over a large area. Surface deformation is most directly related to the subsurface pressure perturbation associated with injection, rather than the distribution of the CO_2 plume. Thus, InSAR is particularly suited to regional monitoring of the pressure plume. Associated geomechanical modelling is required to extract quantitative estimates of stress-strain in the subsurface. Feasibility studies can be conducted with existing historical data to assess the likelihood of successful InSAR monitoring in the absence of installed permanent scatterers, but in rural settings like the Weyburn Field it is likely that deployment of permanent scatterers will be required prior to the onset of CO_2 injection. Depending on the climate and landscape, at least one full season of pre-injection monitoring should be conducted to assess the seasonal variations in ground movements that are unrelated to CO_2 injection operations.

Gravity measurements are another means of regional monitoring. Changes in the gravity field associated with injection should be detectable under two conditions: (1) for large injection volumes at depth, or (2) if significant volumes of CO_2 have migrated to shallow depths so that the CO_2 is no longer supercritical. As with InSAR monitoring, a density model is required to make quantitative estimates of what surface-measured changes in the gravity field imply about subsurface CO_2 distribution. It is feasible that airborne gravity monitoring, as opposed to ground-based measurements, could be applicable in some circumstances. The footprint of satellite-based gravity measurements is generally too broad to be used for individual CO_2 storage projects.

In general, however, regional monitoring methods, such as InSAR and gravity, are best applied in conjunction with other higher-resolution monitoring methods (for example, seismic).

LEERT (long-electrode electrical resistance tomography) currently falls into the category of a research method in regard to deep CO_2 monitoring. At Weyburn, feasibility modelling showed that interwell spacing was a significant limitation. LEERT methods may be most suitable in areas where the overburden is more resistive, and/or the reservoir is significantly shallower than at Weyburn. Alternatively, the method may be better suited to monitoring the overburden for CO_2 leakage. Given these limitations, pre-acquisition modelling is essential in assessing the feasibility of LEERT to specific geological and well geometry in an EOR field.

5.8.3 Downhole Monitoring Methods

Several downhole seismic techniques were tested at Weyburn leading to the following recommendations.

Continuous downhole passive seismic monitoring documents microseismicity associated with the CO_2 injection process. This information is valuable for providing public assurance as well as providing constraints on brittle deformation near the reservoir and in the overburden. The Weyburn results suggest that this monitoring method is not well-suited for directly tracking the CO_2 plume. Although near-surface geophones can be used for passive monitoring, a downhole passive array provides greater sensitivity to small magnitude microseismicity. At Weyburn, moment magnitudes in the range of −3 to −1.5 were detectable only within about 500 m of the array, suggesting that these events would be difficult to detect at the surface.

Downhole arrays should be designed with the monitoring objectives in mind. Where accurate event locations are required, particularly in regard to the depth interval including the reservoir and immediate cap rock, the array should have adequate aperture (or depth extent) and ideally should extend through the reservoir interval. Horizontal location accuracy can be improved by deploying arrays in more than a single well. The use of multiple wells will improve the ability to characterize the source mechanisms, which can be useful for stress evaluation. Velocity logs (both Vp and Vs) should be obtained from the monitoring wells to provide an accurate velocity model for event location. Finally, a geomechanical model is essential for understanding the significance of observed microseismicity.

Active-source downhole seismic methods (crosswell and VSP) provide a means of acquiring subsurface images with resolution on the scale of metres to tens of metres. This makes them particularly amenable for monitoring small injection volumes, or providing geological details in a zone of interest. However, they are less suited to regional monitoring because of their limited areal coverage, and the inherent need for wellbore access. In general, it is recommended that they are used primarily for calibration purposes, complementing surface-based seismic monitoring and downhole passive seismic monitoring.

Downhole electrical methods, as tested in other CO_2 projects, could be considered as an alternative to seismic methods, although experience with electrical techniques for this purpose is very limited.

5.8.4 3D Seismic Methods

Overall, surface-based 3D seismic methods provide the most effective means of monitoring the subsurface distribution of CO_2 over a large area. They provide relatively high resolution at depth (for example, about 10 m at 1450 m depth in the case of Weyburn) and are capable of imaging and hence detecting CO_2. The applicability and depth resolution of 3D seismic methods are dependent on (1) the geological environment, (2) the achievable signal-to-noise levels, and (3) the depth of the target. To date, the effective use of 3D time-lapse seismic monitoring for qualitative mapping of the subsurface CO_2 plume is well demonstrated, whereas semi-quantitative CO_2 saturation estimates are still limited. As demonstrated at Weyburn, discriminating pressure effects from CO_2 saturation is feasible, although the results are highly model-dependent. As noted earlier, the model utilized for this purpose inadequately accounted for the effects of crack- and fracture-related porosity, resulting in pressure estimates that were at odds with calibrated reservoir simulations. This stresses the need to adopt petro-elastic models appropriate to the specific reservoir lithology. Further, seismic inversion results should be clarified by fluid flow simulations at the least, or, optimally, should be formally linked with fluid flow simulations. Acquisition and analysis of multi-component seismic data in the Weyburn project provided information of secondary importance relative to standard vertical-component data.

In considering 3D seismic methods for time-lapse monitoring in a given CO_2-EOR or CO_2 storage project, the seismic survey design used in any seismic survey may be applied. In addition, the following specific aspects should be considered.

Acquisition geometry (stations and acquisition patterns) should be repeated from one survey to the next as closely as possible in order to minimize differences in acquisition-related data. The importance of this will be determined by the magnitude of time-lapse signals that are being sought, as compared to the magnitude of acquisition-related signals.

The near-surface conditions (that is, groundwater levels and ground conditions) should also be as consistent from one survey to the next, again to ensure data repeatability. Obviously, there is limited control over this aspect. This problem can be partially addressed for monitoring surveys by deploying receivers at depth and by acquiring data at the same time of year.

Ideally, for comparing data sets (that is, the baseline survey with subsequent monitor surveys) they should be edited to include only those traces that have common source-receiver pairs. An alternative strategy might be to force agreement between traces that are common to the respective processing flows.

Acquisition of three-component data should be considered if incremental acquisition costs are small (as they often are) and if there is reason to believe that the use of shear wave data will be particularly effective. When good quality shear wave data are available, they can be used to map pressure changes and to monitor fracture systems.

In terms of seismic attributes that can help distinguish CO_2 saturation from pressure-related effects, combinations of the AVA (amplitude-versus-angle) intercept (I) and gradient (G) can be used. The monitoring procedure could be similar to the identification of Class III AVA anomalies. Alternatively, P- and S-impedances can be inverted for using limited offset stacks, and then formally inverted for pressure and CO_2 saturation changes using model-based inversion.

If amplitude-versus-offset/amplitude-versus-angle (AVO/AVA) analysis it to be employed to discriminate CO_2 saturation from pressure effects, the seismic survey must be designed to record high-fidelity data at offsets that are at least comparable to the target reservoir depth. High-fidelity, long-offset seismic data are critical for accurate prestack P- and S-impedance inversion. In particular, the S-impedance inversion is most sensitive to variations observed at the larger offsets. The quality of the longest offset data from the Weyburn data set was too poor to use for prestack impedance inversion, limiting the quality of S-impedance inversion from that area as well.

Detailed analysis of the behaviour of reservoir rocks should be conducted for low effective pressures, where small changes in pressure can produce large effects on seismic data. The petrophysical model used for inversion in the Weyburn project predicted S-impedance changes that are 50% greater than those for the P-impedance, for large pressure changes.

6

HISTORY MATCHING AND PERFORMANCE VALIDATION

Summary

Laboratory-scale and field-scale history matching were used to calibrate and refine reservoir simulation and reactive transport models which are required to forecast the long-term efficacy of CO_2 geological storage.

Reactive transport modelling of CO_2 coreflood experiments showed that faster carbonate mass transfer rates and greater volumes of dissolved carbonates per unit time occurred with homogeneous pore size distributions. Bimodal pore size distribution, with vugs and sub-micron pores, resulted in variable mass transfer rates. Permeability increased drastically when wormholes developed. The results suggest that Darcy flow adequately models reaction fronts for porous carbonate rocks with a uniformly distributed pore space, although it may not hold for heterogeneous carbonate rocks that have preferential flow paths or wormholes.

History matching of produced water compositions and isotopic data at Weyburn require the use of a dual porosity reservoir model. Here, the isotopic data are potentially more useful than major ion concentrations for gaining insights into transport processes during CO_2-EOR.

A Markov Chain Monte Carlo method and reactive transport modelling were used to history match 3D seismic data through refinement of reservoir permeability structure, and to match produced water chemistry through refinement of reservoir mineralogy.

Recommendations arising out this research begin on page 253.

Authors: **J.W. Johnson** (Schlumberger Doll Research, Cambridge, MA, USA) and **D. White** (Geological Survey of Canada, Ottawa, Ontario).

6.1 Introduction

As a matter of convenience, history-matching studies may be considered as those that involve either laboratory-scale or field-scale systems. In laboratory-scale history matching, site characterization is tightly constrained, so that the significant source of uncertainty in field-scale simulations is virtually eliminated. Also, the experiments can be designed to address, individually, many important processes whose interdependencies in nature are often difficult to disconnect.

By essentially removing site characterization uncertainty and permitting focus on specific processes, laboratory-scale investigations enable resolution of prediction/observation discrepancies through model refinement. Thus, they present a unique opportunity to calibrate and improve predictive models.

In this chapter, we present the results of two investigations that addressed laboratory-scale history matching exercises that were well constrained by careful characterization of Weyburn core and formation water data. In the first, the effluent concentrations and porosity/permeability changes seen in CO_2 coreflood experiments were used to calibrate and refine a reactive transport model. In the second, the so-far-uncharacterized fracture transport properties of Marly, Vuggy, and Midale Evaporite (seal) cores, as well as the reaction- and stress-dependent changes of such properties during CO_2 injection, were determined and simulated through an integrated experimental and modelling approach.

In contrast, field-scale simulation efforts face the daunting challenges of typically extreme site-characterization uncertainty and the need to represent all of the relevant integrated processes. Where there is a reasonable degree of confidence in model pedigree (for example, from experimental calibration), field-scale simulations offer the potential for minimizing prediction/observation discrepancies through site characterization refinement.

In this chapter, we present the results of three investigations that addressed field-scale history matching exercises that were well constrained by comprehensive monitoring results (principally, formation water geochemistry and seismic reflection data). In the first, the formation water database was used to constrain field-scale reactive transport modelling of fluid flow and CO_2–formation water–rock transfer in the reservoir. In the second, a novel stochastic inversion method was used to minimize discrepancies between predicted and observed seismic reflection data through refinement of the bulk reservoir permeability field. In the third, the same inversion technique was used to reduce discrepancies between changes in measured and simulated formation water composition through refinement of reservoir mineralogy along inferred flow paths from CO_2 injector to production wells. Both sets of experiments are directly relevant to virtually any CO_2-EOR or CO_2 storage site.

6.1.1 **Context**

History matching involves the use of iterative or stochastic computational methods to minimize discrepancies between the predicted and observed behaviour of laboratory- to field-scale experiments. The process focuses on (1) identifying the hierarchical sensitivity of model results to specific site-independent and site-specific input parameters, and (2) progressively refining key parameters until an acceptable reduction in initial discrepancies between predicted and observed values has been achieved. In the case of subsurface CO_2 emplacement, history matching aspires to achieve accurate forecasting of performance in terms of storage capacity, footprint dynamics, and containment efficacy. These performance parameters are quite different from those of most CO_2-EOR history matching studies, which typically focus on model tuning to achieve accurate simulation of produced fluid volumes and pressure changes.

In carbonate rocks, the key to history matching is strongly dependent on the way porosity and permeability evolve as a function of CO_2-induced mineral dissolution and precipitation. The slow kinetics of these reactions at reservoir temperatures preclude direct field measurement over relevant time-scales. Hence, field-scale history matching must be underpinned by laboratory-scale efforts that achieve accurate simulation of well-constrained batch and flow-through experiments. These can observe compositional and transport consequences of incipient CO_2–formation water–rock transfer at the pore scale, in which the significant impact of reservoir heterogeneity is virtually eliminated. Thus, accurate laboratory-scale history matching is an essential precursor to successful field-scale efforts.

The work reported involved multi-scale history-matching studies, including (1) a broad range of laboratory experiments (coreflood, fracture flow, microbeam), (2) modelling (pore- to field-scale reactive transport modelling), and (3) iterative and stochastic inversion methods (including Markov Chain Monte Carlo).

6.1.2 **Objectives**

The objectives of history matching in the context of CO_2 storage performance are as follows.

1. To use well-constrained experimental results to calibrate and refine reactive transport models. These are needed to predict CO_2–(oil)–formation water–rock transfer and dependent reservoir–seal integrity, CO_2 migration, and containment.
2. To use comprehensive geophysical–geochemical monitoring data and modelling techniques to reduce site-characterization uncertainty with respect to key reservoir transport properties, such as permeability structure and magnitude, and mineralogy.

6.2 Coreflood Experiments

Potential reaction-dependent alteration is of particular concern in highly reactive carbonate reservoirs such as the Weyburn Field. A critical research area for CO_2-EOR and CO_2 storage operations is to develop improved predictive capabilities for CO_2-induced fluid–rock transfer processes and dependent porosity/permeability changes, which may significantly alter storage performance. In this study, a Darcy-scale continuum model was used to simulate effluent chemistry and mineral dissolution effects, documented for two well-characterized coreflood experiments, in which CO_2-equilibrated formation water was injected into Midale Vuggy and Marly core samples under reservoir conditions. The 3D reactive transport model was constructed through X-ray computed microtomography (XCMT) and scanning electron microscopy (SEM) characterization of the Vuggy and Marly samples (highly resolved mineral and pore space distributions), and constrained by experimental formation water data. In this section, we review the principal characterization techniques, coreflood methodology, and Weyburn-specific experimental and simulation results.

6.2.1 Sample Characterization

Accurate pre-flood and post-flood mineralogical and petrophysical characterization of core samples used for CO_2-flood experiments is essential to calibrating reactive transport models. Previous information about the Midale Vuggy and Marly units suggested that they would be difficult to characterize, particularly because the void spaces vary from sub-millimicron to centimetre scale (and larger).

Although X-ray computed micro-tomography captures a range of grey scales in 3D, it does not capture intercrystalline pores and individual minerals that can be seen in 2D back-scatter electron (BSE) images, because the resolution is not sufficient to detect microstructures. Therefore, scanning electron microscopy (BSE mode) was used to calibrate the high-resolution XCMT grey scale, which in turn allowed estimation of micropore connectivity and mineral distribution.

Pre-flood cores were imaged by XCMT at the European Synchrotron Radiation Facility in Grenoble, France, and post-flood cores were imaged at the Advanced Light Source at Lawrence Berkeley National Laboratory, Berkeley, California. Figure 6-1 shows the average initial conditions for the coreflood experiments.

The Vuggy samples display extreme heterogeneity in pore diameters and mineral composition. Dolomite is present as crystal inclusions in the limestone or slightly cemented with calcite (sparse dolomite). Dolomite crystals are <30 μm in diameter. The dolomite-rich regions tend to be more porous and contain well-connected micropores. Fractures, several millimicrons in width, also contribute to this network. Macropores, which account for only 10% of the

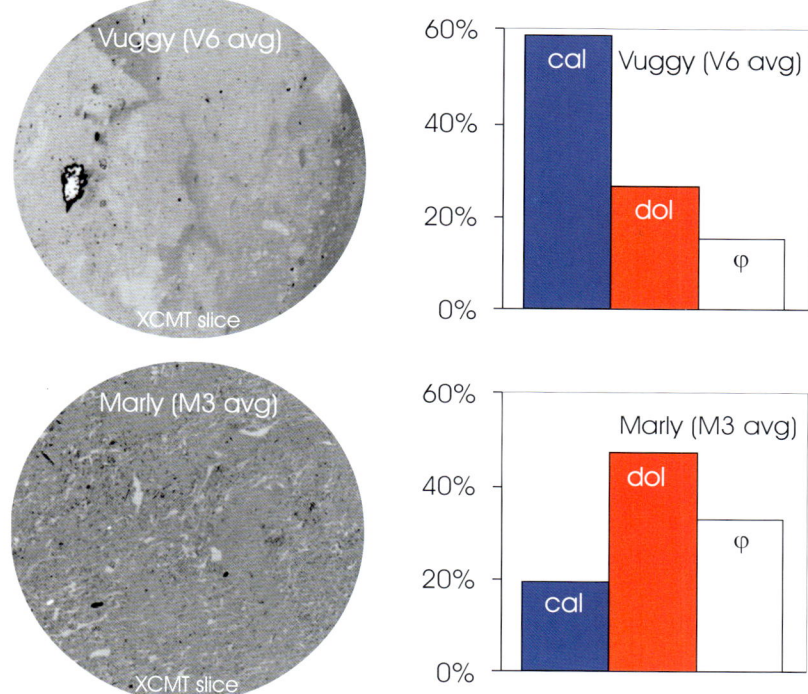

Figure 6-1. *Average mineral volume fractions, and porosity for the Vuggy (V6) and Marly (M3) samples used in the coreflood experiments. Permeability ranges were 0.01–0.03 mD, and 1–2 mD, respectively. The XCMT slice is 30 mm wide.*

porosity, are uniformly distributed among all phases, rarely touching each other and connected only through micropores present in the mineral matrix. In general, the calcite regions are less porous than the dolomite regions.

In contrast, the Marly dolostone samples are quite homogeneous with respect to both mineral and pore-space distribution, primarily consisting of segments that are a mixture of calcite and dolomite with abundant void space and dense calcite with little void space. The porosity is submicroscopic, and unfilled elongated rods are common. The samples generally exhibit very little structure.

6.2.2 Experimental Conditions and Results

All experiments were designed to simulate chemical equilibration with calcite and dolomite at reservoir conditions (60°C, 12.4 MPa), and were run at four CO_2 partial pressures (3, 2, 1, and 0.5 MPa). The composition of the synthetic formation water was based on data from the wellhead fluid sampling program.

The reactor vessel confining conditions were 2.48 MPa and 60°C, with 5–10 mL samples withdrawn for analysis. The system was allowed to equilibrate with CO_2-free formation water before introduction of CO_2-equilibrated formation water.

During the pre-CO_2 flow phase, samples were taken to obtain differential pressure/permeability measurements. Complete analyses were made of the effluent fluid samples, but we report here only calcium, magnesium, total inorganic carbon, and pH. After the introduction of CO_2-saturated formation water, solution pH could no longer be directly measured due to loss of dissolved CO_2 and concomitant pH increase as the sample depressurized at the sampling port. Thus, pH and mineral saturation indices reported for these conditions were calculated using the EQ3/6 geochemical speciation code via charge balance from measured compositions at 60°C and 12.4 MPa.

Figure 6-2 shows the results for the Midale Vuggy beds and Figure 6-3 for the Marly beds. Calcium contents respond quickly to the introduction of CO_2-equilibrated formation water, followed by slow decay to lower steady-state values. For the Vuggy cores, reacted at intermediate values of pCO_2 (partial pressure of CO_2), there were no changes in magnesium levels. This possibly reflects the complex interdependence of carbonate cation concentrations on transport and chemical processes. The Marly experiments were of shorter duration than the Vuggy ones, so the correlations among calcium and magnesium contents are quite striking.

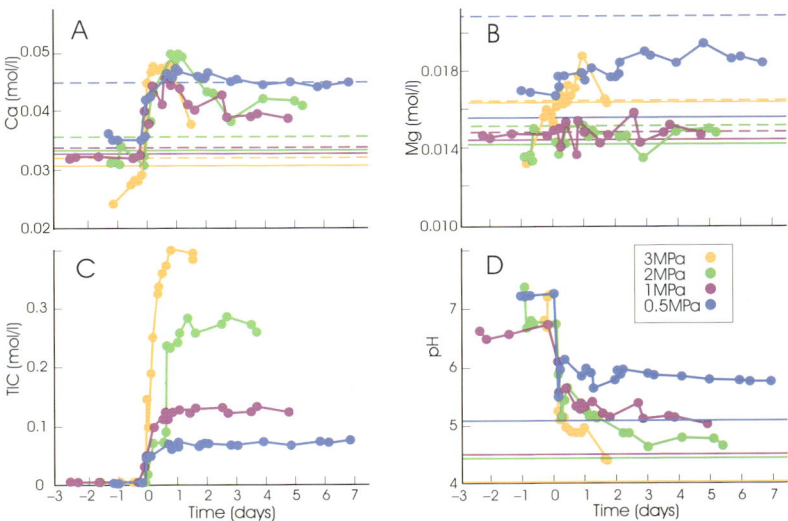

Figure 6-2. *Vuggy coreflood results. Measured effluent for:* **A.** *Calcium.* **B.** *Magnesium.* **C.** *Total inorganic carbon.* **D.** *Estimated pH, for four pCO_2 values (0.5, 1, 2, and 3 MPa). Horizontal lines indicate average input formation water concentrations before (solid lines) and after (dashed lines) each experiment.*

For all pCO$_2$ values, there is also a sharp and rapid increase in calcium content almost immediately upon CO$_2$ introduction. Final steady-state values are higher than those in the Vuggy experiments, and also correlate as expected with pCO$_2$. Although increases in magnesium contents for the Marly experiments are small compared to calcium, all show evidence for consistent steady magnesium increases, in contrast to the Vuggy experiments. These observations agree with the generally smaller dolomite crystals (and thus larger specific surface areas available for reaction) in the Marly compared to Vuggy cores. Measured inorganic carbon values for both sets of experiments show rapid increases and subsequent steady-state behaviour, consistent with their anticipated dependence on pCO$_2$, CO$_2$ aqueous solubility, and carbonate dissolution.

6.2.3 Wormholes

An unstable dissolution front, in the form of a single dominant wormhole (diameter 0.7–2.2 mm), developed in all Vuggy coreflood experiments at all pCO$_2$ values. Figure 6-4 (top figures) shows the pre-flood and post-flood images for the 1 MPa pCO$_2$ experiment. This is contrasted with a post-flood image for the Marly cores (Figure 6-4, bottom figure), which are characterized by relatively stable homogeneous dissolution fronts at all pCO$_2$ values.

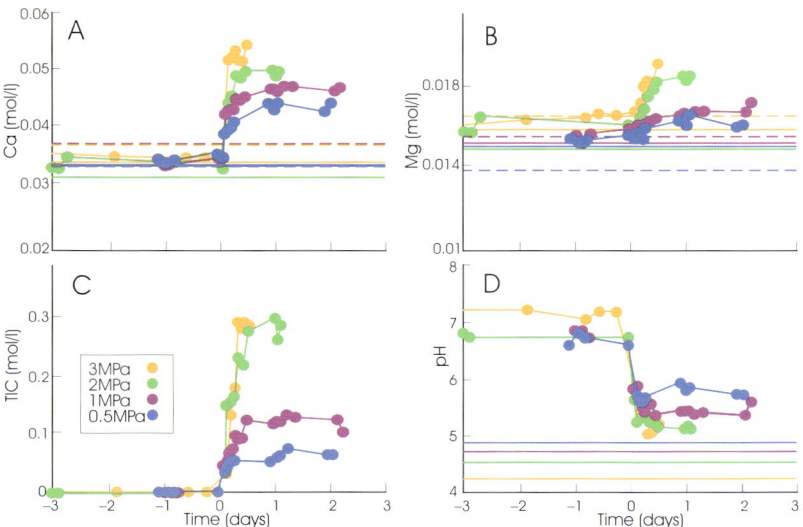

Figure 6-3. *Marly coreflood results. Measured effluent for:* **A.** *Calcium.* **B.** *Magnesium.* **C.** *Total inorganic carbon.* **D.** *Estimated pH, for four pCO$_2$ values (0.5, 1, 2, and 3 MPa). Horizontal lines indicate average input formation water concentrations before (solid lines) and after (dashed lines) each experiment.*

Wormhole development reflects (1) the initial, multi-scale porosity of carbonate minerals that line and bridge the hole, and (2) carbonate dissolution kinetics. Wormholes tend to be narrower in fractures and wider where fluids encounter tight mineral zones, suggesting that they develop through complex interaction of dynamic transport and reaction processes. These observations support a conceptual model that invokes positive feedback between preferential fluid flow to regions of high permeability, followed by enlargement of these regions by rapid dissolution of the matrix mineral, causing the zone to receive even more flow.

6.2.4 Conclusions

Under conditions of reactive flow using formation water equilibrated with varying levels of pCO_2, we can generalize the experimental observations as follows.

1. Faster carbonate mass transfer rates and greater volumes of dissolved carbonates per unit time were observed for the Marly cores, which have more homogeneous pore size distributions and larger reactive surface areas. Solution chemistry evolved toward carbonate saturation levels and remained in this steady state.
2. Variable mass transfer rates (tending towards saturation at early times but sharply decreasing at later times with increasing pCO_2) were observed for Vuggy cores, which possessed a more bimodal distribution of pore sizes, dominated by sub-µm pores and macroscopic vugs. Larger dissolved carbonate volumes occurred for these experiments as a result of their longer experimental times.
3. Combined chemical and tomographic data show that fast pathways developed according to the distribution of pre-existing physical heterogeneities (large vugs, microfractures), and localized dissolution in these areas contributed to rapid and drastic increases in bulk sample permeability.
4. In general, periods when formation water was saturated with respect to both calcite and dolomite correspond to periods of Darcy-like flow (for example, pre-breakthrough times for Vuggy experiments and entire experimental durations for Marly experiments). During these periods, fast carbonate kinetics were not a limiting factor in the evolution of the reacted formation waters.
5. In the Vuggy experiments, where breakthrough of a fast preferential pathway, or wormhole was noted, the inferred reduction in available reactive surface area allowed transport processes to dominate over reaction kinetics, resulting in formation water solutions undersaturated with respect to carbonate minerals.

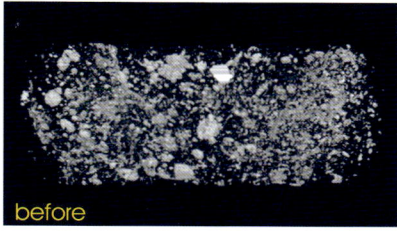
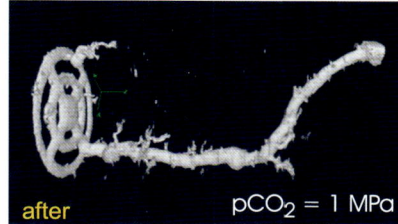
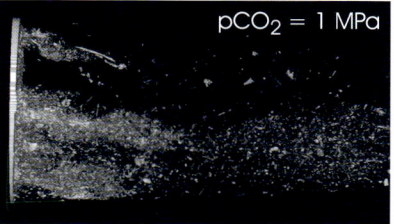

Figure 6-4. Top diagrams. *Pre-flood and post-flood XCMT images illustrating the development of unstable dissolution fronts (wormholes) in the Vuggy core reacted with CO_2-equilibrated formation water at 1 MPa pCO_2. Flow was from left (inlet) to right (outlet), and the post-flood image has been filtered to highlight the development of new porosity only.* **Bottom diagram.** *Post-flood XCMT image illustrating the development of relatively stable, homogeneous dissolution fronts in the Marly core reacted with CO_2-equilibrated formation water at 1 MPa pCO_2. Flow was from left (inlet) to right (outlet).*

6.3 Simulation Model

A Darcy-scale continuum model was used to simulate reactive transport and mineral dissolution processes for two of the coreflood experiments described above, namely, the Vuggy at pCO_2 of 3 MPa and the Marly at pCO_2 of 1 MPa. The model was based on the geochemical and physical characteristics for each unit, and employed scaling parameters that facilitate incorporating the XCMT-resolved spatial distribution of minerals and pore space into the much larger representative element volumes required for the simulations. It is important to note that these models represent preliminary efforts to match the empirical characterization and chemical data. The goal of this work was (1) to exploit these accurately characterized cores, and well-constrained experimental data, and (2) to calibrate and improve reactive transport models that predict CO_2-induced fluid–rock transfer and the dependent evolution of pore space and permeability during CO_2 storage in carbonate rocks.

6.3.1 **Model Overview**

The reactive transport simulations were performed using the Nonisothermal Unsaturated-Saturated Flow and Transport code, NUFT, which is based on an approximation for Darcy flow. NUFT is a highly flexible software package for modelling multiphase, multi-component, heat and mass flow and reactive transport in unsaturated and saturated porous media. An integrated finite-difference spatial discretization scheme is used to solve mass and energy balance equations in both flow and reactive-transport models. The Newton–Raphson method was used to solve the resulting nonlinear equations.

Dolomite was not allowed to precipitate in this study. The rate constants used in the simulations represent a lower threshold needed to develop wormholes in the Vuggy experiments. Lower rate constants failed to generate wormholes, and higher rate constants did not improve the fit. These same rate constants were used in the simulation of dissolution fronts for the Marly experiment. Mineral dissolution and precipitation processes are treated as kinetically controlled, and are represented by a transition-state theory rate law. Because the 3D, 0.5 mm grid-size numerical mesh used for this study is much coarser than the XCMT image resolution (15 μm), an arithmetic averaging procedure was used to convert pixel grey-scale values from XCMT analysis to effective porosity and mineral composition at each NUFT numerical gridblock. Comparison of the

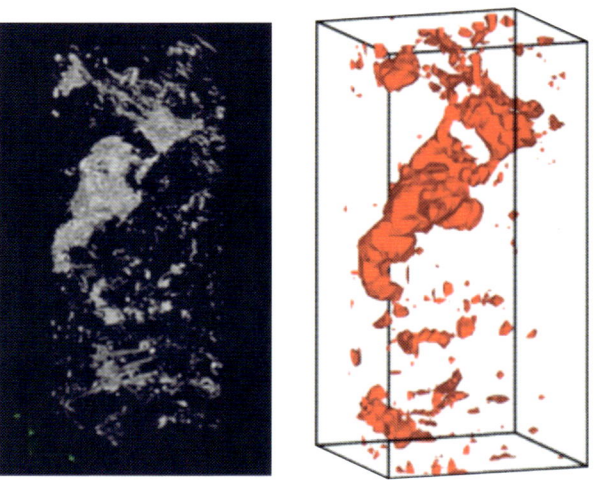

Figure 6-5. *Reactive transport model of the base case pre-flood experiment.* **Left.** *The 15 μm resolution of the XCMT image (0.5 iso-surface threshold).* **Right.** *Averaged values appropriate for 0.5 mm gridblock resolution in the NUFT model (0.5 iso-surface threshold).*

XCMT images and simulated core shows that this averaging strategy captures the macro-scale properties for the initial porosity and mineral composition distributions of Vuggy and Marly cores (Figure 6-5).

The macro porosity–permeability relationship is an important factor for modelling coupled flow, reactive transport, and mineral dissolution processes at the Darcy scale. This relationship is time-dependent, due to strong coupling between flow and mineral dissolution processes. In a reactive-transport model that considers mineral dissolution/precipitation, the traditional Kozeny–Carman correlation is often used to describe this relationship. However, in this study, it failed to account for higher permeability within preexisting or dissolution-induced highly porous and conductive preferential flow pathways. As a result, the initial simulations underpredicted the faster wormhole development observed in the experiments. To correct this, modified K–φ correlations were constructed. These account for enhanced permeability for regions with porosity >0.5 that cannot be captured by the standard Kozeny–Carman relationship, and therefore help maintain a higher flow rate once wormholes develop.

6.3.2 Simulation Results for Vuggy Corefloods

A series of numerical tests was performed to explore the effects of such factors as the chemical reaction rate and permeability–porosity (K–φ) relationship on coupled physical and chemical processes associated with porous media dissolution, and the development of unstable dissolution fronts or wormholes. A side-by-side comparison of the porosity distribution obtained from an XCMT image and that obtained from the NUFT model is shown in Figure 6-5 and Figure 6-6 for before and after CO_2 flooding, respectively.

Despite this reasonable physical agreement, a close inspection of numerical results reveals that in addition to the main conductive porous channel, multiple finger-shaped pore structures developed near the inlet, which are absent from the XCMT image.

The simulated wormhole was not established until 96 h after CO_2 flooding, instead of the breakthrough time (20–40 h) observed from experimental measurement. The underprediction of breakthrough is also reflected in a comparison of pressure and chemistry data. Experimental results imply that the highly conductive wormhole was formed at about 20 h, and was accompanied by a significant pressure-difference decrease. Within about another 20 h, all the mineral residues in the channel dissolved until solution pH reached about 4.3 as a result of loss of mineral buffering. This pH value was the same as that of the injected formation water. Except for the first few hours, the pressure difference predicted by the NUFT model decreases at a much slower pace than that of the experimental measurement. Final breakthrough occurs at around 96 h. Similarly, simulations predict higher pH while the experiment shows a more acid pH after breakthrough. In the model, the input solution continues to be buffered by calcite and dolomite dissolution.

Sensitivity studies showed that when the reaction rate constants were reduced by an order of magnitude, there was (1) somewhat less branching and fingering near the inlet, (2) breakthrough extended beyond 96 h for the development of a preferential flow pathway, and (3) the preferential pathway originated at a slightly different location. A second sensitivity study explored the dependence of wormhole formation on the permeability–porosity relationship, and showed that the dissolution reaction front movement was less aggressive after the preferential pathway was developed.

6.3.3 Simulation Results for Marly Corefloods

For NUFT simulation of the Marly experiment, separate permeability–porosity (K–φ) relations were developed (not shown). Other model parameters are the same as those outlined above. The development of dissolution reaction fronts at discrete time steps is shown in Figure 6-7. Of particular significance is the accurate portrayal of dissolution-front evolution along the dolomite half of this core. The simulations confirm the experimental finding that connectivity of the pore space is more important than overall reactivity of the mineral phases in carbonate rocks. Even though slight fingering is observed at the late stage of the simulation, the overall reaction front propagates more uniformly than in the Vuggy experiments because of the more homogeneous distribution of pore space in the Marly core (at least on the dolomite half).

6.3.4 Conclusions

The goal was to use detailed experimental data to constrain, calibrate, and refine reactive transport models that describe and predict the evolution of pore space and permeability during CO_2 storage in carbonate rocks. The ability to model the evolution of unstable reaction fronts (wormholes) that develop in the highly heterogeneous Vuggy limestone is limited by the ability to properly scale this heterogeneity and to capture the fast channel (wormhole) flow phenomena within the model. The simulated results suggest that Darcy continuum-scale flow can adequately model reaction fronts for highly porous rocks with a uniform distribution of pore space. We are on the verge of fully capturing the development of wormholes by using an empirical relationship that ties mineral dissolution and resulting increases in porosity to increases in permeability. However, comparison of the model output to pressure gradient and solution chemistry shows that the model underpredicts the amount of time needed to generate these features.

The discrepancies between the experimental and simulated results found in Vuggy limestone may be explained as follows.

1. In certain regions, the Darcy flow approximation may not hold.

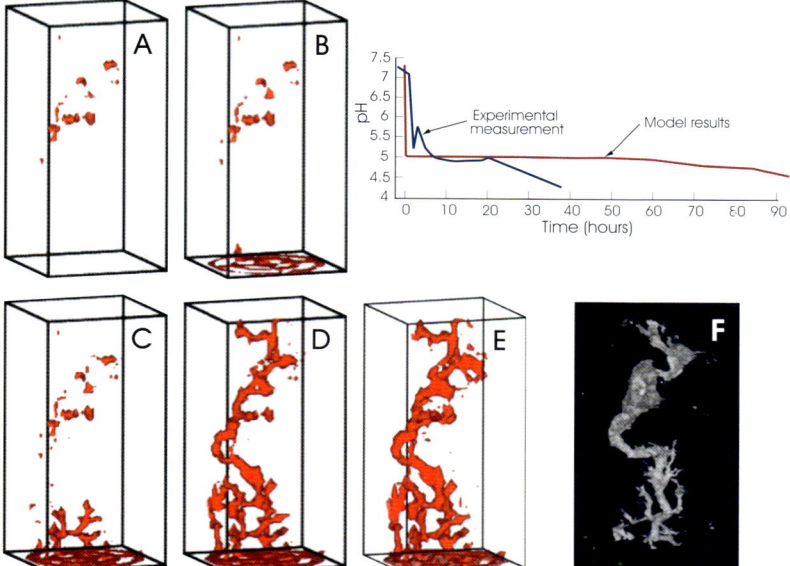

Figure 6-6. *Simulated evolution of porosity (0.6 iso-surface threshold) caused by unstable dissolution (wormhole formation and fingering) during the Vuggy coreflood experiment (pCO$_2$ 3 MPa) at:* **A.** *0 h,* **B.** *24 h,* **C.** *48 h, and* **D.** *96 h, for post-flood porosity distribution (0.5 iso-surface threshold).* **E.** *As predicted by the NUFT simulation.* **F.** *Observed experimentally through XCMT imaging. The graph shows a comparison of effluent pH as calculated using EQ3/6 (blue line) and predicted by NUFT (red line).*

Stokes, or the Darcy–Brinkman models, may better represent flow and transport physics of highly heterogeneous carbonate rocks that result in the formation of preferential flow paths or wormholes.

2. The coarser mesh used in the simulation compared to the actual heterogeneity in the limestone may 'smear' pore-scale heterogeneity during scaling up, and may introduce relatively large numeral diffusion that spreads out the reaction fronts.
3. The permeability–porosity correlations and averaging methods are not well calibrated. Robust calibration would require detailed pore-scale modelling.

The study does, however, capture the mechanism for wormhole evolution in heterogeneous carbonate rocks. The impermeability within the Vuggy core is an artifact of the isolation of macropore features by millimicron-sized pores. The

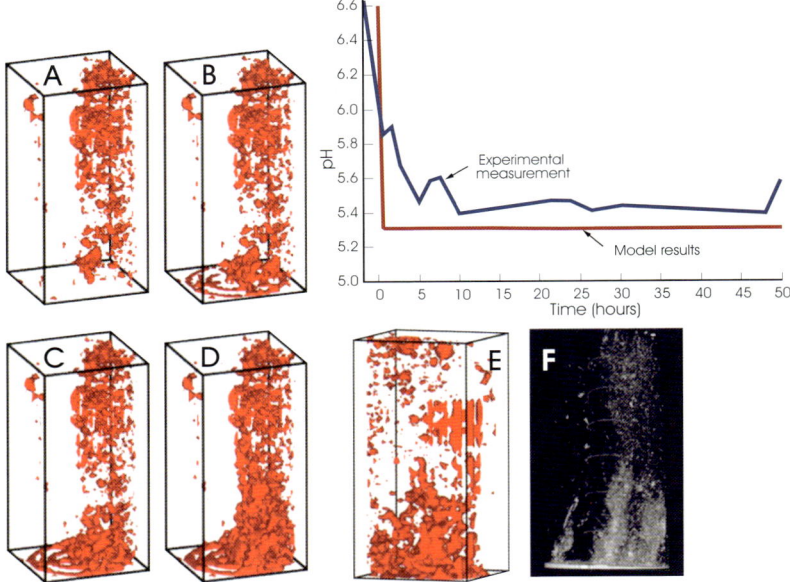

Figure 6-7. *Simulated evolution of porosity (0.45 iso-surface threshold) caused by progressive stable dissolution during the Marly coreflood experiment (pCO$_2$ 1 MPa) at:* **A.** *0 h,* **B.** *12 h,* **C.** *24 h, and* **D.** *60 h, for post-flood porosity distribution (0.45 iso-surface threshold).* **E.** *As predicted by the NUFT simulation.* **F.** *Observed experimentally through XCMT imaging. The graph shows a comparison of effluent pH as calculated using EQ3/6 (blue line) and predicted by NUFT (red line).*

wormhole grows by dissolving the carbonate minerals between the macropores. Observed calcite equilibrium prior to breakthrough supports this observation.

6.4 Fracture Flow Experiments

Injection of CO$_2$ for geological storage invariably imposes transient stresses and chemical disequilibria. In turn, these change deformation and fluid-rock transfer processes, and so modify key petrophysical and transport properties of the reservoir–seal system. Mechanical stresses can deform (even create) local structural features such as fractures, and chemical disequilibria along these features can alter bounding rocks through mineral dissolution/precipitation and associated volume change. The changes in porosity and permeability caused by these processes may result in significant modification (degradation or enhancement) of pre-injection reservoir, seal, and wellbore integrity, thus affecting CO$_2$ containment. Hence, the need to investigate fracture flow, particularly within fractured reservoirs.

Weyburn samples were examined using an integrated experimental and modelling study that focused on quantifying changes in fracture permeability during reactive fluid flow. Representative cores from the Marly and Vuggy reservoir zones and the evaporite seal were characterized before and after being subjected to well-constrained reactive flow-through experiments, which were then simulated using flow and reactive transport models. Particular emphasis was placed on analysing cores from the Vuggy unit, because fractures contribute significantly to flow in this zone, and on the evaporite seal, to assess the potential for developing (or mitigating) leakage pathways during injection. The experimental and modelling results reveal highly contrasting behaviour among the Vuggy, Marly, and evaporite cores, which highlights the importance of developing models that are capable of predicting fracture permeability evolution due to chemo-mechanical deformation in different rock types under varied flow conditions.

6.4.1 Experimental Conditions and Results

A critical aspect of this study is the determination of the roughness (topography) and mineralogy of both fracture surfaces. From this, it is possible to calculate fracture aperture when the two are mated, and to assess possible changes in formation water–mineral reactions.

Fracture aperture is defined as the void space between rough fracture surfaces. Currently, it is impossible to measure fracture aperture in situ on small diameter cores at millimicron-scale resolution using X-ray computed tomography (CT) techniques. Therefore, this information must be obtained indirectly before and after each flow-through experiment. This requires carefully measuring the topography of both rough fracture surfaces and numerically combining them to obtain the fracture aperture field.

A high-resolution NANOVEA ST40 optical profilometer was used, which provides accurate (±1 μm) measurements of fracture surface topography. Each measurement represents an integrated measure of surface elevation over a 16-μm area of the surface. Each fracture surface was scanned at a spatial resolution of 20×20 μm. Numerically mating the measured surfaces requires accurate alignment of the two fracture halves, which was done by using an appropriately fabricated jig to effectively secures the core halves to the profilometer stage. Mating the opposite surfaces of the fracture must be extremely precise. In fact, it was necessary to correct for a minute (6.0×10^{-6} μm/μm) incremental error in stage movement, which would otherwise have imposed a slight mismatch between the surfaces. The correction involved numerically 'stretching' the two surface profiles prior to numerically mating the surfaces.

Scanning electron microscopy (SEM) images of selected core samples were obtained before and after the flow-through experiments to assess the impact of reactive flow on fracture surfaces at the microscopic scale. Energy dispersive spectra (EDS) were also obtained to provide information on changes in the relative proportions of the main chemical elements. Together, the SEM and EDS data yield considerable insight into the reaction-dependent compositional and microstructural evolution that occurred during the experiments.

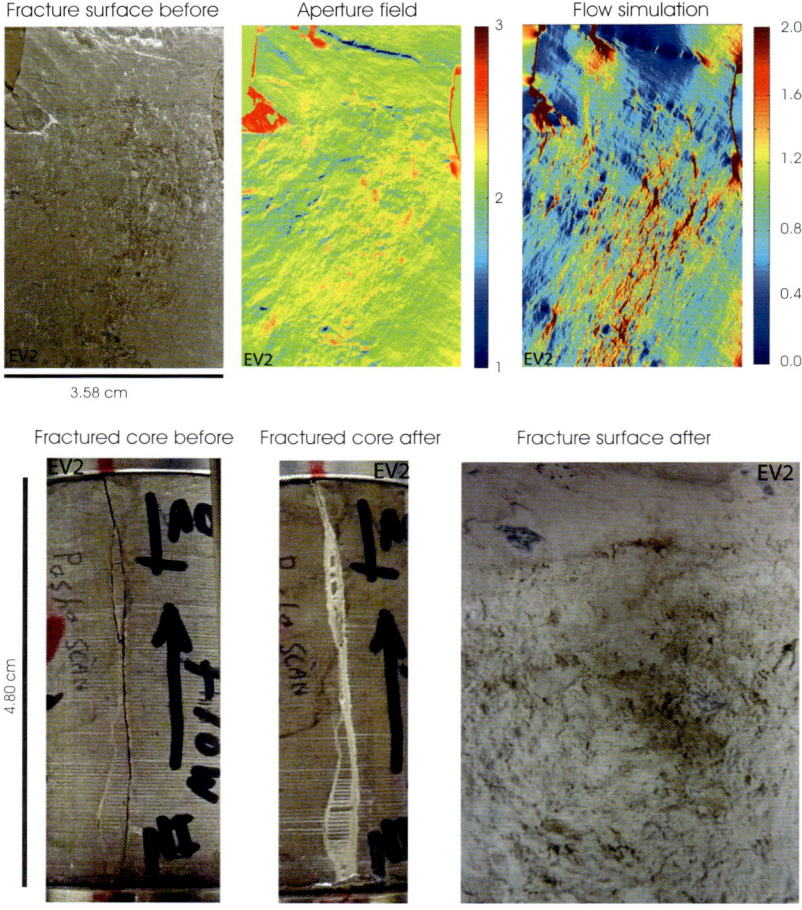

Figure 6-8. *Fracture experiment on the Midale Evaporite using deionized water.* **Top diagrams.** *Pre-experiment optical image of one fracture surface; the aperture field [in log10 (b, μm)]; and the corresponding normalized flow simulation (flux at each point is normalized to total flux across the fracture, and the colour scale is saturated to 2).* **Bottom diagrams.** *Fractured core before and after the flow-through experiment, together with an image of one post-experiment fracture surface.*

Seven reactive fracture flow experiments were conducted using the described system. Two involved deionized water and five used synthetic CO_2-equilibrated formation water. Three samples were fractured Midale Evaporite cores. Others were artificially fractured and re-cored before analysis, and sometimes sand-blasted to introduce heterogeneities to the fracture surfaces. All were jacketed and placed in the reaction vessel under confining pressure and controlled temperature. Two sets of boundary conditions were applied: (1) pressure control at the inlet and flow control at the outlet for the CO_2–formation water experiments, and (2) flow control at the inlet and pressure control at the outlet (open to atmosphere) for the deionized water experiments.

We report here the results from selected experiments on each of the units examined.

6.4.2 Midale Evaporite

Three experiments were conducted on core from the Midale Evaporite seal. In the experiment using synthetic CO_2-equilibrated formation water, the dissolved CO_2 had a negligible influence on anhydrite dissolution, as expected. Consequently, the two other experiments used deionized water as the injection fluid, which accelerated the reactions and simplified the experimental conditions.

Figure 6-8 shows images of the pre-flow and post-flow fracture surfaces, including 'before' and 'after' pictures of the actual core. There was a two-order-of-magnitude decrease in permeability, and about a 300-mm decrease in sample diameter perpendicular to the fracture near the inlet. Both of these changes are consistent with progressive aperture reduction and mechanical destabilization during the experiment. Effluent calcium and sulphate contents also increased, reflecting some anhydrite dissolution. However, these increasing contents never closely approached saturation values. This was because the fracture-lining anhydrite was simultaneously converting to gypsum ($CaSO_4 \cdot 2H_2O$), which is significantly less stable mechanically but 60% larger in molar volume. Together, these coupled processes caused aperture reduction through simultaneous dissolution and gypsum conversion of anhydrite asperities, while the fracture and local matrix diffusion zone became increasingly filled with gypsum. Pre- and post-experimental images of the shear-displaced fracture reveal the effect of anhydrite dissolution and conversion to gypsum, within both the fracture itself and the near-fracture diffusion zone. After the post-experimental core was separated, the thickness of the layer of gypsum constrained between the relatively unaltered anhydrite was estimated to be in the range of 150–1300 mm, narrowing from inlet to outlet.

6.4.3 Vuggy Unit

Three experiments were conducted on cores from the Vuggy unit, all using the

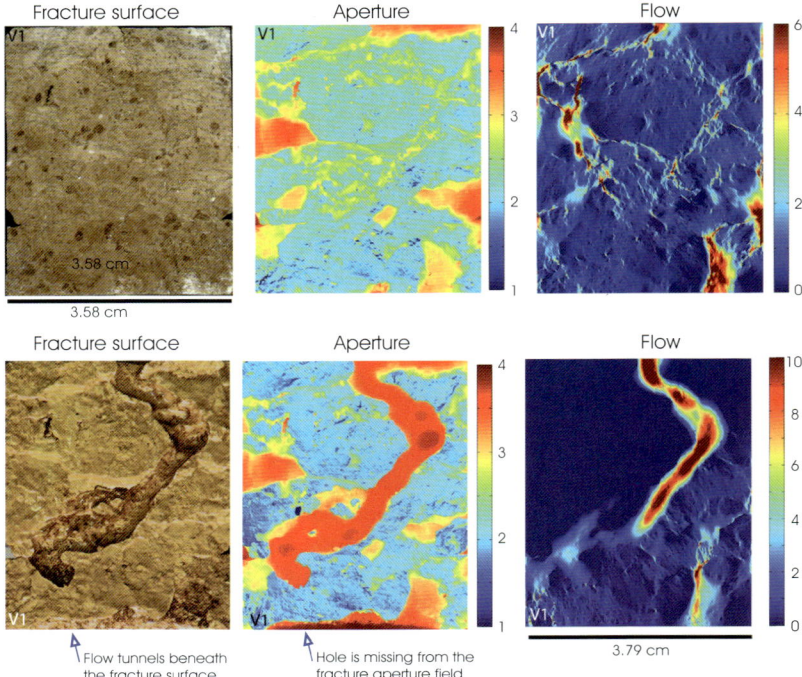

Figure 6-9. *Fracture experiment on the Vuggy unit using synthetic CO_2-equilibrated formation water.* **Top diagrams.** *Pre-experiment optical image of one fracture surface; the aperture field [in $\log_{10}$ (b, μm)]; and the corresponding normalized flow simulation (flux at each point is normalized to total flux across the fracture, and the colour scale is saturated to 6).* **Bottom diagrams.** *Post-experiment analogues of top results. The channel started in the bulk of the Vuggy sample and emerged into the fracture about 0.8 cm from the inlet. The tunnel feeding the fracture channel is missing from the measured aperture field; hence, the channel in the flow simulation is truncated.*

representative synthetic CO_2-equilibrated formation water as the infiltrating fluid. Details about only the naturally fractured Vuggy core are reported here.

Two wormholes were observed at the inlet end of the core and one at the outlet end, only one of which grew until it reached the outlet end of the core. Figure 6-9 shows details of this experiment. The average aperture increased from 414–1004 mm.

A reactive transport simulation was carried out using the initial aperture field for one of the artificially fractured samples, to calibrate the ability of the model to predict the observed dissolution patterns and measured evolution of effluent calcium concentrations. The qualitative match was good between the predicted and observed evolution of the aperture field during reactive flow and dependent carbonate dissolution. A comparison between the measured and predicted calcium content of the effluent suggested that development of the channel occurred during the introduction of the first several hundred millilitres of fluid. These close agreements support the effectiveness of the depth-averaged reactive transport model for predicting both dissolution rates within a fracture, and changes in the dependent porosity and permeability.

6.4.4 Marly Unit

A single experiment was conducted using CO_2-equilibrated formation water on an unfractured core from the Marly dolostone. Both significant mineral dissolution (several wormholes propagated about 1 cm into the core) and mechanical deformation (axial compression of the inlet end of the core) were observed. Comparison of the pre-experiment inlet surface of the core with a near-inlet slice of the post-mortem core revealed the spatial correlation of pre-existing large-pore concentrations (exposed on the inlet surface) with the development of wormholes in the near-inlet core interior. Post-mortem SEM images provided clear micro-scale evidence of incipient dolomite dissolution, as well as the wholesale dolomite dissolution that initiates wormhole development. These details are seen in Figure 6-10.

6.4.5 Conclusions

Flow-through experiments were conducted using representative Weyburn cores and reactive fluids synthesized to represent CO_2-equilibrated formation water. Flow of formation water, equilibrated with supercritical CO_2 through a naturally fractured anhydrite core, resulted in significant increases in permeability, due to increased dissolution-induced porosity (aperture).

Characterization of pre-flow and post-flow fracture surfaces (and reconstructed aperture fields) were used as input to a reactive transport simulator. For the experiments with carbonate cores, the simulations showed reasonable agreement with the measured dissolution patterns and rates. In particular, the simulations reproduced the trends in the observed effluent chemistry under different flow conditions, and qualitatively reproduced the dissolution patterns observed in the experiments. At low flow rates, dissolution caused the growth of wormholes, whereas at higher flow rates, dissolution occurred more uniformly over the surface.

The formation of wormholes due to flow of low-pH fluids has been observed in intact carbonate rocks. These results show that the presence of fractures does not significantly alter wormhole formation, but the fractures do tend to focus the wormholes along fracture planes.

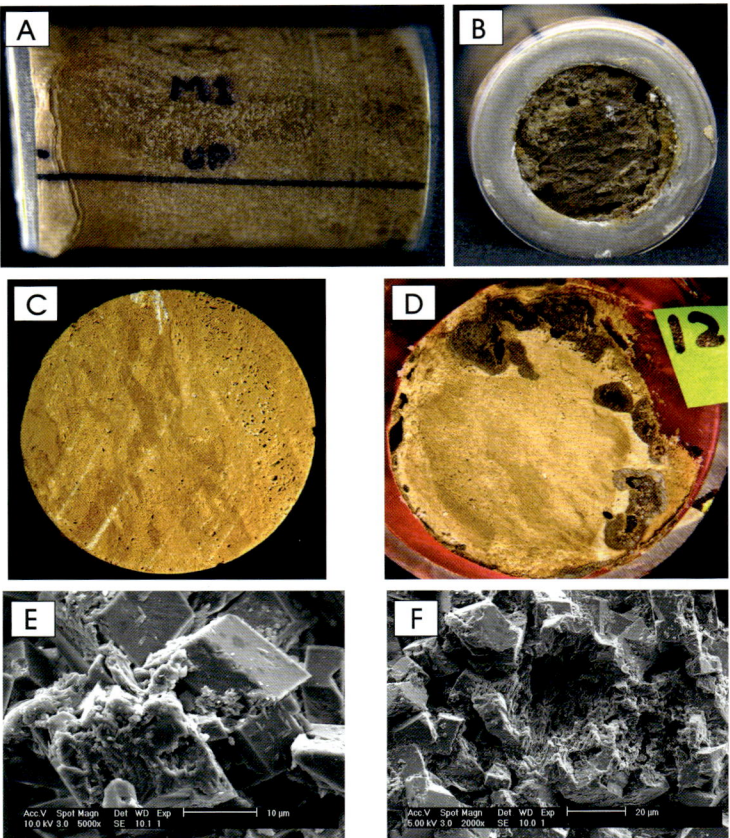

Figure 6-10. *Fracture experiment on the Marly unit using synthetic CO_2-equilibrated formation water.* **A.** *Post-experiment core with significant deformation at the inlet (left edge).* **B.** *Channel formation is readily apparent.* **C.** *Pre-experiment inlet surface.* **D.** *Local concentrations of large pores can be roughly correlated with channeling features observed in this post-experiment slice of the core about 1 cm in from the inlet surface (pink epoxy borders the sample).* **E.** *Post-experiment SEM images that illustrate selective dolomite dissolution (together with minor salt precipitation).* **F.** *Wholesale dolomite dissolution (at the centre of the image) that initiates wormhole development.*

The good agreement obtained between simulation and experimental results for the Vuggy cores suggests the potential utility of the predictive model for exploring the dependence of porosity/permeability evolution on formation water composition, which affects mineral dissolution kinetics and flow rates.

6.5 Field-scale Studies

6.5.1 Introduction

In contrast to the laboratory-scale history matching studies presented above, field-scale simulations face the daunting challenges of both uncertainty of site-characterization, and the need to represent all of the relevant integrated processes. In cases where there is a reasonable degree of confidence in model pedigree (for example, from experimental calibration efforts), field-scale simulations offer at least the potential for resolution of discrepancies between prediction and observation by refining site characterization. In this section, we present the results of these investigations. Each addressed a field-scale history matching exercise that was well constrained by comprehensive monitoring results (principally, formation water geochemistry and seismic reflection data), and is also directly relevant to virtually any CO_2 storage site.

In the first example, the formation water geochemical database, developed by the project, was used to constrain field-scale reactive transport modelling of fluid flow and CO_2–formation water–rock transfer in the reservoir. Formation water composition and isotope data represented unique constraints on reservoir permeability structure and dependent flow dynamics. In the second investigation, a novel stochastic inversion method was used to minimize discrepancies between predicted and observed seismic reflection data, through refinement of the bulk reservoir permeability field. In the third study, this same inversion technique was used to reduce discrepancies between changes in measured and simulated formation water composition, through refinement of reservoir mineralogy along inferred flow paths from CO_2 injector to production wells. While still in the development stage, the conceptual framework underlying the last two of these projects shows great promise, and produced useful insights into reservoir permeability and mineralogy at Weyburn.

6.5.2 Formation Water Chemistry

The modelling work summarized here focuses on history matching the changes observed in produced waters, specifically, conservative and reactive components, and stable isotopes. Conservative components (for example, Na and Cl) are not added to or removed from solution to a significant degree through CO_2–(oil)–formation water–rock reactions. Therefore, changes in their content primarily reflects that of (1) the injection water, (2) the mixing of this water with formation fluids, and (3) the permeability field, which dictates fluid flow direction and magnitude. Changes in the content of reactive components (for example, calcium and bicarbonate) are due to these same factors, plus the contribution of transfers through CO_2–(oil)–formation water–rock reactions. Hence, accurate simulation of the observed compositional signals requires accurate representation of initial reservoir conditions, operational history (in particular, evolving injection-fluid compositions), and multiphase flow and geochemical processes.

At Weyburn, the chemistry of injection water was not documented during waterflooding or subsequent CO_2–WAG operations, and therefore both formation water and injection water compositions are poorly constrained. Further, the nature of reservoir permeability is poorly understood. Phase-1 simulation work adopted the conventional single porosity (unfractured) reservoir model, but there is good evidence that the reservoir is indeed fractured. This means that a dual porosity model is more appropriate. Moreover, although it is possible to model fluid production history without including the influence of fractures, accurate simulation of aqueous geochemical signals appears to require explicit representation of dual porosity flow.

Because the dual porosity model provides a better representation of permeability structure and fluid transport in the Weyburn reservoir, it was used to identify the more rapidly transmitted short-term chemical signals, which bring the predicted and observed behaviour into closer agreement. This improved representation of fluid transport is even more essential for modelling the behaviour of reactive tracers, particularly those associated with slower chemical reactions, because these signals indicate the contact time between aggressive solutions and minerals.

Traditional reservoir simulation work is usually calibrated against pressure and fluid production history because they represent the most commonly available relevant field data. The emergence of more sophisticated reactive transport simulators (for example, GEM®), and the availability of additional relevant field information, such as formation water composition and isotope data, together provide a unique opportunity to more rigorously calibrate and refine both reservoir modelling and characterization efforts.

6.5.3 Modelling Approach

The original concept of representing a fractured reservoir as the superposition of two porous media (Barenblatt et al., 1960) was further developed by using an idealized model of such reservoirs, based on the mathematical concept of superposition. In fractured reservoirs, fluid exists in both porosity systems (that is, within matrix blocks and their bounding fractures). This is represented by either of two modelling conventions: dual porosity or dual permeability. In dual porosity models, which are in fact dual porosity–single permeability systems, fluid transport takes place only through the connected fracture system, while the matrix blocks act as source regions for this flow. In contrast, in dual permeability models, which are true dual porosity–dual permeability systems, both the matrix blocks and fractures contribute to fluid transport, and there is direct flow communication between neighboring matrix blocks.

The dual porosity reservoir model developed for the project was calibrated using fluid production information dating from initial development of the field, through secondary waterflood, and 3 years of CO_2-EOR. Fracture density and permeability were the principal model parameters adjusted during model calibration. After a match was obtained of field production rates comparable to that achieved during Weyburn Phase 1, non-reactive tracers were then used to predict both the age and source distribution of waters produced at two specific vertical wells and one horizontal well.

While flow patterns and age distributions of produced waters can be calculated using the single porosity model, the predicted distributions are overwhelmingly more uniform than expected, given the variability observed in produced water chemistry. For example, consider the change in composition of Na and Cl in water produced at a single well within the Phase 1A area (Figure 6-11). As can been seen, the measured values fluctuate by >20% between many sampling trips. This variation is almost certainly due, at least in part, to permeability structure and dependent flow dynamics within the reservoir. Both Na and Cl are conservative elements, so these fluctuations are attributed to changes in the proportions of differently sourced waters. In principle, it should be possible to predict these proportions using a reservoir model.

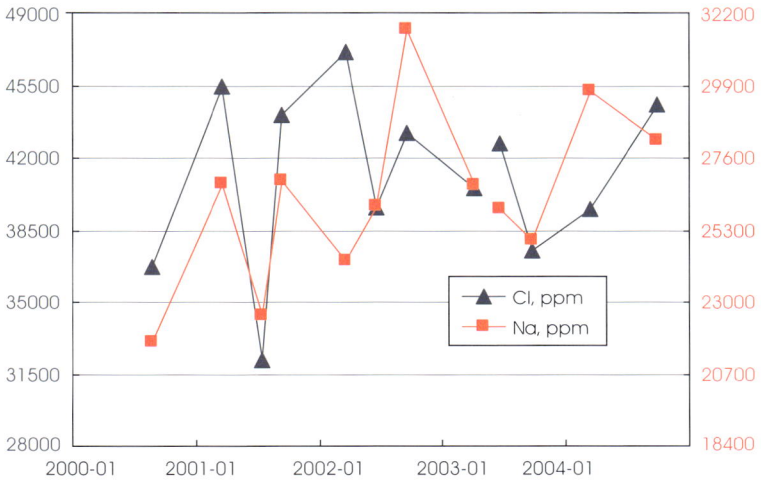

Figure 6-11. *Measured changes of conservative components, Cl and Na, in a vertical well in the Phase 1 area, Weyburn Field.*

The simulations that were run addressed the transport of conservative tracers that were present either in initial formation water or those injected during secondary and tertiary EOR. Two types of tracers were considered.
1. Age-based tracers with a distinct composition, injected through all injection wells during each of four successive 10-year periods.
2. Source-based tracers with a distinct composition, injected through each injection well during the entire 40-year injection period.

Age-based tracers are used to introduce compositional variations into the model that might be typical of those induced by changes in the source of water injected during EOR. Source-based tracers demonstrate the impact that changes in flow patterns (associated with water injection rates) may have on produced water chemistry.

6.5.4 Single Porosity Model

Figure 6-12 shows the changes in age-based tracer distribution (top graphs) and source-based tracer distribution (bottom graphs) for the single porosity model at a vertical injection well (left column graphs) and a horizontal injection well (right column graphs).

The top graphs show the traces of five lines, each representing the fraction of water recovered that was either initially in the reservoir (initial) or injected during one of four 10-year periods. The simulation results indicate that the produced water is very heavily weighted to 'old' water, with 40–60% derived from the formation source, and >90% being older than 1984. Further, the age distribution varies very smoothly with time. This result indicates that the fluid transport involves a great deal of mixing, as expected for single porosity (matrix only) flow, which smoothes out (homogenizes) any rapidly transmitted compositional signals (read fractures) that are likely the source of observed fluctuations in Na and Cl signals.

Simulated variations in age-based tracer distributions at a horizontal production well show very little short-term variability. However, in contrast to the vertical producer, they are characterized by a predominance of contributions from injected (versus initial formation) waters. This principally reflects the proximity of the horizontal producer to certain injection wells, and the lack of such proximity for the vertical producer.

For the vertical well, results from the source-based tracer calculations indicate that, as in the case for age-based tracers, (1) 40–60% of the produced water is initial formation water, and (2) the source distribution varies smoothly with time due to the homogenizing nature of single porosity fluid transport.

Simulated variations in source-based tracer distributions at a horizontal production well show very little short-term variability in the calculated evolution of tracer distributions. However, in contrast to the vertical producer, they are characterized by a predominance of contributions from injected (versus initial formation) waters. This contrast principally reflects the proximity of the horizontal producer to certain injection wells and the lack of such proximity for the vertical producer. As with the age-based simulations, the source-based tracers show that fluid transport into this producer is very uniform over time, with most of the well-mixed waters being derived from nearby wells.

In summary, the age-based and source-based tracer distributions calculated for both vertical and horizontal producers vary smoothly when the reservoir description uses a single porosity model. Significant changes occur only on a scale of several years. In contrast, for short terms of a few months, field observations of produced fluid chemistry show significant noise. So, although the model

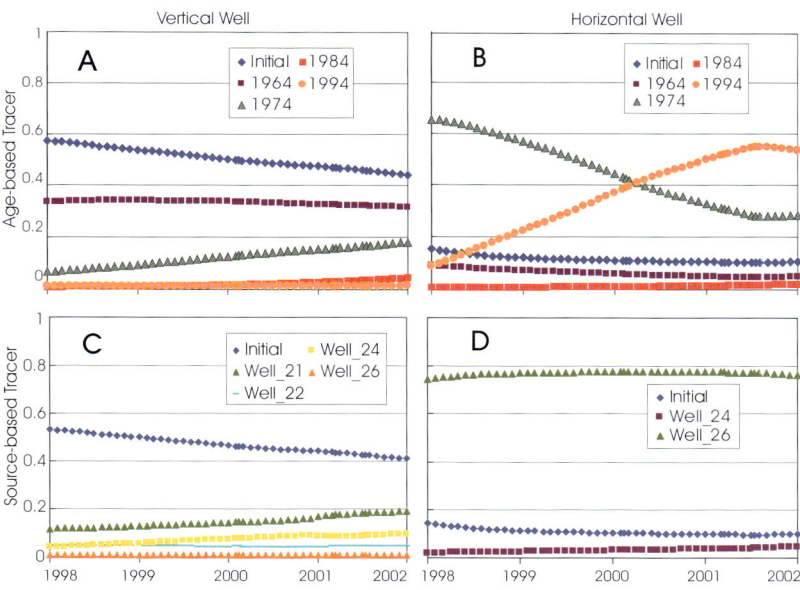

Figure 6-12. *Simulations using the single porosity model.* **Top graphs.** *Age-based tracer distribution.* **A** *and* **C** *at a vertical production well,* **B** *and* **D** *at a horizontal production well. Both wells occur within a pattern that contains 10 injectors. Water sources are the initial formation fluids (initial) and injected waters from 1964–1974 (1964), 1974–1984 (1974), 1984–1994 (1984), and 1994–2004 (1994).* **Bottom graphs.** *Source-based tracer distribution.* **C.** *Water sources are the initial formation fluids (initial) and injected waters from four source-predominant injectors.* **D.** *Water sources are the initial formation fluids (initial) and injected waters from two source-predominant injectors.*

accurately predicts overall fluid production rates, it does not appear to accurately represent fluid transport through the reservoir. This limitation requires that the geological description of the reservoir must be adapted in order to generate flows that are compatible with the geochemical signals. This conclusion provided the impetus to convert the model to a dual porosity or dual permeability formulation.

6.5.5 Dual Porosity Model

As was done for the single porosity reservoir representation, the dual porosity model was adapted to incorporate age-based and source-based tracers (Figure 6-13). There are several significant differences between results from the single- and dual-porosity models. First, there is a clear decrease in the amount of 'old' water produced, with most of the produced water having been injected after 1974

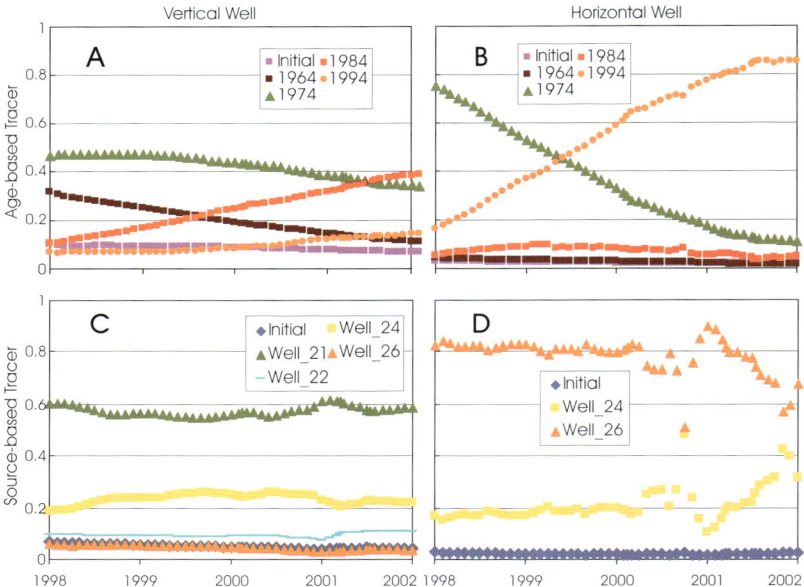

Figure 6-13. *Simulations using the dual porosity model.* **Top graphs.** *Age-based tracer distribution.* **A** *and* **C** *at a vertical production well,* **B** *and* **D** *at a horizontal production well. Both wells occur within a pattern that contains 10 injectors. Water sources are the initial formation fluids (initial) and injected waters from 1964–1974 (1964), 1974–1984 (1974), 1984–1994 (1984), and 1994–2004 (1994).* **Bottom graphs.** *Source-based tracer distribution.* **C.** *Water sources are the initial formation fluids (initial) and injected waters from four source-predominant injectors.* **D.** *Water sources are the initial formation fluids (initial) and injected waters from two source-predominant injectors.*

and 1994. This age reduction implies that the amount of mixing between potential water sources is decreased in the dual porosity model, and this decreased mixing results in increased compositional variability of the produced water.

In addition, the age distributions change more rapidly and erratically with time in the dual porosity simulations than in the original single porosity runs. This is most evident in results for the horizontal well, which show a very significant change in the age distribution from 1998 to 2001. While less pronounced for the vertical well, the rate at which age fractions change does vary noticeably over the 4 years shown. This is in contrast to the single porosity model, where the age fractions evolve at very constant rates. Fractures in the reservoir provide highly permeable conduits for fluid transport, which result in rapid breakthrough and sudden changes in the produced water compositions over a short time period. For example, in 1998 roughly 80% and 20% of the waters produced from the vertical well are derived from those injected during 1974–1983 and 1994–2003, respectively; however, by 2002 these proportions are reversed.

As with the single porosity model, tracers were also used to identify the source distribution of produced waters. Simulation results for the same vertical and horizontal wells addressed with the unfractured model, demonstrate that the predicted fraction of original water produced is significantly smaller in the fractured model. Aside from this difference, the waters produced at both wells display similar signatures to those obtained from the single porosity model. Somewhat similar conclusions can be drawn about the produced tracer distribution at the horizontal well, which shows that most (about 80%) of the water produced is derived from effectively only one well, with the remainder almost entirely initial water.

6.5.6 Conclusions

This study demonstrated that the observed variability of produced water compositions at Weyburn for non-reactive tracers is greater than that obtained from simulations carried out using a single porosity model. Modelling using a dual porosity representation of the reservoir produces waters that are less thoroughly mixed – and therefore intrinsically more variable – than those produced by the single porosity model. As a result, for age-based and source-based tracers, the dual porosity model yields predictions that more closely represent the actual variability observed in the field.

GEM® simulations also proved quite useful in interpreting selected isotope evolution during CO_2-EOR at Weyburn. In some respects, the isotope data are of potentially more value than the major ion compositions because of the relatively low variability of baseline samples across the field. Although the potential to model isotopic processes with reservoir simulation tools appears to be great, it is important to note restrictions that limit the capabilities of the current simulator. There are difficulties in calculating isotope ratios when there is extensive depletion of component concentrations of the isotopically distinct molecules used to model the fractionation. Further, the modelling is hindered to some extent by a lack of equation of state properties for isotopically distinct components.

Despite these shortcomings, very promising results were obtained from simulations of the isotopic evolution of light alkanes. A potential mechanism exists to fractionate isotopically distinct molecules – by preferentially mobilizing the lighter molecules in the gas phase. There is clear field evidence that some process associated with CO_2 flooding is fractionating these molecules within the Weyburn Field. Simulations incorporating this mechanism failed to generate any historical variation using the reservoir description adopted during the original Phase 1 study. However, signals similar to those seen in the field were obtained when the reservoir description incorporated the dual porosity model. This observation suggests that calibrating reservoir models against isotopic signals can provide further insights into reservoir transport properties processes during CO_2-EOR.

The success of a model in replicating field observations gives confidence that the numerical representation of operative processes and the spatial domain are realistic – acknowledging that 'realistic' is not necessarily accurate. If the model also yields realistic representations of data that were not used in its initial calibration, then it is more likely accurate than if it does not. The observation that the original, single porosity model smoothes out most compositional variations implies that this model does not provide a good representation of actual fluid transport in the reservoir. While the fractured (dual porosity) model presented here does not necessarily provide a better representation of the actual flow fields, it at least provides a more realistic proxy for 'true' reservoir behaviour.

6.6 Seismic Simulation and History Matching

In conventional oil and gas production, 'history matching' entails matching field production history, predicted by numerical fluid flow simulations, with actual production history. Typically, model parameters (primarily permeability) are iteratively modified until the predicted history emulates the actual production history. For the purposes of CO_2 storage, the resultant model can also be used to forecast CO_2 injectivity and migration. Unlike typical oilfield operations, where there is abundant well-based production and engineering information, CO_2 storage operations have a greater reliance on remote monitoring data (for example, time-lapse seismic) for history matching. Thus the geological model used for simulations is iteratively modified to history match not only available injection–production data, but also remote monitoring data.

For the purposes of CO_2 storage monitoring in the Weyburn Field, there are two regional monitoring data sets that are most suitable for history matching. They are the geochemical fluid sampling data and the 3D time-lapse seismic results. The 3D time-lapse monitoring data are well-suited to tracking CO_2 migration within the reservoir, whereas the changes of fluid chemistry in the reservoir informs us of the ongoing chemical interaction between the pore fluids and rock matrix. Here, we describe attempts to utilize 3D time-lapse monitoring data to constrain the history matching process. Figure 6-14 shows a schematic flow chart for history matching information, including seismic data.

As described, the parameters of a numerical geological model are iteratively modified until the resultant flow simulations produce fluid distributions that are consistent with the production data and the time-correlative seismic data. 'Consistent,' in this sense, means that the seismic properties determined numerically from the geological model for the resultant pore fluid distribution (specifically CO_2 saturation) match those that are derived from the seismic data.

This comparison of predicted versus measured seismic properties can be done in a highly qualitative manner or semi-quantitatively. Further, the iterative history matching process can be applied in a trial-and-error manner, or ideally can be formalized and applied in a semi-automated fashion. Both approaches are reported here, although the latter approach has only been partially implemented.

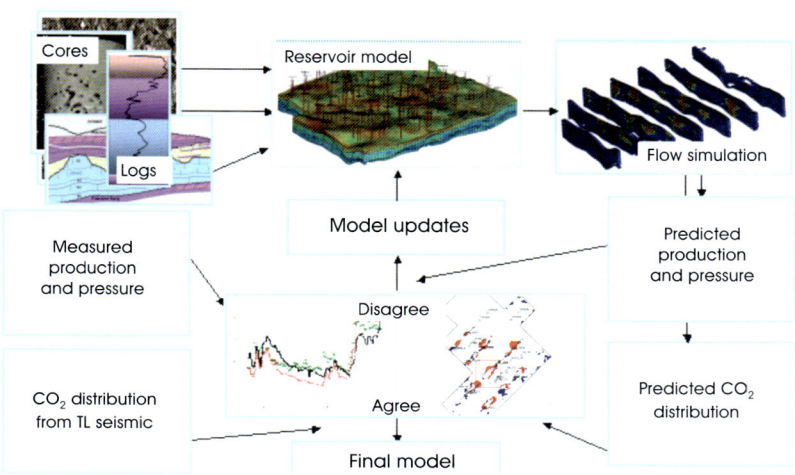

Figure 6-14. *Schematic flow chart for history matching, including the direct use of the seismic information.*

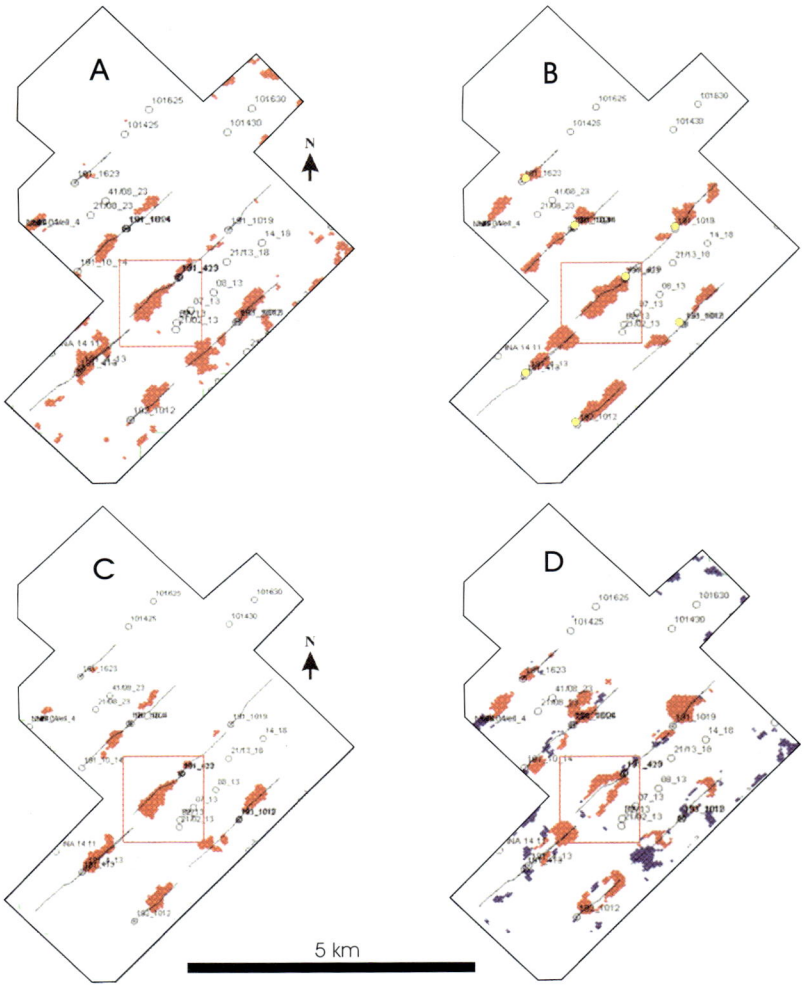

Figure 6-15. *Distribution of CO_2 at the time of the 2002 seismic survey as predicted by* **A.** *the P-wave amplitude difference map, and* **B.** *the starting reservoir simulation model. In A, the CO_2 distribution is determined by applying a threshold of 1.5 standard deviations of the amplitude map. In B, the CO_2 distribution represents areas where CO_2 saturation exceeds 10 molar per cent.* **C.** *and* **D.** *show where the two estimates of CO_2 distribution agree or disagree. In C, red zones indicate areas of agreement, and in D, red zones are where the simulator predicts CO_2 saturation that is not observed in the seismic map and blue zones are where the seismic predicts CO_2 saturation that is not observed in the simulator CO_2 distribution map.*

6.6.1 Trial-and-error History Matching

An iterative process of reservoir simulation, comparison of simulated and seismic-derived CO_2 distribution, and modification of the reservoir permeability model, was conducted using the P-wave time-lapse seismic results (Figure 6-15).

To facilitate this process, two simplifying assumptions were made. First, the areal extent of the Marly amplitude anomalies was used as a proxy for the presence of CO_2 in the Marly unit because the amplitude anomalies determined at the Marly horizon are particularly sensitive to the presence of CO_2 in the Marly unit. Second, the pressure contribution to the Marly amplitude differences was not considered.

Figure 6-15 shows a comparison of the CO_2 areal distribution within the Marly unit predicted by the seismic data at the time of the 2002 survey, compared to the distribution from the reservoir simulator obtained using an original reservoir model. There is relatively good agreement in the immediate vicinity of the horizontal injection wells, but there are also significant areas of disagreement. In particular, and compared to the seismic CO_2 areal distribution, the reservoir simulator predicted larger areas of gas saturation in the Marly around the majority of horizontal injection wells.

Careful examination of the injection pressure histories suggested that there may be a vertical permeability pathway, in the area of the CO_2 injection well, associated with either fractures or cross-flow between existing vertical wells that spanned the Marly and Vuggy zones. On the basis of this observation, the vertical permeability was increased near the injection well, resulting in a reduced volume of CO_2 in the Marly unit, and leading to an improved match between the simulated production and the production history (compare curves in Figure 6-16). This demonstrates the potential for improving the reservoir model using the additional constraints provided by seismic.

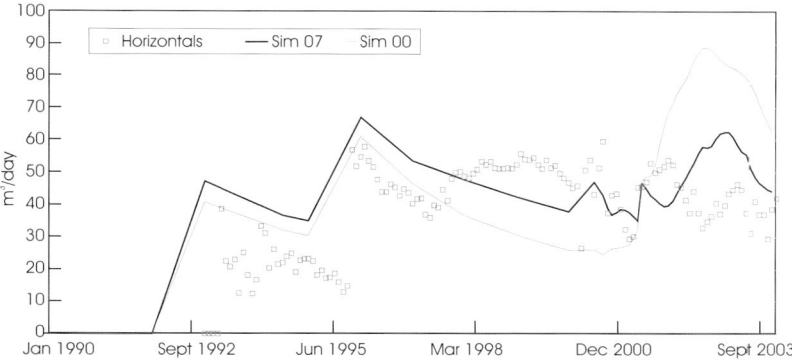

Figure 6-16. *Production rate at horizontal wells before and after update of the model.*

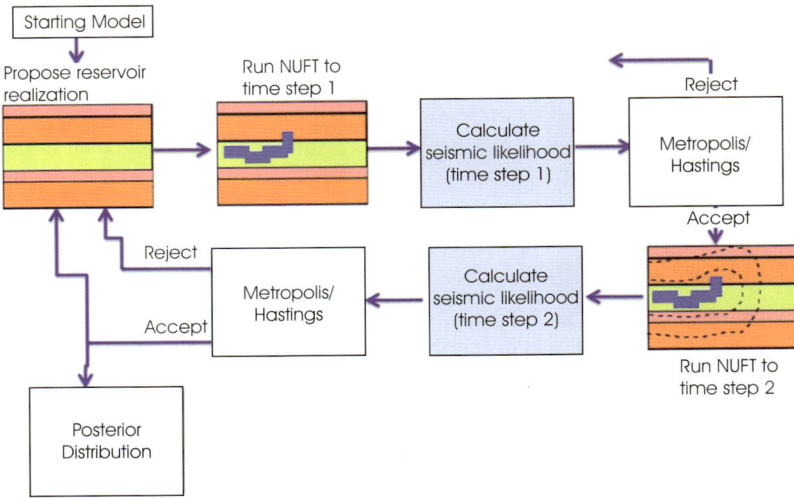

Figure 6-17. *Flow diagram for the first permeability-seismic inversion. Note that the process will use two or more seismic data sets simultaneously.*

6.7 Markov Chain Monte Carlo Stochastic Inversion

Ideally, it would be desirable to achieve results comparable to the history matching described in the previous section, with less direct intervention by the geophysicist, geochemist, or reservoir engineer. Further, it would be advantageous to allow a more quantitative comparison of the predicted simulation results and the observed data with fewer simplifying assumptions.

One potential means of doing this is through stochastic inversion. Accordingly, a stochastic inversion algorithm was developed that aims to identify reservoir permeability models that optimize agreement between the predicted and observed storage performance, the latter consisting of production data and the observed seismic or geochemical response. A Markov Chain Monte Carlo (MCMC) methodology was applied to use existing data to refine the spatial distribution of reservoir permeability, which effectively defines the spatial framework of CO_2 migration. The adopted MCMC approach accounts for uncertainties associated with both limited sensitivity and measurement error through the use of a Metropolis–Hastings algorithm (Metropolis et al., 1953) as described by Mosegaard and Tarantola (1995). This approach is similar to classical deterministic inversion, except that the deterministic model updating scheme is replaced with random perturbations of the model. In both cases, an initial model is chosen for which simulated monitoring data are calculated.

The simulated data are compared to the observed monitoring data, and the differences are utilized to choose an updated model. This iterative process is continued until some convergence criterion is achieved. The two approaches differ fundamentally in terms of the nature of the final result. Specifically, MCMC produces a probability distribution that describes likelihoods for a wide sampling of models.

6.7.1 Stochastic Inversion of Time-lapse Geophysical Data

Figure 6-17 is a flow diagram indicating how this process works, specified in this diagram for time-lapse seismic data. The algorithm was tested on a single production area in the Weyburn Field, for an injection scenario comprising 2 years of water injection accompanied by 1.3 years of CO_2 injection which started 0.7 years after the start of water injection. The resultant pressure and CO_2 saturations were used as the basis to calculate the change in P-wave velocity, which was then used to calculate the associated seismic reflection response. These seismic data were adopted as the observed seismic data. The stochastic inversion algorithm was then run, using the observed seismic data, to see whether the true permeabilities could be successfully estimated. After 70 iterations, the recovered permeability is very similar to the true model. In subsequent iterations, although the gross patterns of the permeability models are qualitatively similar to the true model, the inverted values are generally higher than in the true model.

The utility of seismic-constrained history matching has been demonstrated above by application of a trial-and-error process to modify model permeabilities, so that the simulated CO_2 saturations within the storage reservoir are more consistent with the seismic monitoring data.

A more objective procedure for applying seismic-constrained history matching utilizes a Markov Chain Monte Carlo stochastic inversion. The ultimate advantage of this method is that it should require less time spent by the reservoir engineer doing trial-and-error model updating. This method works with a synthetic example, and has the further advantage of providing an assessment of the variability in models that are capable of explaining both the production data and the seismic monitoring results. It has also been applied with actual seismic data from a single production pattern, with limited success. This procedure as implemented is very computationally intensive. Ongoing research efforts are being conducted to make the algorithm more robust in real data cases.

6.7.2 Stochastic Inversion of Formation Water Data

The specific approach taken with the formation water and reservoir mineralogy was to employ a reactive transport model to represent the flow path, populated by a set of mineral phases that characterize the reservoir (calcite, dolomite, anhydrite, and K-feldspar), which were assigned appropriate thermodynamic and kinetic properties. The target parameters for the inversion are the mineral volume fractions along the flow path and the composition of the effluent formation water.

The complex requirements of a full forward reactive transport model depend upon a wide range of parameters related to myriad coupled processes. In contrast to the seismic inversion, the geochemical inversion entails multiple input parameters (for example, the kinetic parameters for each reactant mineral) and output values (for example, the contents of calcium and magnesium that reflect the extent of mineral dissolution). The degree to which different output values must be weighted in compiling an overall goodness-of-fit is not known *a priori*. Further, samples drawn from a given well represent a spatial average of conditions across an unknown reservoir source region, where the well intersects relatively permeable materials – a situation made more complex for horizontal wells. For these and other reasons, and to reduce the computational burden, the initial tests were limited to a 1D representation.

The Midale Vuggy and Marly units consist primarily of dolomite and calcite, with lesser amounts of aluminosilicate minerals such as feldspars, illite, and kaolinite. To posit a plausible geochemical model of the reservoir for geochemical parameter inversion, the formation water composition data (which include pre-CO_2 baseline measurements and subsequent monitoring data), were reconciled with the inferred volume fractions and presumed reactivities of key mineral phases.

Reactant minerals – calcite, dolomite (disordered), anhydrite, and K-feldspar (representing a silicate phase most likely to undergo dissolution in response to a pH decrease) – are distributed along the column, each with a fixed intrinsic dissolution rate and mineral surface area. Running the inverse problem entails conducting multiple forward reactive transport simulations for the 1D column using the software PHREEQC (Parkhurst and Appelo, 1999), with the tacit assumption of single-phase flow.

Comparisons of formation water composition changes generated by the best likelihood proposal for the inversions are shown in Figure 6-18. Information includes monitoring data for two wells and for calcium, magnesium, pH, and total dissolved CO_2. In general, the agreement with the monitoring data is reasonably good. Measured concentrations presumably reflect multiple factors, including periodic injections of imported water, non-steady CO_2 injection, and the influence of CO_2 injection from other sources. Nonetheless, the modelled magnitudes of changes in each of the parameter values over time are consistent with the observations. Particularly noteworthy are the inversion results for dissolved CO_2 because total dissolved CO_2 values were not used to calculate the likelihood values that guided the MCMC search. The inverted result agrees well with the trend in the observations, which suggests the inversion is producing internally consistent results.

6.7.3 Conclusions

Constrained by the computational burden imposed by a full 3D model, and limited spatial resolution inherent in the monitoring data, the MCMC algorithm

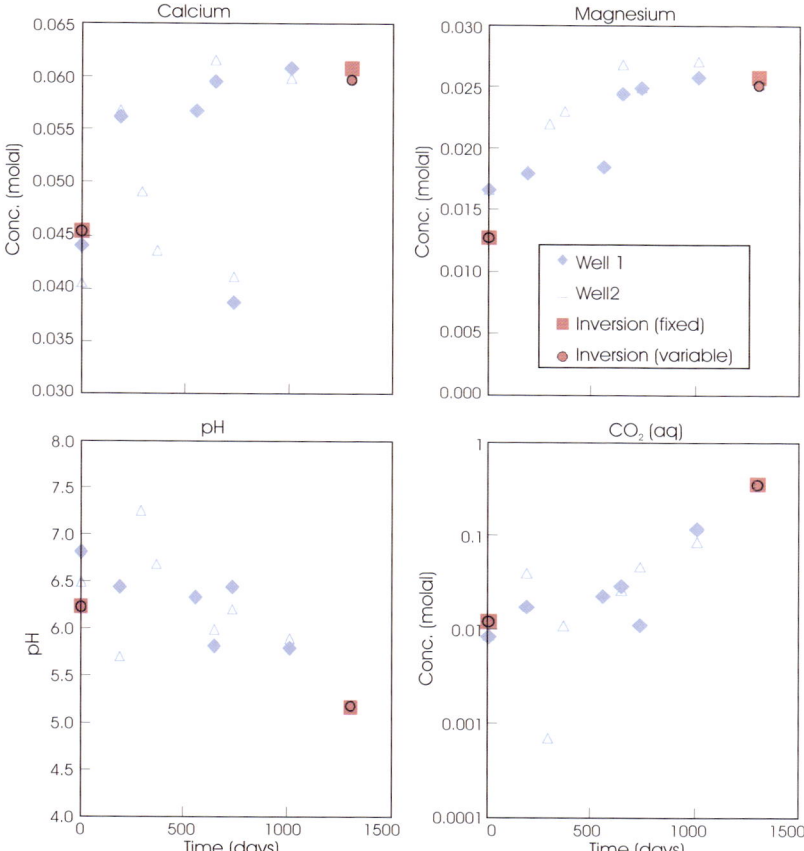

Figure 6-18. *Comparison of calcium, magnesium, pH, and total dissolved CO_2 measured in two wells with the modelled values – based on the best likelihood proposal generated by the MCMC inversions using different mineral dissolution kinetic models (that is, fixed versus variable intrinsic dissolution kinetics). The inversion values at time t = 0 represent the initial condition prior to the introduction of CO_2-rich formation water into the flow path. Note that total dissolved CO_2 was not included among the parameters used to compute composite likelihood functions used to rank proposals.*

was applied to a simplified, 1D idealization of a reactive transport in the Pattern 16 area to demonstrate applicability of the method and to provide some degree of constraint on mineral volume fractions. It is important to recognize that the inversion approach, as employed here, is highly idealized, not only in assuming 1D flow but also in assuming that fluid flow is single-phase. While the single-phase model can capture salient features of the CO_2–(oil)–formation water–rock interactions by assuming the influx of CO_2-enriched formation water, it neglects the multiphase physics of the flow system. Nonetheless, inversion of the inherently low-resolution Pattern 16 geochemistry data indicates a reactive flow path between the injection well and the production well, that is characterized by interactions with carbonate minerals, predominantly dolomite-rich, near the injection well and progressing to more calcite-rich farther downgradient. This observation is consistent with the initial interaction of the CO_2-rich formation water with the dolomite-rich Marly in contact with the injection well, followed by subsequent interaction with the calcite-rich Vuggy along the flow path.

In summary, the MCMC approach for inversion of reactive transport for the geochemical conditions specific to the Weyburn-Midale reservoir appears capable of yielding plausible results that are consistent with data and constrained by independent parameter estimates. Application is tempered, however, by computational constraints, the complex operational history and, significantly, by the limited spatial resolution associated with formation water samples, as well as the undocumented nature of injection-water compositions during both waterflood and subsequent water-alternating-gas (WAG) CO_2-EOR.

6.8 Summary of Results
6.8.1 Coreflood Experiments

Provided the experimental conditions emulate reactive flow using realistic pCO_2 values, with natural or made-up formation water, coreflood experiments can provide valuable information about potential field situations.

For Weyburn carbonates, faster carbonate mass transfer rates and greater volumes of dissolved carbonates per unit time were observed for the more homogeneous pore size distributions and larger reactive surface areas. A more bimodal pore size distribution, with vugs and sub-micron pores, resulted in variable mass transfer rates.

Bulk sample permeability was increased drastically when fast pathways developed as a result of pre-existing heterogeneities, such as large vugs and microfractures. In some cases, wormholes developed, and it was inferred that available reactive surface area had allowed transport processes to dominate over reaction kinetics, resulting in formation water solutions undersaturated with respect to carbonate minerals.

When the formation water was saturated with respect to both calcite and dolomite, Darcy-like flow was observed. In this situation, fast carbonate kinetics were not a limiting factor in the evolution of the reacted formation waters.

6.8.2 Simulation Model

Results from coreflood experiments can be used to validate flow models that simulate reactive transport and mineral dissolution processes. Understanding these processes is one important aspect of determining storage complex integrity.

The simulated results suggest that Darcy continuum-scale flow can adequately model reaction fronts for porous carbonate rocks with a uniformly distributed pore space. However, flow may not be Darcy-like in very heterogeneous carbonate rocks, which results in the formation of preferential flow paths or wormholes.

The permeability–porosity correlations and averaging methods used in the model are not well calibrated, and a more robust calibration is required for detailed pore-scale modelling.

The study did, however, capture the mechanism for wormhole evolution in heterogeneous carbonate rocks. Wormholes grow by dissolving the carbonate minerals between the macropores. Observed calcite equilibrium prior to breakthrough supports this observation.

6.8.3 Fracture Flow Experiments

Injection of CO_2 for geological storage invariably imposes transient stresses and chemical disequilibria, which can result in fractures. Chemical disequilibria along the fractures can alter bounding rocks through mineral dissolution/precipitation with associated volume change. These changes may result in significant modification (degradation or enhancement) of pre-injection reservoir, seal, and wellbore integrity, thus affecting CO_2 containment. Hence, there is a need to investigate fracture flow, particularly within fractured reservoirs. A critical aspect of this study is the determination of the roughness (topography) and mineralogy of both fracture surfaces. From this, it is possible to calculate fracture aperture when the two are mated, and to assess possible changes in formation water–mineral reactions.

Flow-through experiments using a naturally fractured anhydrite core resulted in significant increases in permeability, due to increased dissolution-induced porosity (aperture). These results show that the presence of fractures does not significantly alter wormhole formation, but the fractures do tend to focus the wormholes along fracture planes.

6.8.4 Field-scale Studies

Where there is a reasonable degree of confidence in laboratory-scale experiments, field-scale simulations offer the potential for resolution of discrepancies between prediction and observation by refining site characterization. Although a single porosity model accurately predicts overall fluid production rates, it does not properly represent fluid transport through a fractured reservoir, and so the model was converted to a dual porosity or dual permeability formulation. The dual porosity model was tested with formation water data and 3D time-lapse seismic results.

For formation waters, the software GEM® showed that isotope data are potentially more useful than major ion composition for simulating some processes in the reservoir. There were difficulties in calculating isotope ratios when there is extensive depletion of component concentrations of the molecules used to model the fractionation. This software is also useful for simulations of the isotopic evolution of light alkanes. Thus, calibrating reservoir models using isotopic signals can provide further insights into transport processes during CO_2-EOR.

Changes of fluid chemistry in the reservoir are related to ongoing chemical interaction between the pore fluids and rock matrix. In contrast, because 3D time-lapse seismic monitoring data are able to track CO_2 migration, the parameters of a numerical geological model could be iteratively modified until the resultant flow simulations produced fluid distributions consistent with the production data and the time-correlative seismic data. This comparison of predicted versus measured seismic properties was done both in a trial-and-error manner and in a semi-automated process. The results showed that there is potential for improving the reservoir model by using the additional constraints provided by 3D time-lapse seismic.

6.8.5 Markov Chain Monte Carlo Stochastic Inversion

History matching as described above for formation water and 3D time-lapse seismic requires both direct intervention by the geophysicist, geochemist, or reservoir engineer, and requires many simplifying assumptions. Accordingly, a stochastic inversion algorithm was developed which aimed to identify reservoir permeability models that optimized agreement between the predicted and observed storage performance. The latter consisted of production data, and the observed geochemical or seismic response. A Markov Chain Monte Carlo method was applied to existing data to refine the spatial distribution of reservoir permeability, which effectively defines the framework of CO_2 migration.

Formation water composition and reservoir mineralogy data were used in a reactive transport model to represent the flow path, which was populated by a set of mineral phases that characterize the reservoir. The resultant mineral volume fractions along the flow path, and the composition of the effluent formation water were the parameters that were matched to field data. Running the inverse problem entailed conducting multiple forward reactive transport simulations, which for simplification were treated as single-phase flow in a 1D column. Agreement with the monitoring data is reasonably good, bearing in mind that not all variables were taken into account. The inverted results agreed well with the trend in the field observations, which suggests that the inversion is producing internally consistent results.

The algorithm for the stochastic inversion of the time-lapse geophysical data was tested on a single production area, and for a specific injection scenario of water and CO_2. After a few tens of iterations, the recovered permeability was very similar to the true model. For the real seismic data, trial-and-error seismic-constrained history matching was demonstrated whereby model permeabilities were manually modified so that the simulated CO_2 saturations within the storage complex are more consistent with the seismic monitoring data. Using a Markov Chain Monte Carlo stochastic inversion on the actual observed data had only limited success. This procedure, as implemented, is very computationally intensive, and ongoing research efforts are being conducted to make the algorithm more robust in real data cases.

6.9 Recommendations

Laboratory-scale history matching provides a unique method for calibrating and refining the reservoir simulation and reactive transport models required to forecast the long-term efficacy of CO_2 geological storage. It also enables detailed measurement and prediction of the site-specific CO_2–(oil)–formation water–rock transfer and reaction-dependent reservoir–seal and porosity–permeability changes that control dynamic reservoir–seal integrity, and therefore CO_2 migration and containment.

Reactive transport modelling of CO_2 coreflood experiments that use core and fluid samples from a potential storage complex can provide considerable insight into anticipated reactivity and long-term isolation performance. Tuning the model to achieve close agreement between experimental and simulation results also provides the necessary point of departure for addressing the long-term, field-scale predictive efforts that may be required for future regulations. Given the relatively low cost of such laboratory-scale studies, they are highly recommended for any potential project.

Although challenged by (1) the uncertainty of site characterization, (2) complex process coupling, and (3) various scaling-up issues, field-scale history matching plays a critical role in CO_2 storage projects. First, any potential project requires an initial long-term forecast of CO_2 migration, mass partitioning, and containment, in order to determine appropriate lease requirements. Once a project is underway, continuous feedback between monitoring, site characterization, and modelling efforts is required to ensure continued accuracy of the forecasts, regulatory compliance, and optimal deployment of the monitoring tools. In this regard, field-scale history matching – especially in the context of a model tuned to site-specific properties through comprehensive experimental calibration – can provide well-informed insight into strengths and shortcomings of the geological model of the site, especially with respect to permeability structure and magnitude. Again, the relatively low costs involved make well-designed field-scale history matching efforts well worth the investment.

7

WELL INTEGRITY

Summary

The purpose of the study was to identify issues related to well integrity that may arise in CO_2 storage projects. Although most of the work was based on a synthesis of relevant open literature, regulations, and a survey of CO_2-EOR and acid gas-EOR operators, a well integrity field test was also made.

Aspects evaluated included (1) the identification of the potential for leakage of individual wells through a regular assessment program, and (2) methods to be considered for securing well integrity when constructing new wells, remediating and converting existing wells, or abandoning wells. For all these situations, the most critical factor was the condition of the cement in the well. A variety of well logging techniques, pressure testing methods, and other tests were evaluated to check on their efficacy for ensuring the integrity of the well.

In the field test, the following suite of logs and other methods were used. Some or all of them should be considered when evaluating the integrity of a well. They included: Schlumberger Isolation Scanner, Schlumberger Sonic Scanner, Baker Atlas High-Resolution Vertilog™, Schlumberger Reservoir Saturation Tool (RST), and the Expro borehole camera to image 'perforations' and cement sampling holes, in order to assess the geometry of these penetrations. A built-for-purpose Pressure Transient Tool was used to quantify the hydraulic properties of the cemented annulus at selected intervals within the regional seal, and the MaxPERF® tool was used to create a series of holes, which established the hydraulic communication at each of the zones of pressure communication. The cement-filled annulus was penetrated and sampled, undamaged, using a cement coring tool (CemCore) developed for this testing program based on modifications to the MaxPERF® tool.

Recommendations arising out this research begin on page 284.

Authors: C. **Hawkes** (University of Saskatchewan, Saskatoon, Saskatchewan) and C. **Gardner** (Chevron, Houston, TX, USA).

7.1 Introduction

The purpose of this study was to identify issues related to well integrity that may arise in CO_2 storage projects, and to consider means of assessing and, as necessary, dealing with these issues. Guidance is provided on measures that can be taken to assure the hydraulic isolation of CO_2 in the injection zone. Much of the information presented here is based on a synthesis of relevant open literature, regulations, and results of a survey of CO_2-EOR and acid gas-EOR operators. Results of this synthesis indicate that an essential first step in an assessment of well integrity is to identify wells that may be exposed to CO_2 or acidified formation fluid, or affected by injection-induced pressure increase, and rank them in order of importance for investigation.

Key aspects of assessing existing wells include the depth and condition of cement plugs (if present) and the cement sheath (for cased wells). For new wells in a prospective storage site, probably the most important aspect of wellbore construction is the design, execution, and post-job assessment of the cementing program. Implementing a monitoring program is also important, and regardless of the monitoring methods planned, developing a program to obtain baseline measurements prior to CO_2 injection is essential. During well abandonments, the positioning and placement of plugs is of utmost importance – squeezing perforations in conjunction with abandonment yields superior results compared to plugs alone.

This chapter also presents information from a field testing and sampling program that was conducted prior to the abandoning of an oil-producing well that was drilled in the Weyburn Field in 1957. Key logging-related information includes the following: (1) the combined use of sonic and ultrasonic logging tools was beneficial because it enabled a broad range of depths of investigation, yielding key information on cement sheath conditions, and (2) the electromagnetic casing inspection tool used was effective for characterizing casing conditions. The combined analysis of casing and cement evaluation logs confirmed the commonly-held belief that a well-placed cement sheath provides effective protection against corrosion. Cement sampling was conducted using a newly developed tool, capable of drilling small sidewall core samples in 139.7-mm (5.5 in.) casing. This confirmed the log-based interpretation that a good cement sheath was present in the test well. Results obtained from pressure transient testing suggest that tight hydraulic conditions can exist even in a 1950s vintage well. Two key practical facts emerged from the testing program: (1) stringent depth-control measures should be used during specialized downhole testing, and (2) it is important to take real-time bottomhole pressure measurements during cement-sheath pressure transient tests.

7.1.1 Purpose

The specific purpose of the study was to provide information on measures that should be taken to assure the hydraulic isolation of CO_2 in the injection zone. For storage sites with many wells penetrating the storage zone (for example, CO_2-EOR sites), the assessment of well integrity is perhaps the most important part of assessing the containment potential of the site.

7.1.2 Scope

The project was not involved in the operation of the wells in the Weyburn or Midale fields. Therefore, the information presented here is based largely on a report prepared by a well integrity consultant that included a review of relevant open literature and regulations, and a survey of CO_2-EOR and acid gas-EOR operators. Information from the report has been supplemented by (1) results obtained in a field testing and sampling program that was conducted in a soon-to-be-abandoned oil well drilled in the Weyburn Field in 1957, (2) wellbore database and simulation projects conducted by researchers at the University of Alberta, and (3) a casing corrosion literature review.

7.2 Well Integrity Assessment

Prior to developing a well integrity assessment program, the area must be identified in which wells may potentially be contacted by CO_2 or otherwise affected by injection operations (for example, by induced elevated pressures). The methods used to define this area will vary depending on the nature of the project, but will generally involve modelling and (at some stage) monitoring methods described in earlier sections of this book. A safety margin should be incorporated into the assessment area, to allow for errors in the prediction of its extent. Over time, the plume location can be verified by project monitoring, and the assessment area can be updated as required.

Wells in the assessment area should be evaluated prior to the start of injection to determine wellbore integrity. This is done to mitigate the potential for CO_2 or pressurized fluids to leak past the cap rock and other reservoir boundaries and to potentially impact other permeable zones. This is a particularly important consideration where aquifers with potable water may be jeopardized by leakage (Celia et al., 2008).

Ultimately, the integrity assessment should answer seven key questions for wells in the assessment area.

1. Where is the well?
2. What is the status of the well?
3. What is the expected future function of the well as part of the storage project?

4. Does the well penetrate the storage or EOR reservoir?
5. Is the well sealed across the reservoir–cap rock interface?
6. Does the well have a high probability of uphole well leakage?
7. Will the well need to be remediated to ensure integrity?

The integrity assessment of existing wells should be conducted systematically, where the first step is to identify all wells by location, depth, and trajectory, and collect data relevant for the integrity assessment. Once the area of assessment has been determined, all oil and gas wellbores, stratigraphic test holes, and water wells in the assessment area must be identified. Depending on the jurisdiction of operation and the regulatory framework that previous oil and gas operations have operated under, this first step may be quite time consuming.

The easiest wellbores to identify and assess will generally be those where oil and gas exploration and production is fairly recent. Oil and gas wells and stratigraphic test holes should be well-documented and their locations easily accessible through the regulator, with depth, design and history information on newer wells readily available, through the regulator or company files.

Wellbore information in jurisdictions where operations have been ongoing for many decades, or where historical regulations and reporting requirements were not as stringent as they are today, may be more difficult to obtain. Identification of wells in these areas may require investigation in the field, in addition to review of available records. This process may involve the following stages.

1. Review of historical aerial photographs.
2. Review of historical information that may be available through local museums and historical societies.
3. Interviews with long-term residents.
4. Interviews with service providers such as water well drillers and oil and gas service companies.
5. Field investigations such as electrical conductivity surveys which can identify near-surface saline water (for example, due to leakage up wells).
6. Field investigations such as magnetic surveys which can identify well casings below ground surface.
7. Field investigations such as soil gas migration surveys or vegetation impact surveys which can identify wells leaking gas to the surface.

Once all of the wellbores have been identified, they should be ranked in order of importance for investigation. All wellbores should be evaluated for their potential for leakage in the event that the cap rock is breached through any wellbore, and therefore wellbores that penetrate the cap rock over the injection zone should be investigated first. Wells that do not penetrate the cap rock may also provide part of a leakage pathway to surface in the event of a cap rock failure (for example, due to faults or fractures), or leakage through a well that penetrates the cap rock combined with cross-flow through a shallower formation.

7.2.1 Evaluation of Drilled and Abandoned Wells (Uncased)

Drilled and abandoned wells are those that were drilled, evaluated, and abandoned without production casing being run in the well. As indicated from logs or tests, these wells were typically not commercially productive for a variety of reasons.ABandonment procedures typically include setting cement plugs across porous intervals in the wellbore (thereby isolating individual aquifers), and a cement plug set across groundwater intervals and the surface-casing shoe. A key point to remember is that a plug which may have been adequate for isolating a specific interval may not be suitable if CO_2 injection results in greater differential pressure across the plug than was present at the time of abandonment.

Two situations may affect wellbore integrity in abandoned wells, and hence could allow the cross-flow of CO_2 or pressurized fluids.

1. Lack of a cement plug across the injection zone (reservoir)–cap rock interface.
2. Lack of an effective cement plug and lack of zonal isolation due to poor mud removal. Factors which can contribute to poor mud removal include (1) excessive filter-cake buildup, (2) hole washout and enlargement, (3) loss of circulation during cementing, (4) well deviation, (5) poor mud properties, and (6) insufficient hole conditioning prior to cementing.

Lack of a cement plug across the reservoir–cap rock interface may allow injected CO_2 or pressurized fluids to migrate up the wellbore until they encounter a hydraulic barrier, usually a cement plug set higher in the hole. If there are porous intervals between the injection zone and the base of the next higher cement plug, cross-flow can occur into these intervals. Carbon dioxide and other pressurized fluids may then migrate into such zones until they encounter another wellbore which, if also leaky, could provide a conduit to potable groundwater or the atmosphere, regardless of whether or not that wellbore penetrates the cap rock.

Factors that affect the success of plug cementing are similar to those which affect primary cementing success, but plug setting is made more difficult because the relatively small volume of cement slurry used increases the likelihood of the plug being contaminated with drilling fluid. Much has been written concerning plug cementing success (for example, see Calvert et al., 1995).

Factors that may affect wellbore integrity or cause concern due to unknown conditions include the following.

1. Cement of unknown quality and type.
2. Unknown mud properties (such as oil-based muds with no spacers run to protect cement from contamination).
3. Cement top location not confirmed.

Further to the seven key questions listed in the preceding section, the evaluation of abandoned wells which penetrate the cap rock should answer the following six questions:
1. What was the condition of the hole prior to setting cement plugs, especially across the reservoir–cap rock interface?
2. Where are the cement plugs located in the wellbore?
3. How were the locations of the cement plugs confirmed?
4. What fluid is between the plugs?
5. Is the well deviated? Where is the kick-off point?
6. Is there any 'junk' in the hole?

Assuming that records exist, the answers to these questions can often be found in the well drilling tour reports, deviation surveys, and open-hole logs. Where records do not exist, the assessment may assume that wells drilled in the same era by the same operator were abandoned with a similar program, but should be flagged as potential problems due to missing information.

Drilling tour reports should provide the following information.
1. Mud type and properties prior to running cement plugs.
2. Cement plug locations based on top and bottom locations of drill pipe.
3. Cement type used (plug cement slurry details are often not recorded).
4. Plug location method (tagged with drill pipe, density logged, etc.).
5. Hole problems while drilling and the depth of these problems (for example, lost circulation, sloughing, tight hole, 'junk', kicks).
6. Kick-off point.

After the cement plug locations have been identified, the open-hole logs should be analysed to determine the top of the injection–disposal zone and the top of the cap rock at the particular location. The cement plug location should be overlaid on the open-hole logs and the following questions asked.
1. Does a cement plug extend across the interface between the injection–disposal zone and the cap rock?
2. What is the condition of the hole where the cement plug is located? For example, is there filter-cake buildup or are there hole washouts and deviations?

3. Is there another cement plug across the top or through the cap rock?
4. Where is the next zone that could be infiltrated by leaking CO_2 or pressurized fluids?

The answers to these questions will identify the wells that are most at risk for leakage, and will help determine the most likely pathway and destination for potential leakage. Should leakage occur, knowing the physical location of the well (and its physical relation to the injector well) can lead to an accurate assessment of the right remediation program. In some instances where remediation costs are excessive, this information may lead to relocation of the injection well.

7.2.2 Evaluation of Cased Wells

Cased wells are wells that have been drilled, evaluated, and cased for the purpose of production, injection, or observation. After evaluation, steel casing will typically be run in the hole. Often mechanical devices will be run on the casing, including centralizers to assist in centralizing the casing within the wellbore, and scratchers to assist in the removal of filter cake. These devices are often only run across the interval that is targeted for production or injection at the time the well was drilled. After casing is run, the mud in the hole will be conditioned prior to cementing. Mud conditioning changes the rheological properties to allow for better mud displacement during cementing.

A review of well history can provide valuable insight into the possible risks to wellbore integrity, and it is recommended that a thorough review be conducted by staff that have adequate experience and training to identify possible problems.

An evaluation of well attributes specific to cased wells may assist in identifying wells that have a relatively high probability of leakage from any zone. This type of information is listed below and is sometimes available in electronic form.

1. Depth of cement top.
2. Location (areal coordinates).
3. Year well was drilled.
4. Number and depth of perforated intervals.
5. Abandonment date (if applicable).
6. Hole size and well deviation.

This information may sometimes be mined from the databases of regulators or operators, thereby facilitating the process of prioritizing wells for in-depth analysis of leakage potential. Such an approach is illustrated in Watson and Bachu (2008). In-depth assessment of cased wells includes the following.

1. Cement evaluation log analysis.
2. Casing corrosion log analysis.

3. Well history information, such as casing failures, surface-casing vent flow or gas migration, cross-flow, fish in the hole, drilling events such as lost circulation, kicks, and doglegs.
4. Production information, such as a history of broken sucker rods at a particular depth (possibly indicating a dogleg and lack of centralization leading to inadequate cementation).
5. Tubular information that could indicate a problem for corrosion, such as steel grades or dissimilar metals.
6. Abandonment techniques.
7. Further assessment through logging and integrity testing.

Evaluation of existing logs (in particular cement evaluation logs) may provide important information regarding the condition of the zonal isolation between the injection zone and the cap rock. The quality of the zonal isolation or the cement sheath may be compromised due to completion, production, and injection activity. The evaluation of existing logs is important, and may give critical clues to where problems may exist, or could develop, because zonal isolation seldom improves over time (unless remedial measures are taken).

Note that because many cement evaluation logs are run prior to completion to ensure zonal isolation before stimulation (for example, fracturing and perforating), care must be taken to determine if this information is current and, if not, how applicable it is to the current condition of the wellbore.

If a well is abandoned, the only easily obtained information will be the historical well file and any existing electronic data. The method of abandonment is a significant factor in the determination of a well's long-term integrity. Historically, many jurisdictions allowed the use of mechanical, cast iron bridge plugs that use elastomer sealing elements. A common, additional requirement when using mechanical plugging devices, such as a bridge plug, has been the emplacement of a cement cap on top of the plug. Perforations or an open hole provide access to the wellbore below the plug, and in the event that the original completed interval is in the CO_2 injection zone, it is expected that CO_2 as a gas or in aqueous phase (carbonic acid) will enter the wellbore – easily corroding cast iron, elastomers, and typical casing materials (Schremp and Roberson, 1975). Where these types of abandonment exist, and the cement plug has been installed by dump-bailing, there is a high probability of the well being leaky due to a failed bridge plug and cement plug. Dump-bailing of cement often does not yield adequate cement plugs, and both laboratory tests and field observation confirm this (White et al., 1992).

Very old wells may have been abandoned with other methods that will not provide isolation if reservoir pressures are increased. The use of dense mud, wooden plugs, burlap, rocks, wire rope, or other materials that may have been used in the early days of oil and gas development (generally prior to 1950) will not provide adequate wellbore integrity over the long-term.

7.2.3 Compilation of Well Data

A database of wellbore information for 183 wells in the Phase 1A area of the Weyburn Field was compiled. Data were collected from two main sources: (1) older well data, obtained from historical well files and transferred to microfiche (records stored in a database managed by the Saskatchewan Ministry of Energy and Resources) and (2) newer well data, obtained from internal well files (in paper format and electronic), provided by Cenovus. In general, the same information was found in both places but substantially more detailed well records were available in the newer, electronic records. Data existing for wellbores drilled in the last decade were more detailed and complete than the data available for those wells drilled in the early development of the Weyburn Field.

Information on wells drilled in the 1950s was generally limited to details about the geometry of the wellbore (diameter, length), casing (diameter, length, grade), and cement type and volume. Information for a typical wellbore perforated in the 1990s was very detailed and complete. Cementing data include records of cement type, volume, depth set, circulation pressures, spacer volume, casing reciprocation, stroke length, and time. Casing data contain casing length, grade, accessories, number, and spacing. Daily drilling reports provide a detailed progress of perforation with information like mud density, volume, pressure, and daily drilling advance. Figure 7-1 illustrates a range of typical wellbore construction practices used in the field, as identified from the database. Figure 7-2 illustrates selected statistics for the wells within Phase 1A of the project, extracted from the database.

7.3 New Well Design Considerations

Wellbore construction practices depend in large part on the composition of the injection stream, the content of CO_2, contaminants, and moisture content. This information is necessary to design the tubing and packer system that are to be employed. The design and execution of an effective, quality cement job is the most important aspect of wellbore construction. Poor cementing leads to loss of zonal isolation, external casing corrosion, casing failure, inability to perform effective kill operations, inability to seat tools within the casing, and many other problems.

7.3.1 Surface Casing

In addition to its drilling-related functions (such as the installation of blow-out prevention equipment), surface casing provides additional protection over the life of the well. Most surface casing is set across useable or potable water

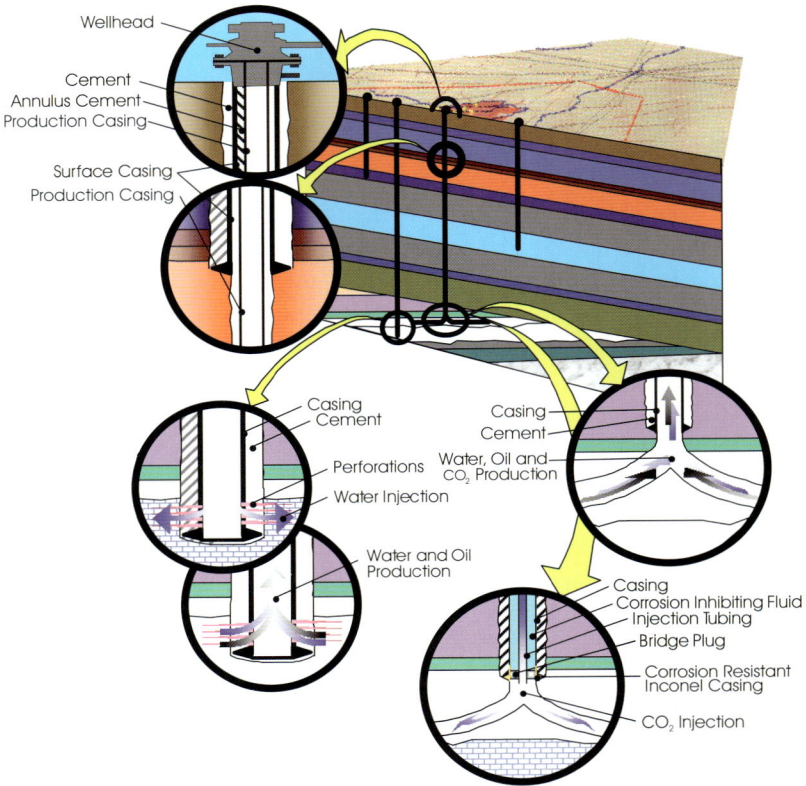

Figure 7-1. *Range of well types in the Weyburn Field.*

aquifers, and provides an effective second barrier to subsurface flows that may occur along the wellbore. Surface casing effectively protects groundwater from deeper hydrocarbon flows or, in the case of CO_2 storage, from CO_2 or saline water flows. Regulations usually require that surface casing is set and effectively cemented below any groundwater aquifers that must be protected. Further, because of the important role that surface casing and its cement play over the life of the well, careful attention should be given to this aspect of wellbore design.

After the main hole has been drilled (and during the logging of the main hole), it is recommended that the surface casing be logged for cement evaluation. This evaluation will indicate not only how well the cement job is performing but will also indicate if any damage was imposed on the cement sheath by the drilling operations. If zonal isolation is compromised behind the surface casing, serious consideration should be given to remedial measures if groundwater aquifers are at risk.

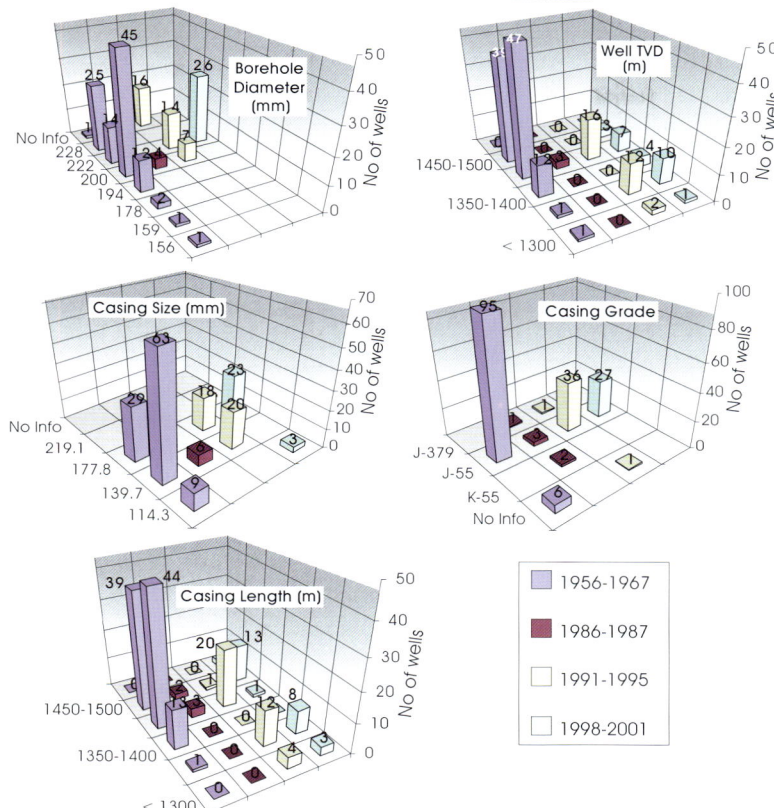

Figure 7-2. *Wellbore statistics for wells within the Phase 1A area of the Weyburn Field.*

7.3.2 Well Trajectory

Previous analysis showed that well trajectory can be an important factor in the incidence of wellbore leakage (Watson and Bachu, 2009). Wells that are slant-drilled from the surface have a high incidence of leakage due to poor centralization of casing. This can leave a channel on the low side of the casing due to inadequate mud removal during cementing, and a channel on the high side due to solids settling and free water. Where wells are drilled directionally, attention must be paid to the depth interval(s) over which the angle is built. Ideally, measures should be taken to obtain a vertical well trajectory over a significant proportion of the cap rock. High-angle trajectories through cap rock formations jeopardize the integrity of the cement zonal isolation behind casing due to poor centralization.

Accumulation of water in tubulars due to condensation in low-lying intervals of the wellbore can occur and can cause corrosion which may lead to failure, even for very low humidity fluid streams (Gunaltun, 1991). This means that well angle can contribute to corrosion of tubing and casing, in addition to contributing to poor zonal isolation. Careful analysis should be conducted to evaluate temperature as a function of depth, well angle, and dogleg severity, in order to ascertain whether condensation is likely to be a problem.

7.3.3 Casing Considerations

The casing above the anticipated packer landing depth and the top of the injection zone need not be specifically designed for CO_2 injection, unless the operator wants to allow for the possibility of later perforating a shallower portion of the injection zone and moving the packer up-hole. This casing must be designed to be competent in the event of a failure by the tubing or packer to withstand the anticipated injection pressures. Packer fluid should be chosen to lessen the short-term degradation of casing when in contact with CO_2. The annulus between the tubing and the casing should be continuously monitored during injection operations to ensure that injection ceases immediately in the event of a failure, and repairs made.

Casing below a packer should be designed to withstand any anticipated corrosion effects by the injection stream. Although the injection stream may be dry (dehydrated), eventual kill operations may require the use of water or brines that could induce corrosion.

7.3.4 Cementing

As noted previously, probably the most important aspect of wellbore construction is the design and execution of the cementing program. Much work has been done to determine the long-term effects of CO_2 on typical Portland-based oilfield cements, and many industry service providers can formulate cement blends that have good CO_2 resistance. In general, commonly used oilfield Portland-based cements will perform adequately in the presence of CO_2, a fact demonstrated in the cement sampling program undertaken in this project and by various researchers (Duguid, 2008; Kutchko et al., 2008; Strazisar et al., 2008). The ultimate goal should be to achieve a cement sheath that does not have channels, micro-annuli, cracks, or gaps. With this in mind, cement design and placement should focus on the wellbore stresses over the life of the well to ensure that the cement sheath will perform under all anticipated conditions.

Wells for which cementing best practices (Nelson and Guillot, 2006) have been applied should have the following characteristics.

1. The drilling mud is fully displaced.

2. The casing is well-centralized throughout the length of the well, especially through the cap rock and overlying aquitards (secondary seals).
3. There is effective removal of filter cake prior to cementing.
4. Adequate cement velocities are achieved during placement.
5. Pipe reciprocation or rotation is used where possible.
6. Adequate spacers are pumped.
7. The cement formulation is correct for the pressures and temperatures expected in the wellbore.

For wells that are to be used as CO_2 injectors, CO_2-resistant cement formulations may be used across the CO_2 injection zone. Such formulations include Pozzolan–Portland cement blends as well as proprietary formulations. Although dry CO_2 is unlikely to significantly impact cement, there will be occasions during the life of a well when it may be necessary to utilize water-based fluids for stimulation or kill operations. Introduction of water-based fluids may create an acidic environment and so produce negative impacts on the cement sheath adjacent to the CO_2 injection interval.

7.3.5 Surface Equipment

After the casing is set and cemented, seals and the wellhead will be installed. The design of the wellhead equipment is often subject to regulatory approval.

Figure 7-3. *Surface-casing vent installed on an existing well.*

It is strongly recommended that a surface-casing vent and valve be installed, as shown in Figure 7-3, and that the vent be left open at all times. Where regulation requires that vents be closed, continuous monitoring of the surface-casing vent is required to assist in early detection of casing failure or CO_2 leakage.

7.4 Well Remediation and Conversion

7.4.1 Remediation

Assessment of the integrity of pre-existing wells will indicate which wells require remediation prior to the commencement of CO_2 injection. Remediation may be deferred depending on the time it takes for the CO_2 plume to reach the wellbore. Although the integrity assessment will indicate which wells have the highest potential to leak, these wells may in fact be secure. A risk assessment should be made on each well identified as having the potential to leak, in order to determine which is more cost-effective, pro-active repair or remediation, should a leak be detected in the future.

Several factors need to be evaluated to determine which wells should be repaired. Key questions to be asked are as follows.

1. What is the likely fate of leaked CO_2?
2. Would CO_2 only migrate to a deep saline aquifer, or could it migrate to the useable or in-use groundwater intervals?
3. What would the impact be on the reputation of the storage operator and regulator, should a leak occur?
4. Are the injection zone and the potential leak areas in a populated area?
5. What are the chances of a successful repair?
6. What other risks may be imposed due to the repair, such as risk of blow-out, sour gas release, worker injury, public impact due to noise, land disturbance, increased anxiety, odours, and traffic?

Where repair is to be deferred, or not undertaken unless a leak has been confirmed, additional monitoring may be required to determine when measures should be taken.

7.4.2 Conversion

In many reservoirs that are appropriate for CO_2 storage, a number of wellbores will penetrate the injection zone. Many of these wells could be used as injectors if they have been adequately constructed to meet the regulatory requirements, or have been remediated to appropriate standards. In many cases, older wells which may be candidates as injectors will not be cemented to the surface. It is recommended that if these wells are to be used for injection, remedial operations focus on the injection zone only.

Remedial cementing to cover up-hole porous zones or non-saline aquifers introduces potential failure pathways in the wellbore. Where cement does not meet the regulatory requirements, monitoring of the uncemented annulus between the production casing and the surface casing can provide important information regarding the integrity of the zonal isolation. However, having methods in place at the surface to detect and bleed off increasing pressures is better than having a poor remedial cement job – because with only a poor remedial cement job, leakage problems will be masked, with the possibility of longer term leakage to other formations or wellbores.

Wellbore failure of acid gas injection wells indicated that failed zonal isolation was not the cause, but rather failure modes common to all well types. For the acid gas wells studied, there was no leakage of acid gas to the surface or to other downhole formations. Rather, it was the casing that failed due to external corrosion, coupled with surface-casing vent flow, both typically occurring as a result of uncemented intervals far above the injection zone. Failures of a packer or tubing could result in acid gas or CO_2 in the tubing–casing annulus, and were part of the operational problems and planning for these operations (Bachu and Watson, 2008).

Isolation of the injection zone should be confirmed through logging (which may include radioactive tracers, temperature logging, and other methods) that can detect fluid or gas movement behind pipe. If needed, remedial operations should be conducted at the injection zone–cap rock interface, and remedial perforations should be placed so that they are above the injection packer, so that any failure is immediately evident through annular pressure monitoring. The tubing–casing annulus must be continuously monitored, and where injection pressures are below hydrostatic pressure, positive pressure should be maintained on the annulus so that a failure of the casing, packer, or remedial perforation is immediately evident.

Prior to selecting a well for conversion, zonal isolation integrity must be confirmed. Casing inspection logs should be run the full length of the wellbore to provide baseline information. This logging should provide information on both the internal and external surfaces of the casing. Where flow rates are too low to detect with other methods or are intermittent, the casing inspection logging can be helpful over time to detect leakage behind pipe. Baseline cement evaluation should also be conducted the full length of the wellbore, to determine if there are other risks that may make the wellbore an unsuitable candidate for conversion.

7.5 Well Abandonment Considerations

In most jurisdictions, regulatory requirements for wellbore abandonment will already be in place, but they will likely be minimum requirements. Most jurisdictions created wellbore abandonment requirements prior to the advent of CO_2

storage or CO_2-EOR. Many requirements do not take into consideration the possibility of reservoir pressures increasing above virgin pressures or normal hydrostatic gradients, or that a reservoir may at some time contain a corrosive substance. It is important for operators to take these possible future uses and liabilities into account, and abandon wells with a view to long-term containment.

7.5.1 **Open-hole Abandonment**

As noted above, non-productive or uneconomic wells are abandoned without production casing being set. Typically, this involves setting cement plugs at selected intervals, although occasionally the entire borehole is filled with cement. Regardless of which method is used, the most important task is to ensure that a competent plug is set across the reservoir–cap rock interface and across shallow aquifers. It is also highly desirable to set plugs across intermediate porous zones, taking care to ensure that these plugs cover the interfaces between these zones and the non-porous zones (seals) adjacent to them.

Well logging to assist in designing open-hole abandonment plans should at least include a gamma ray–neutron log or an electric log with a caliper – possibly both. Information from these logs will assist in determining (1) the position of lithological boundaries, (2) where increased cement volumes may be needed to accommodate for enlarged holes, and (3) where filter cake may impede an effective hydraulic seal between the cement and the formation.

Prior to setting cement plugs, mud should be conditioned to reduce cement channeling. If there is a large density difference between the cement and the mud, high density pills should be placed to keep the cement from falling away at the bottom of the plug or roping down-hole. Appropriate cement blends, designed for the plug location temperature, mud type, and pumping time must be used. Cement quality should be closely monitored during mixing and pumping to ensure consistent cement density.

Care must be taken to avoid disturbing the plug before it has set, otherwise mixing of mud and cement may occur, especially if circulation is too close to the plug top after emplacement. If this happens, it may result in the top of the plug being washed away. Logging plugs using tools that are run through the plug to detect density differences (rather than waiting for the cement to set, then tagging it with pipe) can result in mixing of the mud and cement.

7.5.2 **Cased-hole Abandonment**

Evaluation of the cement sheath behind the casing is the first step in determining the requirements for abandoning a cased hole. In general, tools that assess conditions around the full circumference of the casing are preferred over those that generate averaged readings representing the entire circumference. Further, tools that can adequately distinguish low-density cements should be used. An experienced log analyst can interpret the log to determine areas where zonal isolation is compromised. This cement evaluation log, along with open-hole logs to determine porosity, lithology, and permeability, will guide requirements for remedial cementing.

Once the evaluation is completed, there are several options to consider for remedial action – if such action is deemed necessary. Access to the casing-hole annulus can be gained through perforating, abrasive jetting, section milling, or some other cutting mechanism. It is important to choose the appropriate method. For example, section milling of the casing is not appropriate where the casing is not well-cemented above and below the section, because it may make future remedial operations below the milled section impossible or extremely expensive. It is recommended that methods be used that leave many future options available in the event that the initial remediation fails.

It is also recommended that operators squeeze all perforated intervals when abandoning any well. The squeeze method will vary, depending on pressure, feed rate, equipment, and other factors. Squeezing perforated zones may isolate sources of fluid or gas which could migrate through pathways along cement cracks or micro-annuli induced by operations such as perforating, stimulation, and pressure cycling due to production. It is recommended that previous abandonment should be pressure-tested before moving up-hole to conduct further operations.

After perforated intervals have been squeezed, other pathways for potential leakage into and through the wellbore should be sealed. Although the cement sheath in a well-cemented casing–open-hole annulus provides a barrier to vertical flow of gas or fluids, leakage may occur even in well-cemented wellbores. This may occur through discrete gaps, small channels, cracks, or micro-annuli that allow reservoir fluids and gases to contact the external surface of the casing. Over time, the corrosion of casing can allow fluids and gases to enter the wellbore. Because of this, it is desirable to follow a plugging procedure internal to the casing, by setting plugs across intervals corresponding to those where open-hole abandonment plugs would be set. Heavy, non-corrosive mud should be left in place between plugs.

Where casing is free to the surface from an external cement top, consideration should be given to cutting and pulling casing, then following open-hole abandonment procedures for this upper portion of the well. Data from Alberta suggest that wells abandoned open-hole have fewer incidences of leakage than cased wells. If this option is viable, it is important to ensure that all operations conducted below the cut point have been successful in creating isolation.

7.6 Well Integrity Monitoring

Monitoring CO_2 storage projects is multifaceted and must consider (among other important aspects) the monitoring and establishment of baseline norms for (1) wellbore integrity, (2) plume dispersion, (3) cap rock integrity, (4) groundwater, and (5) soil gas. A broader discussion of monitoring methods and baseline surveys is given elsewhere in this book. Here, only monitoring aspects specific to wellbores are discussed.

Wells near a proposed storage project can provide important information about pre-existing leak paths and historical information regarding leakage. If a well has not been abandoned, cut and capped, it should be tested for surface-casing vent flow, and gas samples analysed. Soil gas migration testing should be conducted near the well, and the source zone determined if leakage is detected. This may be accomplished through gas analysis, cement evaluation logging, analysis of open-hole logs, and identification of wells with external casing corrosion. If a pre-existing leakage source has been identified, future monitoring of surface-casing vent flow and gas migration changes will help to distinguish leakage of injected CO_2. This evaluation will also identify casing failures.

Baseline information to help identify deep zonal isolation issues in the future may be secured by analysing gases and liquids from porous zones above and below the CO_2 injection zone in wells in the vicinity of the injection well.

Once construction of a new injection well is complete, baseline information should be gathered. This should include logs for zonal isolation, cement evaluation, and casing inspection. Tests should include a soil gas migration test and surface-casing vent flow test both for hydrocarbons and CO_2. Gas analysis done on samples obtained should include determination of carbon isotopes on all gas components that contain carbon.

The cased-hole logging suite will provide information for initial zonal isolation. A cement evaluation log should be run after (1) the wellbore has been initially pressure tested, (2) the mud or displacement fluid has been replaced with clean fluid (similar to the proposed injection packer fluid), and (3) the casing has been scraped to remove any residual mud, cement or mill-scale. Note that both the casing pressure test and the displacement to lighter completion fluid can have negative impacts on the cement evaluation logs due to the formation of micro-annuli. The cement evaluation log(s) should be run initially under normal hydrostatic pressure, then a comparison log run at pressures that accurately reflect the pressure that the well would have experienced during cement curing and setting. The application of this pressure when running the log will indicate the potential presence of micro-annuli along the casing–cement interface. A casing inspection log should then be run that clearly identifies external and internal defects. Because new casing should not be corroded, this log can be used later to identify other defects of milling, casing attachments, and tong damage for comparison to future inspection logs.

Temperature and radioactive tracer logs, or possibly oxygen activation logs, should be run to confirm initial zonal isolation prior to injection. The well should be logged in a static state prior to completion, to determine if there is cross-flow due to natural pressure drive and poor cementing, and then logged under injection conditions. After the well has been completed and stimulated, a comparative suite of zonal isolation logs should be run. If the well is to inject only CO_2, then the logs should be run using CO_2 as the test gas/fluid.

Monitoring the cap rock or the zone immediately above it is strongly recommended. The drilling of dedicated observation wells, or the use of pre-existing observation wells, may be used for this purpose. These wells will need to have similar logging and testing conducted to ensure that only the monitored zone is in communication with the wellbore. Also, the placement of monitoring wells will depend on the area that has the highest risk of leakage, perhaps due to the density of pre-existing wells or known cap rock faults. The number of these wells should be carefully managed because each wellbore creates a new potential pathway for leakage.

Monitoring of the wellbore for potential leakage during the life of the CO_2 injection project is highly dependent on the duration, frequency, and quality of the pre-injection data. Monitoring programs should closely follow the initial testing that was conducted, so that meaningful comparisons can be made. Monitoring at the leading edge of the CO_2 plume and where significant pressure increase has occurred should be a priority. Because pre-existing wellbores may be leaking through existing pathways in the casing or cement sheath, this leakage should become more pronounced as fluids in the near wellbore area become more acidic and more highly pressured.

Injection well monitoring should include continuous tubing and annuli monitoring for pressure and temperature. Injection well annuli should have some positive pressure applied so that failures can be easily identified if they occur across sub-pressured zones. At regular intervals (for example, every 5 years) a new suite of comparison logs should be run for casing inspection, cement evaluation, and zonal isolation. Many regulatory jurisdictions dictate the minimum requirements for logging and monitoring. These logs should be run in similar conditions to the original logging suites so that easy comparisons can be made. With the development of tools that can detect smaller variations in casing defects and cement quality, care must be taken to ensure that the change that is noted is actually due to a change in the wellbore condition and not due to enhanced technical capability. Similarly, observation wells should have comparison log suites run at regular intervals.

Monitoring post-closure will depend greatly on the project size, depth of the storage reservoir, and proximity to populations. It is expected that dedicated observation wells will be left in place for a period of time, probably of the order of 20–30 a. These wells will be a combination of deep observation wells that monitor the cap rock and the zone immediately above the injection zone, and groundwater monitoring wells. To date, there are no projects in the post-closure stage or nearing closure, so there is no pertinent information.

7.7 Well Integrity Field Testing
7.7.1 Need for Field Testing

Prediction of potential leakage rates from CO_2 storage zones requires knowledge of the potential transport properties of wellbores. This may entail the prediction of effective wellbore permeability as a function of, or the potential for, geomechanical or hydrochemical damage to the hydraulic barrier components of the well.

Phase 1 of the project included an assessment of the methodology of wellbore integrity, from the perspective of hydraulic and transport properties. The objective was to identify key leakage paths within a wellbore system, and the processes or mechanisms that influence the characteristics of these leakage paths. The wellbore system was chosen to include all the major components of a wellbore: internal abandonment configuration (bridge plug and cement), casing corrosion, annular cement (between casing and formation), excavation damage zones in the near-wellbore formation region, and all the interfaces that exist between these components.

A good understanding of these mechanisms can only be obtained following a comprehensive review of the wellbore environment during the different stages in the life of the well. The main stages, from a hydraulic integrity point of view, can be classified as drilling, completion, well history, and abandonment. The influence of each of these stages on the final hydraulic integrity of the well can be further classified into four major groups: geomechanical damage, hydrochemical damage, mud removal, and cement deterioration. Geomechanical damage encompasses any stress-induced changes in the hydraulic conductivity properties of materials within the wellbore system. Hydrochemical damage refers to any alteration of hydraulic conductivity in the near-well formation region – typically called formation damage. The third group, the efficiency with which mud is removed from the annulus during cementing operations, addresses the development of mud channels and their impact on hydraulic integrity of the wellbore system. Cement deterioration refers to porosity alterations due to geochemical processes under in-situ conditions and in the presence of CO_2.

A modelling workflow that accounted for the above-noted factors was undertaken. Representative results are presented in Figure 7-4. Though due

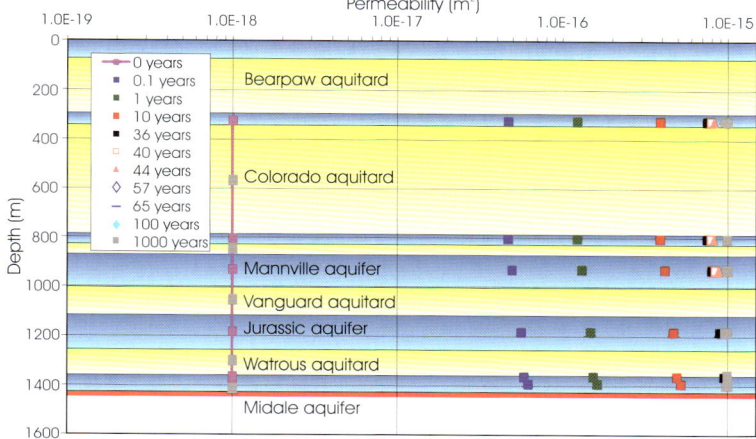

Figure 7-4. *Permeability of the wellbore system at various times, predicted for several sealing zones overlying the Midale Beds, based on modelling conducted in Phase 1 of the project.*

diligence was followed during this work, it was clearly evident upon its conclusion that field measurements were required in order to calibrate these model predictions.

7.7.2 Weyburn-Midale Field Test – Well Description

The well investigated in the Final Phase of the project was drilled as an oil producer in 1957, and is located at the corner of an area that had been under CO_2 flood for approximately five years before testing. Surface casing was set at 122 m depth, and then a 222-mm (8.75-in.) hole was drilled to 1471 m depth and cased with 139.7-mm (5.5-in.) casing.

The well was suspended in 2009, at which time a retainer was set in the Midale Evaporite, approximately 5 m above the top of perforations (1440 m). Due to active water and CO_2 injection in offset wells, elevated pressures occur beneath the retainer. To mitigate operational risks during the course of the well integrity testing program, the retainer was left in place and used to run a cement squeeze into the Midale Beds prior to commencement of the testing program.

7.7.3 Logging Program

A suite of wireline logging tools was run as part of the well integrity field program. One purpose for these logs was to identify suitable intervals for the destructive tests to follow (see below); another purpose was to provide results to compare against the quantitative information obtained from the destructive testing. The following logs were run.

SCHLUMBERGER'S ISOLATION SCANNER was used to evaluate the casing, cement, and cement interfaces. It was used to create maps and estimates of the integrity of the casing and the cement behind the casing. The isolation scanner works using a pulse-echo and flexural mode to radially measure the surface of the inside, the surface of the outside, and the thickness of the casing. It was also used to create solid–liquid–gas maps of the material behind the casing and identify any contamination effects or defects present within the cement sheath. Lastly, the isolation scanner was used to detect the third interface, in this case, the annulus–formation contact, in order to assist in the selection of testing and sampling depths in which the casing was centralized.

SCHLUMBERGER'S SONIC SCANNER is a sonic logging tool which provides information on both the integrity of the well and the properties of the surrounding formation. The sonic scanner was run to collect acoustic data that were used to create cement bond logs and variable density logs comparable to those collected from a cement bond tool. Simultaneously, the sonic scanner acquired data that were used to calculate formation compressional and shear wave slowness, hence enabling the calculation of rock mechanical properties.

BAKER ATLAS' HIGH-RESOLUTION VERTILOG™ is an electromagnetic casing inspection tool which generates a 360-degree defect map and accurately pinpoints the location, size, and shape of casing defects, whether internal or external.

SCHLUMBERGER'S RESERVOIR SATURATION TOOL (RST) is a pulsed neutron tool which was run to obtain an indication of the fluids in the near-well area, and to provide an indication of formation porosity and clay content.

EXPRO'S BOREHOLE CAMERA was used to image 'perforations' (see below) and cement sampling holes to assess the geometry of these penetrations.

7.7.4 Cement Sheath Pressure Transient Testing

A built-for-purpose Pressure Transient Tool (PTT), developed by Opsens Solutions Inc., was used to implement this task, which was conducted to quantify the hydraulic properties of the cemented annulus at selected intervals within the regional seal (Watrous Formation). The PTT incorporates downhole sealing elements (packers), and tubing and cable bypass devices, allowing for isolation between the zones of interest. The tool consists of four hydraulically inflatable packing elements, packer inflation lines, injection ports, pressure and temperature sensors, and the coiled tubing string to lower it to the test depth.

Each test involved the creation of two zones of hydraulic communication between the well and the cement sheath, one near the top and one near the base of the test interval. One packer each at the top and bottom of the tool was inflated to isolate the entire test interval from the rest of the well. The two zones of communication were then hydraulically isolated from one another, using a pair of packers, with a pressure gauge located between them to monitor for leakage at the packer–casing interface. Water was then injected into the upper packed-off zone, and the pressure response monitored at the lower zone.

The MaxPERF® tool, supplied by Penetrators Canada Inc., was used to create a series of holes that established the hydraulic communication at each of the zones of pressure communication. Each hole was drilled to a sufficient depth to fully penetrate the cement-filled annulus and to penetrate several millimetres into the surrounding formation. A sufficient number of holes (16) was drilled to emulate the hydraulic behaviour of a slot. The benefits of drilling the 'slots' include good control on depth of penetration, and the creation of a clean penetration (compared, for example, to conventional perforating technologies), with no damage to the adjacent cement sheath.

7.7.5 Cement Sampling

When this project was initiated, there was no technology for obtaining sidewall cement cores in 139.7-mm (5.5 in.) casing. A cement coring tool (CemCore) was developed for this testing program, based on modifications to the MaxPERF® tool. For these coring operations, the modified MaxPERF® tool was run on jointed tubing in order to achieve better depth and steering control. At each of the four sampling depths, the tool was landed at the zone of interest and locked into place. The tool first milled through the casing, then drilled a core of the cement. These cores measured 9.5 mm (3/8 in.) in diameter and approximately 10 mm (0.4 in.) in length.

7.8 Conclusions

7.8.1 Synthesis of Existing Practices, Modelling and Laboratory Testing

The following is a summary of important conclusions drawn from the review of existing literature, regulations, and operational experience in CO_2-EOR and acid gas injection projects.

An essential first step in a well integrity assessment is to identify wells that will potentially be exposed to CO_2 or acidified formation fluid, or affected by injection-induced pressure increase, and rank them in order of importance for investigation.

Numerous key aspects of well construction and well history have been identified, many of which focus on the location and condition of cement plugs (if present) and the cement sheath (for cased wells).

For new wells in a prospective CO_2 storage site, probably the most important aspect of wellbore construction is the design and execution of the cementing program. Assessment, including logging of the cementing program performance itself, and a continuous monitoring program (for example, surface-casing vent flow or annular pressure buildup), are important follow-up tasks.

Wellbore failure in acid gas injection wells has generally not been due to zonal isolation failure, but rather to failure modes that are common to all well types (for example, packer or tubing failures), which can be addressed through operational planning and remediation programs.

The positioning and placement of abandonment plugs is of utmost importance, and squeezing perforations in conjunction with abandonment gives superior results, compared to plugs alone.

Regardless of the monitoring methods planned (for example, surface-casing vent flow, sustained annular pressure, soil gas monitoring, logging programs), the importance of obtaining baseline measurements prior to CO_2 injection cannot be overestimated.

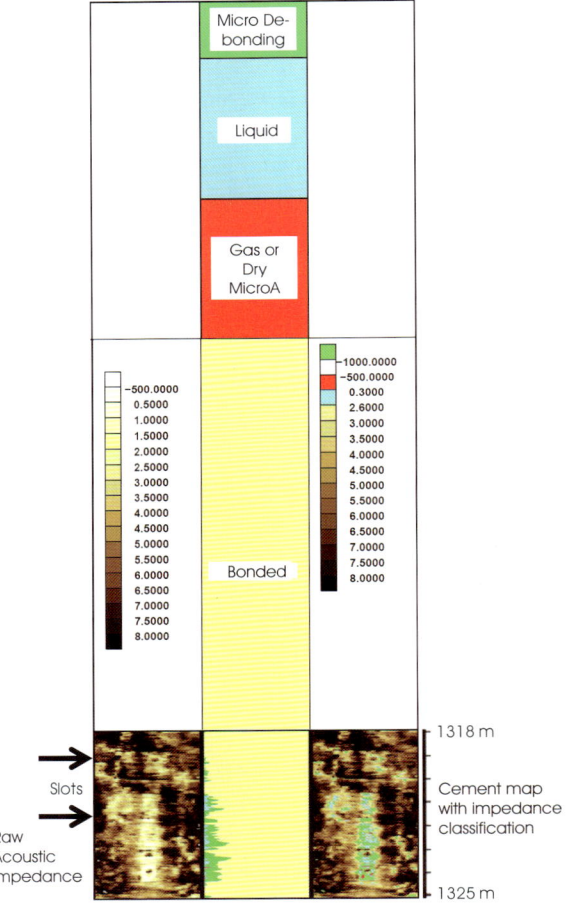

Figure 7-5. *Solid–liquid–gas map of the material behind the casing generated using the Schlumberger Isolation Scanner.*

7.8.2 **Field Testing**

The most significant information based on the well integrity field testing program is given below.

Outputs generated from the ultrasonic tool used for this work (Schlumberger Isolation Scanner) that were particularly useful were the solid–liquid–gas maps of the material behind the casing (Figure 7-5), and the imaging of the third interface (that is, the cement formation, Figure 7-6). Both of these outputs were used to select the depths for pressure transient testing and cement sampling.

The data generated by the sonic tool (Schlumberger Sonic Scanner) were useful for interpreting the top of the cement-filled annulus (approximately 1135 m, see Figure 7-7). However, a focused tool (such as the Weatherford Sector Bond Log, which had been run by Cenovus prior to suspending the well in 2009), could have been more useful (for example, for identifying channels in the cement sheath). Use of the sonic scanner may be advantageous in cases where sonic velocity measurements for the formation are more important than focused measurements of the cement sheath.

The combined use of sonic and ultrasonic logging tools for casing and cement evaluation was beneficial because it enabled a broad range of depths to be investigated. The electromagnetic casing inspection (Vertilog™) tool used for this work proved effective for accurately characterizing the location, size, and shape of casing defects, both internal and external. The results obtained from this tool suggested that numerous corrosion-induced flaws existed on the outer surface of the casing above the cement top but, most importantly, the casing condition below the cement top was very good (Figure 7-8).

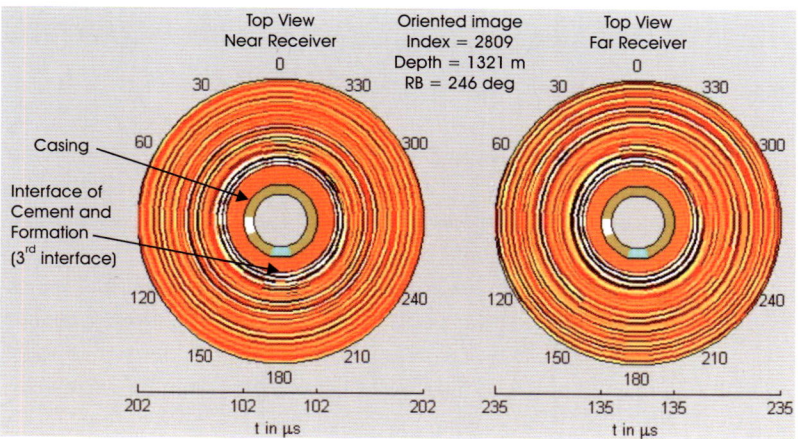

Figure 7-6. *Third interface map generated using the Schlumberger Isolation Scanner.*

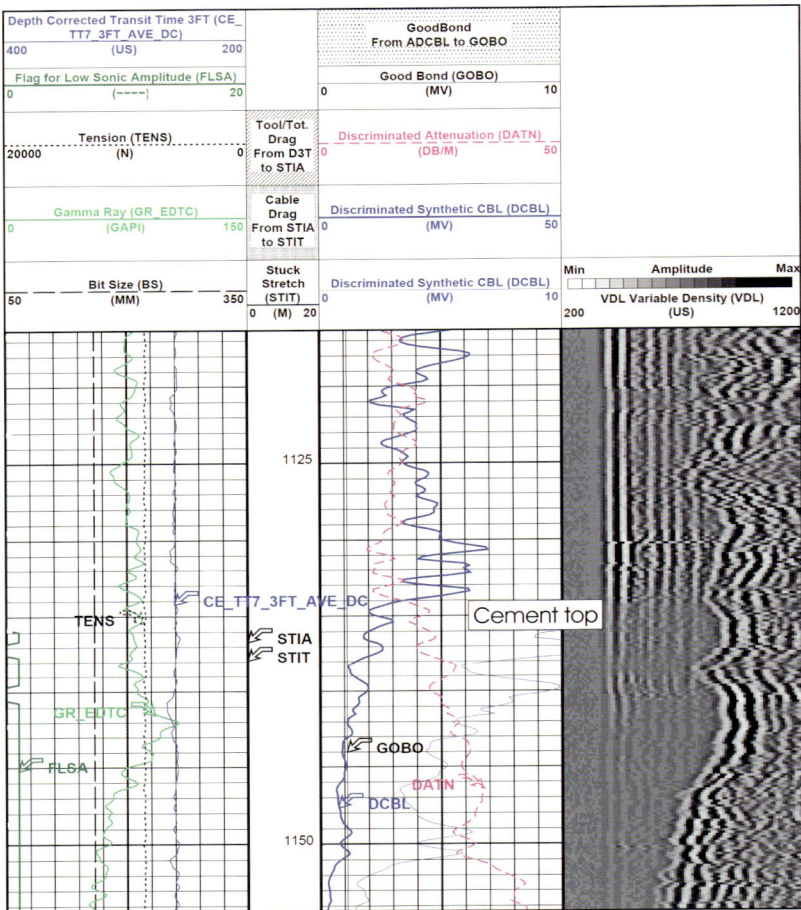

Figure 7-7. *Cement top identified using the Schlumberger Sonic Scanner.*

The combined analysis of casing and cement evaluation logs confirmed the commonly-held belief that a well-placed cement sheath provides effective protection against corrosion. The vast majority of casing defects that were logged were found above the cement top.

The pulsed neutron reservoir saturation tool (RST) was useful for demonstrating the absence of gas outside of the casing immediately above the retainer (Figure 7-9). Discrepancies were noted between the lithologies interpreted by the RST tool and lithologies interpreted by geologists from offset well core samples. In particular, the RST data suggested significant anhydrite contents in zones where actual anhydrite contents are believed to be low or non-existent, a discrepancy which has not yet been resolved. Porosity values interpreted by this tool provided useful data for the selection of porous zones suitable for cased-hole fluid sampling that was conducted in support of hydrogeological characterization activities.

The borehole camera was useful for confirming the location and conditions of the drilled perforations (Figure 7-10); however, it was only effective above the liquid level in the well after it had been partially blown dry. Multi-finger calipers also proved to be effective for locating the drilled perforations, regardless of the type(s) of wellbore fluids that were present, and should generally be sufficient for similar projects in the future.

The MaxPERF® tool proved to be a viable means of drilling damage-free perforations, with acceptable control of vertical position (that is, depth), radial position, and radial depth of penetration. A direct comparison of the transient data obtained through drilled perforations and conventional perforations was not possible in this test program, but is recommended as a follow-up activity.

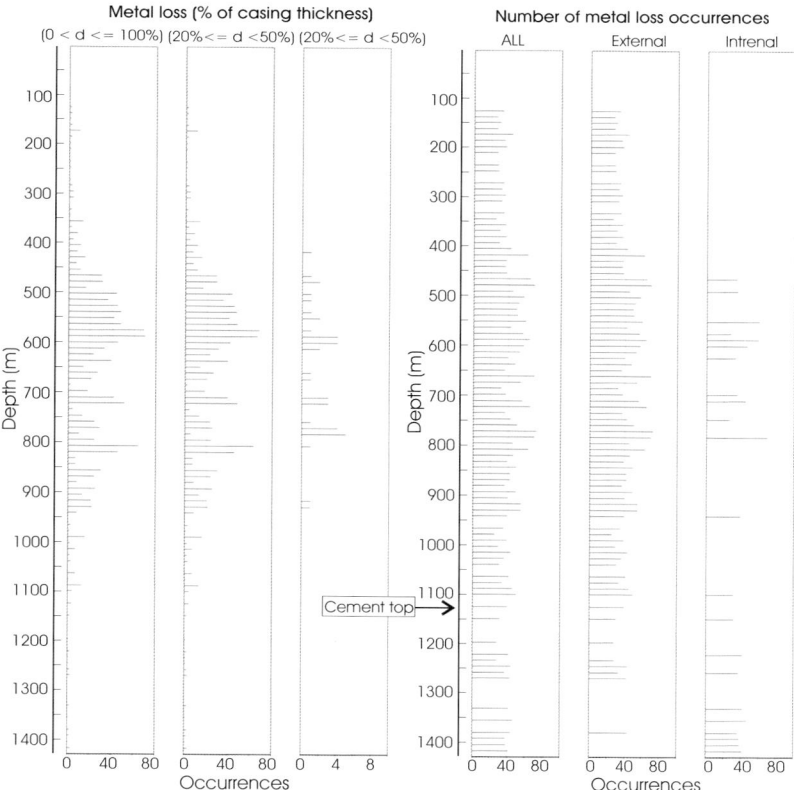

Figure 7-8. *Metal loss occurrences identified using the Vertilog.*

Pressure transient testing using the PTT was effective in one interval that was logged as free of casing and cement defects, and the results suggest that tight hydraulic conditions exist. The pressure transient testing demonstrated the tenuous balance that exists between tool complexity and tool reliability. Experience gained with the PTT (and other tools that were later used to execute modified pressure transient tests on different casing intervals) underscores the importance of real-time bottomhole pressure measurement, and the use of robust packers implemented in the simplest configuration possible.

The modified MaxPERF® tool proved to be a viable means of obtaining small-diameter cement samples, and for directly confirming the presence of cement behind casing. Remaining challenges include the need to achieve larger sample-size capabilities, and designing a better means of extruding samples from the core bit.

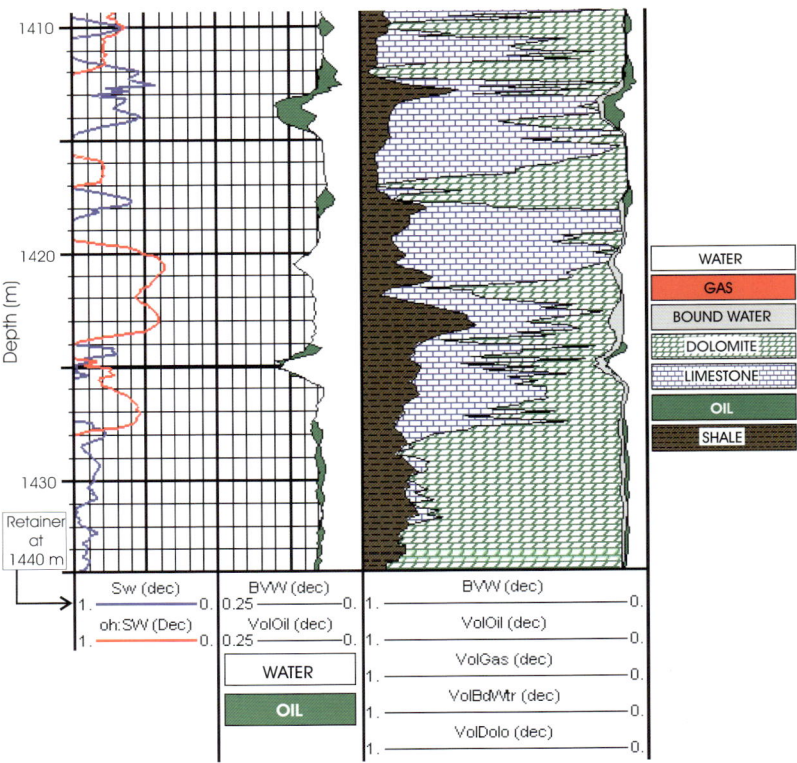

Figure 7-9. *RST log showing the absence of gas above the retainer.*

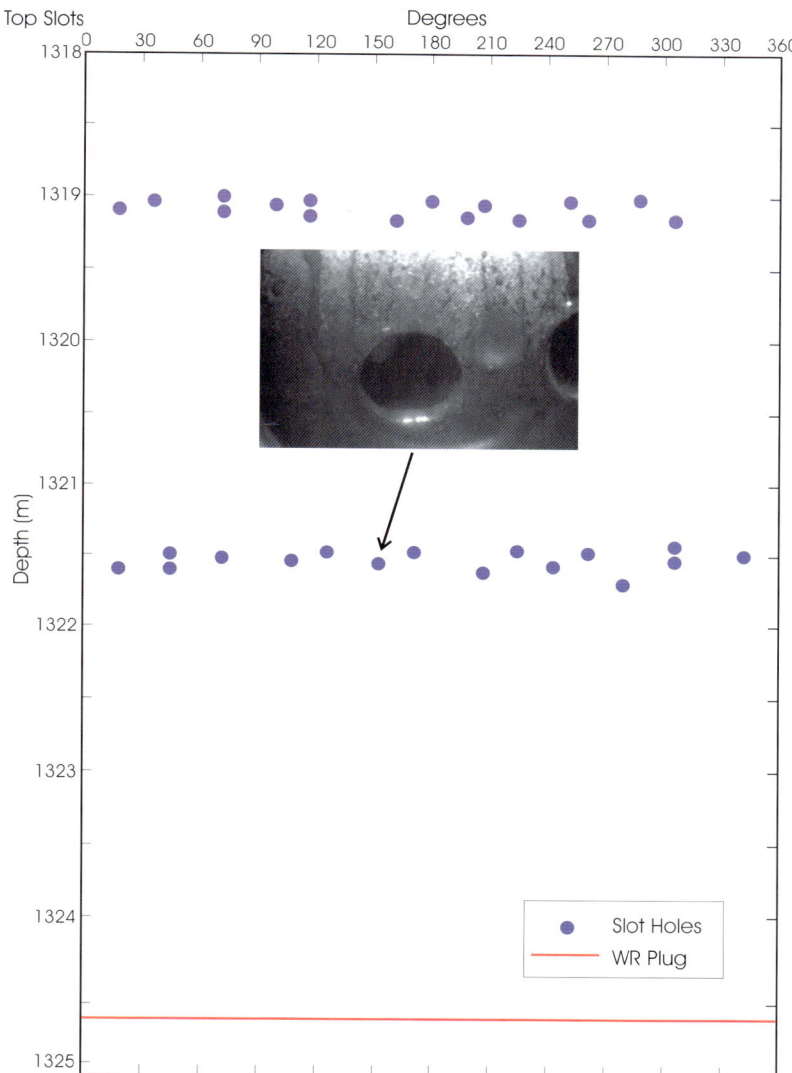

Figure 7-10. *Positions of two 'slots', each consisting of 16 perforations that were drilled through casing and cement using the MaxPERF® tool. Blue dots show drill locations, as interpreted from Weatherford multi-finger caliper. Inset shows an image acquired using the Expro borehole camera.*

A challenging aspect of virtually all downhole operations conducted during this program was accurate depth control. Depth control measures that proved effective included the setting of wireline retrievable plugs as reference markers, running casing collar locators, and using gamma ray traces on wireline logs for depth-correlation purposes. Problems were encountered when rig personnel failed to fully understand (or adhere to) the prescribed operational procedures, and when errors were made in the scale settings selected by logging personnel when providing field prints.

7.9 Recommendations

The purpose of the study was to identify issues related to well integrity that may arise in CO_2 storage projects. Much of the information presented here is based on a synthesis of relevant open literature, regulations, and a survey of CO_2-EOR and acid gas-EOR operators. The essential first step in an assessment of well integrity is to identify wells that may be exposed to CO_2 or acidified formation fluid, or affected by injection-induced pressure increase, and rank them in order of importance for investigation.

For existing wells, key information relates to the condition of cement plugs and the cement sheath (for cased wells). For new wells, probably the most important aspects of wellbore construction are the design, execution, and post-job assessment of the cementing program. Implementing a monitoring program is also important, and regardless of the monitoring methods planned, developing a program to obtain baseline measurements prior to CO_2 injection is essential. During well abandonments, the positioning and placement of plugs is of utmost importance. Squeezing perforations in conjunction with abandonment yields superior results compared to plugs alone.

A field testing and sampling program showed that the key logging-related information included the use of combined sonic and ultrasonic logging tools, and an electromagnetic casing inspection tool. These confirmed the commonly held belief that a well-placed cement sheath provides effective protection against corrosion. Two key practical facts emerged from the testing program: (1) stringent depth-control measures should be used during specialized downhole testing, and (2) it is important to take real-time bottomhole pressure measurements during cement-sheath pressure transient tests.

7.9.1 Well Integrity Assessment

Prior to development of a well integrity assessment program, the area must be identified in which wells may potentially be contacted by CO_2 or otherwise affected by injection operations. Wells should be evaluated prior to the start of injection, and the information should include, but not be limited to, the status of the well, its future function, and its relation to the storage complex (for example, does it penetrate the reservoir–cap rock interface?). Sources of this information may include historical aerial photographs, and interviews with such service providers as water well drillers and oil and gas service companies. They may also include such field investigations as electrical conductivity and magnetic surveys. Soil gas migration or vegetation impact surveys can identify wells leaking gas to the surface.

Once all of the wellbores have been identified, they should be ranked in order of importance for investigation, with wellbores that penetrate the cap rock over the storage complex assigned top priority. For all abandoned wells, the most important considerations relate to the status of existing cement plugs and sheaths. Clues to the status may be provided in the databases of regulators or operators, well drilling tour reports, deviation surveys, and open-hole logs.

7.9.2 New Well Design Considerations

Wellbore construction practices depend in large part on the composition of the injection stream, the content of CO_2, contaminants, and moisture content. This information is necessary to design the tubing and packer system that are to be employed. The design and execution of an effective, quality cement job is the most important aspect of wellbore construction. Surface casing provides additional protection over the life of the well, especially with respect to potable groundwater resources, and should be logged for cement evaluation.

Well trajectory can be an important factor in the incidence of wellbore leakage, due to poor centralization of casing. For directionally drilled wells, ideally, measures should be taken to obtain a vertical well trajectory over a significant proportion of the cap rock. Accumulation of water in tubulars, due to condensation in low-lying intervals of the wellbore, can occur and can cause corrosion which may lead to failure. Consideration of the relation of temperature as a function of depth, well angle, and dogleg severity, will help to determine if condensation is likely to be a problem.

Casing must be designed to be competent in the event of a failure of the tubing or packer, and so the annulus between the tubing and the casing should be continuously monitored during injection operations to ensure that injection ceases immediately in the event of a failure. Casing below a packer should be designed to withstand corrosion effects by the injection stream.

For wells used as CO_2 injectors, CO_2-resistant cement formulations should be used across the CO_2 injection zone. Although dry CO_2 may not significantly impact cement, water may already be present in the storage complex or may be introduced, thus creating an acidic environment with consequent negative impact on the cement sheath adjacent to the CO_2 injection interval.

It is strongly recommended that a surface-casing vent and valve be installed, and that the vent be left open at all times. Where regulation requires that vents be closed, continuous monitoring of the surface-casing vent is required to assist in early detection of casing failure or CO_2 leakage.

7.9.3 Well Remediation and Conversion

Rather than drilling new wells, pre-existing wells may be remediated or converted for CO_2 injection purposes. Following a risk assessment, a well might require remediation prior to the commencement of CO_2 injection, or remediation may be deferred depending on the time it takes for the CO_2 plume to reach the wellbore. Additional well monitoring may be required depending on the specific situation. Many older wells may be candidates as injectors but will not be cemented to the surface. It is recommended that if these wells are to be used for injection, remedial operations focus on the injection zone only.

A study of acid gas injection wells showed that it was the casing that failed due to external corrosion (rather than any other cause), usually due to uncemented intervals far above the injection zone. Therefore, isolation of the injection zone is critical, and should be confirmed through logging that can detect fluid or gas movement behind pipe. If needed, remedial operations should be conducted at the injection zone–cap rock interface, and the tubing–casing annulus continuously monitored. For injection pressures below hydrostatic pressure, positive pressure should be maintained on the annulus so that any failure of the casing, packer, or remedial perforations is immediately evident.

7.9.4 Well Abandonment Considerations

As with well remediation and conversion, cementing is critical. It is important for operators to take all possible future uses and liabilities into account, and abandon wells with a view to long-term containment.

For open-hole abandonment, the most important task is to ensure that competent plugs are set across the reservoir–cap rock interface and across shallow aquifers. It is also highly desirable to set plugs across intermediate porous zones, taking care to ensure that these plugs cover the interfaces between these zones and the non-porous zones (seals) adjacent to them. Prior to the setting of cement plugs, mud should be conditioned to reduce cement channeling. A gamma ray–neutron log or an electric log with a caliper (possibly both) will assist in confirming cement placement.

Evaluation of the cement sheath behind the casing is the first step in determining the requirements for abandoning a cased hole. Tools that assess conditions around the full circumference of the casing are preferred, as are those that can distinguish low-density cements. Access to the casing-hole annulus can be gained through a variety of methods. It is recommended that methods be used that leave many future options available, in the event that the initial remediation fails. Operators should also squeeze all perforated intervals when abandoning any well. It is recommended that previous abandonment plugs should be pressure-tested before moving up-hole to conduct further operations. Ideally, plugs should be set across intervals corresponding to those where open-hole abandonment plugs would be set, and heavy, non-corrosive mud should be left in place between plugs. Where casing is free to the surface from an external cement top, consideration should be given to cutting and pulling casing, then following open-hole abandonment procedures for this upper portion of the well. Data from Alberta suggest that wells abandoned open-hole have fewer incidences of leakage than cased wells. If this option is viable, it is important to ensure that all operations conducted below the cut point have been successful in creating isolation.

7.9.5 Well Integrity Monitoring

Wells near a proposed storage project can provide important information about pre-existing leak paths and historical information regarding leakage. If a well has not been abandoned, cut and capped, it should be tested for surface-casing vent flow, and gas samples analysed. Soil gas migration testing should be conducted near the well, and if leakage is detected, the source zone should be determined. This may be accomplished through gas analysis, cement evaluation logging, analysis of open-hole logs, and identification of wells with external casing corrosion. If a pre-existing leakage source has been identified, future monitoring of surface-casing vent flow and gas migration changes will help to distinguish leakage of injected CO_2. This evaluation will also identify casing failures.

Baseline information should be gathered from new injection wells, including, but not limited to, (1) logs for zonal isolation, cement evaluation, and casing inspection, (2) tests for soil gas migration, and (3) a surface-casing vent flow test, both for hydrocarbons and CO_2, with carbon isotopes on all gas components containing carbon. Temperature and radioactive tracer logs, or possibly oxygen activation logs, should be run to confirm initial zonal isolation, prior to injection. The well should be logged in a static state prior to completion, to determine if there is cross-flow due to natural pressure drive and poor cementing, and then logged under injection conditions. After the well has been completed and stimulated, a comparative suite of zonal isolation logs should be run. If only CO_2 is to be injected, then the logs should be run using CO_2 as the test gas/fluid.

Monitoring the cap rock or the zone immediately above it is strongly recommended. The drilling of dedicated observation wells, or the use of pre-existing observation wells, may be used for this purpose.

Monitoring of the wellbore for potential leakage during the life of the CO_2 injection project is highly dependent on the duration, frequency, and quality of the pre-injection data. Monitoring programs should closely follow the initial testing that was conducted, so that meaningful comparisons can be made. Monitoring at the leading edge of the CO_2 plume and pressure pulse should be a priority.

Injection well monitoring should include continuous tubing and annuli monitoring for pressure and temperature. Injection well annuli should have some positive pressure applied, so that failures can be easily identified if they occur across sub-pressured zones. At regular intervals (for example, every 5 years) a new suite of comparison logs should be run for casing inspection, cement evaluation, and zonal isolation. These logs should be run in similar conditions to the original logging suites, so that easy comparisons can be made. With the development of tools that can detect smaller variations in casing defects and cement quality, care must be taken to ensure that the change that is noted is actually due to a change in the wellbore condition and not due to enhanced technical capability. Similarly, observation wells should have comparison log suites run at regular intervals.

Post-closure monitoring will depend greatly on the project size, depth of the storage complex, and proximity to populations. It is expected that dedicated observation wells will be left in place for a period of time, probably of the order of 20–30 a. These wells will be a combination of deep observation wells that monitor the cap rock and the zone immediately above the injection zone, and groundwater monitoring wells. To date, there are no projects in the post-closure stage or nearing closure, so there is no pertinent information.

8

RISK ASSESSMENT

Summary

Risk analysis is the process of identifying, evaluating, and mitigating risks. Well-structured, systematic methods for achieving credible analyses have been developed and continue to evolve. The techniques of risk analysis have advanced substantially during the more than 10 years that this project has been in operation, and so have the data that we have to work with.

In the case of long-term CO_2 storage, the risks to be assessed include long-term areas of concern, such as human health and safety, the environment, and the security of the biosphere, as well as elements related to system performance. Also evaluated were factors that might affect future operations (including continuing availability of CO_2, reservoir capacity, and changing economic considerations).

It is impossible to conduct detailed performance modelling for every conceivable risk, but great attention should be paid to those risks which might pose the most significant consequences, as well as those that pose the greatest uncertainty to the project being able to achieve its objectives.

Formal analysis techniques were used in evaluating data. The data included information acquired from the project itself, the input of experts, and consultation with landowners and other stakeholders. Stakeholders involvement early in the process is very important.

Recommendations arising out this research begin on page 319.

Author: R. Chalaturnyk (University of Alberta, Edmonton, Alberta).

8.1 **Introduction**

8.1.1 **Overview of Risk Assessment**

Risk is a condition resulting from the prospect of an event occurring, and the magnitude of its consequences. Therefore, risk is an intrinsic combination of (1) the likelihood of an event and its associated consequences occurring (this incorporates consideration of the frequency of the event and the probability of the consequences occurring each time the event occurs) and (2) the magnitude of potential consequences of the event.

Risk analysis is the iterative process of identifying, evaluating, and mitigating risks. It consists of several steps, including risk assessment, risk management, and risk communication (Rechard, 1999; IEA Greenhouse Gas R&D Programme, 2009). Risk assessment embodies the overall process of risk identification, risk analysis, and risk evaluation.

Risk identification seeks to identify the risks to be managed. Comprehensive identification using a well-structured systematic process is critical, because a potential risk that is not identified at this stage will be excluded from further analysis. Identification should include all risks, whether or not they are under the control of the project operator. The aim is to generate a comprehensive list of events which might affect each element of the storage project. These are then considered in more detail to identify what can happen. Having identified a list of events, we then need to consider possible causes and scenarios. There are many ways an event can be initiated, and it is important that no significant causes are omitted. Approaches used to identify risks include checklists, judgements based on experience and records, flow charts, brainstorming, systems analysis, scenario analysis, and systems engineering techniques. The approach used depends on the nature of the activities under review and the types of risk.

Within the context of the Weyburn project, risk analysis for CO_2 geological storage involved the systematic use of project information (1) to identify sources and magnitudes of potential CO_2 leaks and (2) to estimate the probability of their occurring, and to assess the magnitude of adverse impacts or consequences due to the leaks. Some of the potential consequences that are associated with loss of CO_2 storage integrity are related to (1) public safety and health, (2) environmental (ecosystem) safety, (3) greenhouse gas emissions to the atmosphere, (4) damage to natural resources (for example, water and hydrocarbons), and (5) financial loss for investors or insurers.

Risk evaluation examines the acceptability of these risks with respect to the needs, issues, and concerns of stakeholders. Consequently, risk analyses conducted early in the project focused on assessing storage system performance. This risk assessment process will ultimately mature into a framework that considers (1) the social, economic, and political factors associated with geological storage, (2) the specific risks associated with the geological storage reservoir, and (3) the effectiveness of remedial actions that might be taken to minimize both near-term and long-term probabilities, and consequences arising from CO_2 leakage. Equally important, this process will provide the basis for communications about the existence, nature, form, magnitude, and acceptability of risks associated with the geological storage of CO_2.

8.1.2 Background

The challenge in reporting on a project that has spanned more than a decade is that over that period, advances in approaches and technology have most certainly progressed. What was relevant 12 years ago is potentially less relevant today, and the body of literature upon which to draw in crafting research programs has changed. Not surprisingly, the world of risk assessment grew substantially over the period from 2000–2012, and research programs undertaken within the risk assessment theme reflect this rapid growth. The risk assessment activities undertaken in Phase 1 of the project led the industry in risk assessment for the geological storage of CO_2. The activities were built on risk assessment methods developed in the nuclear waste storage industry, and slowly evolved into techniques relevant to the requirements of the geological storage of CO_2 and to the specifics of the project area. The research undertaken in the Final Phase of the project was able to leverage and integrate much of the work completed in Phase 1, but also to take advantage of the significant international research that had been conducted in the area of CO_2 geological storage over the latter part of the project timeline.

To provide a perspective of the evolution of risk assessment over the 12-year period of the project, Compendex was searched for a subject/title/abstract using 'risk assessment' as a keyword. Compendex is the most comprehensive bibliographic database of scientific and technical engineering research available, and covers all engineering disciplines. More than 18 000 references were found for the period 1998–2004, and 37 500 for the period 2005–2011, an increase of 100%. But using a combination of 'risk assessment' and 'geological storage of carbon dioxide' as search keywords, resulted in an increase of >400% in the number of publications dedicated to this topic between the period 1998–2004 and the period 2005–2011. The breadth of knowledge in risk assessment methods clearly evolved over the last decade, but risk assessment applied to the geological storage of CO_2 experienced an even greater pace of development.

It is important, therefore, for the reader of this chapter to place the following discussion of risk assessment research conducted within the project over the period 2000–2012, in the context of these rapid changes in the world of risk assessment.

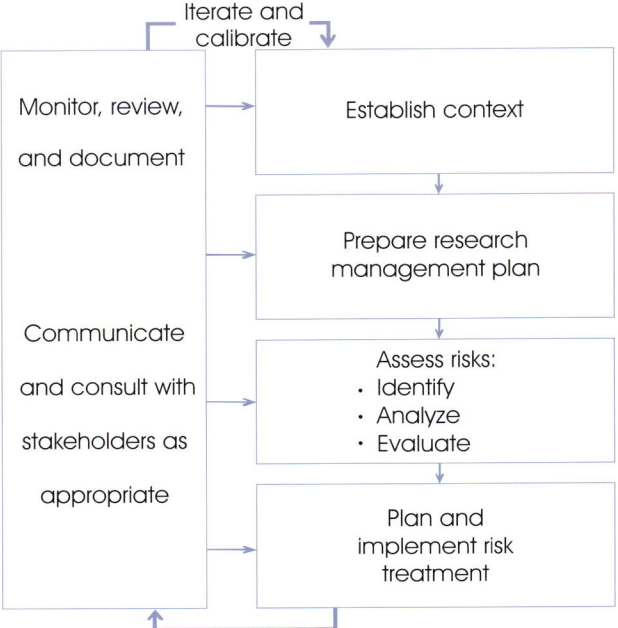

Figure 8-1. *Schematic diagram of the risk management process for CO_2 geological storage projects (CSA Z741-11).*

8.1.3 Risk Management Process

Risk assessment research for the project is described within the framework of the risk management process outlined in the impending release of the CAN/CSA Z741-11 Standard on the Geological Storage of Carbon Dioxide. The risk assessment research activities within the project were primarily associated with three elements of the risk management process outlined in Figure 8-1, namely, (1) establish the context, (2) assess the risks, and (3) consult with stakeholders.

The research reported here was directed at the geological storage of CO_2 within the framework of an operational CO_2-EOR project. The goal was to inform CO_2 storage proponents at CO_2-EOR projects of the risk assessment practices utilized during the Weyburn project. Therefore, it was assumed that an operational hazard assessment and risk management plan already existed for the operational aspects of any CO_2-EOR project, because this would almost certainly be a regulatory requirement in any jurisdiction. It was also assumed that the CO_2-EOR risk assessments focused on operational aspects of the project (for example, wellheads, compressors, facilities, and risks to personnel around facilities and wells). Experience in the CCS community to date suggests that this 'operationally' focused risk assessment may not be sufficient to meet the regulatory requirements for long-term CO_2 storage at a CO_2-EOR site. It should be emphasized, again, that the risk assessment research program at Weyburn focused on the long-term storage of CO_2 (that is, for a CO_2-EOR project that eventually ceases commercial activities and converts to a storage project with the intention of transferring responsibility and liability to a designated authority). The research did not examine issues associated with storage credit application for CO_2-EOR, where there is a difference between CO_2 injected and CO_2 recycled.

8.1.4 Context for Risk Assessment of CO_2 Storage at Weyburn

Establishing the context for any risk assessment involves (1) describing the objectives of the project, (2) developing a conceptual model of the system used for CO_2 storage, and (3) indicating elements of concern for the project, including human health and safety, the environment, and elements related to system performance such as injectivity, capacity, and containment. The following section describes the objectives of the risk assessment research, the scope and parameters of the study, the identification of the data sources used, the approaches utilized in the Weyburn project, and the criteria against which risks are to be evaluated.

For the Weyburn project, the temporal context for risk assessment activities was an exercise for the future, since the CO_2-EOR operation was a commercial project throughout the study and remains so today. Consequently, it was assumed that at some point in the future, CO_2 storage considerations would be applied to the project, either for CO_2 remaining in the reservoir or for direct conversion to a CO_2 storage project.

8.1.5 Objectives

Research on the application of risk assessment methods (qualitative, semi-qualitative, and quantitative) related to the geological storage of CO_2 within a CO_2-EOR project has been a key theme. The goal was the completion of a credible assessment of the permanent containment of injected CO_2 as determined by formal risk analysis techniques, including long-term predictive reservoir simulations. The outcome from this work was intended to help answer questions by regulatory bodies and community stakeholders as to the security of large volume CO_2 storage in the Williston Basin.

At the outset, deterministic and probabilistic risk analysis techniques were employed to achieve two objectives: (1) to identify and evaluate the risks associated with geological storage of CO_2, and (2) to assess the performance and ability of the Weyburn reservoir to securely store CO_2.

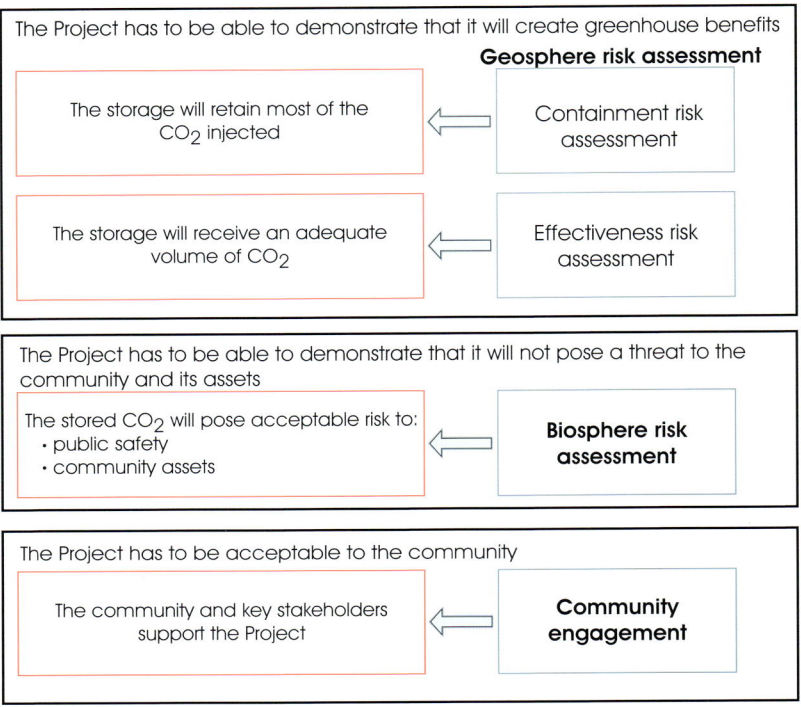

Figure 8-2. *Components and key performance indicators developed for risk assessment in the Final Phase of the project.*

These early assessments were followed by an integrated geosphere/biosphere risk assessment that formally incorporated the consequences of CO_2 leakage. This was done by utilizing the RISQUE modelling method (Bowden and Rigg, 2004), which has been applied to assess CO_2 risk at a number of sites in Australia and Algeria. Figure 8-2 shows the main key performance indicators that were set for the full field risk assessment for the Final Phase of the Weyburn project.

Geosphere Risk Assessment Objectives

1. Assess the risk of CO_2 escaping from the storage complex and reaching the biosphere (termed 'containment risk').
2. Assess the effectiveness of storing the target volume of CO_2 within the confines of the Weyburn unit (termed 'effectiveness risk').
3. Identify further research that is required to address areas of high uncertainty.
4. Identify controls required to reduce risk and areas where further research is required to develop the controls.

Biosphere Risk Assessment Objectives
1. Identify likely movement of CO_2 within various parts of the biosphere once it reaches the biosphere.
2. Determine the potential impacts on flora, fauna, people, and property.

8.1.6 Scope and Parameters

As with many engineered and natural systems, the designated space for CO_2 storage is very complex, and henceforth is simply referred to as the system. A system model is defined below, with three main components.

BIOSPHERE is defined for this project as the air, soils, rivers, lakes and groundwater, and everything contained therein, that lie above the base of the Cretaceous–Tertiary aquifers. It is where the interaction of CO_2 with potable groundwater and biota, and human health risks are assessed. The base of the biosphere occurs at a defined depth of 300 m in the project area.

GEOSPHERE is defined for the project as all the rocks and contained fluids (oil, gas, formation water) that occur below the biosphere. UPPER GEOSPHERE is the zone above the reservoir (that is, above the Midale Beds of the Charles Formation, and below the biosphere). LOWER GEOSPHERE is defined as all strata below the upper geosphere, and the wells therein, which consist of the wellbore, the annulus (at least partially cement filled), the cement seal plug, and the steel casing.

Figure 8-3 is a schematic representation of the elements contained within the system model, which was subdivided into seven aquitards and six aquifers.

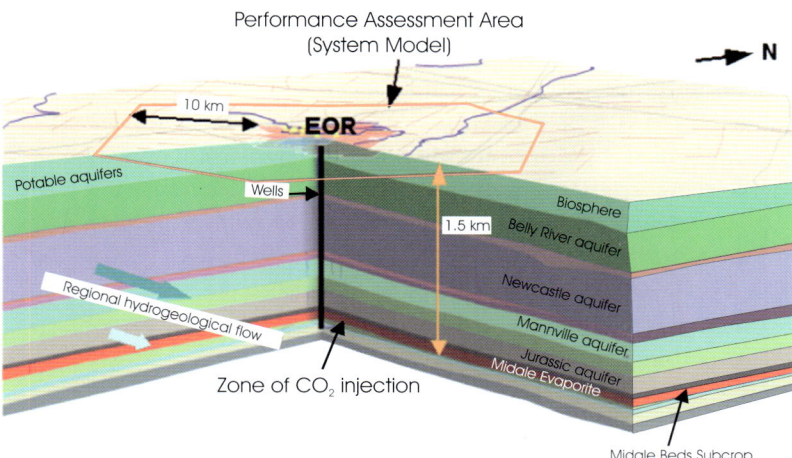

Figure 8-3. *Diagrammatic representation of the system model consisting of the Weyburn EOR reservoir (showing the zone of CO_2 injection), and overlying and underlying geological layers, wells and the biosphere.*

There is an important difference between CO_2 storage in strata with saline formation water and storage in oil or gas reservoirs, such as at Weyburn. In the former, there may be few wells, whereas in the latter many wells are more likely. For example, when the planned 75-well pattern for the Weyburn CO_2-EOR operation is completed, it is expected that there will be about 1070 wells in the EOR project area. In the 10-km zone surrounding the EOR project, the number of wells of all types will total more than 2200, with the majority being vertical wells.

The risk assessment research carried out in the project, related to potential migration along wells, was focused only on the potential for migration along abandoned wells. So for the purposes of the study, the wellbore system was chosen that included all the major components of a wellbore.

1. Configuration of the internal abandonment (bridge plug and cement).
2. Corrosion of the casing.
3. Presence of annular cement (between casing and formation).
4. Excavation of damage zones in the near-wellbore region.
5. All interfaces between these components.

A good understanding of these mechanisms can only be obtained by following a comprehensive review of the wellbore environment during the different stages in the life of the well. The main stages, from the point of view of hydraulic integrity, can be classified as drilling, completion, well history, and abandonment. The influence of each one of these stages on the final hydraulic integrity of the well can be further classified into four major groups.

1. Geomechanical damage: any stress-induced changes in the hydraulic conductivity of materials within the wellbore system.
2. Hydrochemical damage: any alteration of hydraulic conductivity in the near-well formation region – typically called formation damage.
3. Mud removal: the efficiency with which mud is removed from the annulus during cementing operations, specifically addressing the development of mud channels and their impact on hydraulic integrity of the wellbore system.
4. Cement deterioration damage: porosity alteration due to geochemical processes under in situ conditions and in the presence of CO_2.

8.2 Risk Evaluation Criteria
8.2.1 Geosphere

Key assumptions pertinent to the geosphere risk assessment were as follows.

1. A total of 30 Mt of CO_2 is to be stored in the reservoir over the project life.
2. The project life is 1000 years, following the conclusion of EOR.

3. Expected minor losses from primary containment through the seal, fractures, lateral migration, and wells modelled during Phase 1 were considered as losses with 100% likelihood of occurring.

The following criteria were used to evaluate the acceptability of risks for the assessment of the geosphere.

Containment Assessment
1. 80% confidence that 99% of the CO_2 will be contained after 1000 years.
2. <5% chance that one single feature, event, or process, will cause >1% to escape from the geosphere over 1000 years.

Effectiveness Assessment
1. <20% chance that the Weyburn unit (the Midale Beds and the Marly unit in the Frobisher Beds within the extent of the unitized Weyburn Field) will not be able to store 30 Mt.
2. <5% chance that one single feature, event, or process, will prevent the Weyburn unit from accepting 30 Mt of CO_2.

8.2.2 **Biosphere**

Biosphere risk assessment addressed the requirement that the project would not pose a threat to the community, or to community assets in the biosphere, including the risk to public safety and to community assets. Technical disciplines including hydrogeology, hydrochemistry, hydrology, air modelling, human health risk assessment, soil science, agriculture, economics, biology, and ecology were engaged to assess biosphere surface and near-surface impacts. Biosphere assessment also requires a strong understanding of social and community values, needs, and expectations, and potential impacts on lifestyles, recreation, public amenity, heritage, First Nation assets, infrastructure, industry, and local government.

The method adopted to assess biosphere risk was specifically developed for the project. However, the approach can be readily applied to projects elsewhere. It is effectively the same as the approach used to assess environmental impacts of projects or activities, and it involves the identification and understanding of the scale of predicted and potential effects of an activity on the wider environment (that is, the natural environment, communities, and economy). Figure 8-4 is a flowchart that summarizes the process.

Technical Assessment

The technical assessment consists of identifying where and how much CO_2 could enter the biosphere via the various pathways, and evaluating the potential

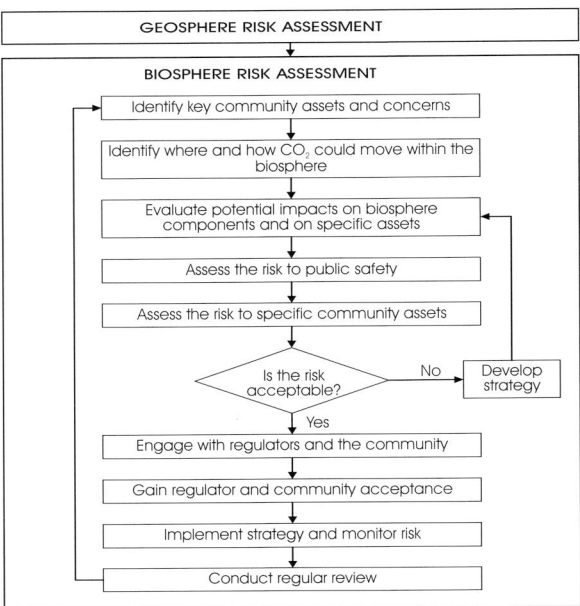

Figure 8-4. *Biosphere risk assessment and management approach.*

impacts on specific assets. Semi-quantitative risk analysis was applied to determine the highest risk pathways, the components of the biosphere that were the main contributors to the risk, and the assets that were most at risk.

Determination of the risk to public safety and to community assets was a key focus for the biosphere risk assessment. This included evaluation of the potential loss of life and/or serious injury due to events related to the CO_2 storage project, and was concerned with (1) the potential pathways of CO_2 movement into environments occupied by people, (2) the potential release rates, (3) the number of people who could be injured, and (4) the severity of potential impacts.

Asset risk assessment evaluated the potential impacts on assets that were of value to the community. These assets are typically livestock, agriculture, lifestyle, industry, wildlife, lakes, rivers, groundwater, native animals, and vegetation. The process evaluated (1) the potential pathways of CO_2 movement, (2) the potential contact with specified assets, and (3) the type and scale of potential impacts on the assets.

Asset risk assessment may be considered from two perspectives. First, by identification of the pathways which pose the highest risk to assets. This type of cause-effect assessment is usually of most interest to project proponents, who use the results to develop strategic plans to reduce risk to assets. Second, by identification of the assets that are most at risk from the proposed activities. This

Environmental Components	Consequences to Assets
Souris River	Damage to property and infrastructure
Other rivers	Environment – damage to ecosystem function, habitat, community assemblages, species
Permanent lakes, intermittent lakes, and dug-outs	Social – damage to amenity, recreation, sensory perception, quality, heritage, and First Nations heritage
Soils	Economic – damage to agriculture, oil and gas, tourism
Aquifers (especially with potable groundwater)	Impact on public health and safety, causing illness, injuries and fatalities

Table 8-1. *Classification of community assets.*

type of assessment is usually of most interest to those who need to understand the potential impact of a project on assets that are of specific value.

A stakeholder consultation meeting, held in the town of Weyburn, was used to identify the social and environmental assets of value and interest to the local community. Concerns that the stakeholders had about future CO_2 storage at Weyburn were also voiced. Table 8-1 summarizes the classification of environmental components and consequences to assets that was developed from the outcome from the meeting and formed the basis of the biosphere risk assessment.

For those changes arising from CO_2 flows that were considered significant, the consequences associated with the flows were assessed using the semi-quantitative risk assessment matrix presented in Table 8-2. In Table 8-2, only two Consequence Levels are shown to illustrate the method. Three other Consequence Levels are as follows: Social (*Amenity/Recreation, Amenity/Perception, Heritage, First Nations Heritage*), Economic (*Agriculture, Oil and Gas, Tourism*) and Public Health and Safety (*Minor Injury/Illness, Major Injury/Illness, Fatality/Disability*). The matrix was originally designed with the assistance of a range of experts from outside the project, with expertise in social, environmental, public safety, and economic impacts, so that the descriptions for each level of consequence represent similar levels of severity or impact. Advice from these experts indicated that the descriptions of changes to the consequences across each level represented approximately one order of magnitude change in severity of impact. Each level of consequence is, therefore, represented as one order of magnitude greater than the previous level in the consequence matrix. For the consequence descriptors below 'Extreme', the level of consequence can be represented as a 1 (or 0.1, 10, etc.) or a 3 (or 0.3, 30, etc.). This enables a distinction between consequences within one category which sit at the lower or higher end of the scale.

Consequence Level	Negligible	Minor
	Minimal impact on some communities. Some impact on small number of individuals.	Low level of impact on some communities, or high impact on small number of individuals.
	0.1–0.3	1–3
Property/Infrastructure		
Repair/replace cost	Up to $0.1 million	$0.1 million to $1.0 million
Environmental		
Ecosystem Function	Change within natural variability.	Measurable changes but no change in function. Recovery in less than 1 year.
Habitat	Disturbance within natural variability. Less than 1% affected.	1% to 5% affected in major way. Re-established in less than 1 year.
Species	Not detectable outside natural variation.	Detectable changes but populations remain viable. Recovery in less than 1 year.

Table 8-2. *Semi-quantitative Biosphere Consequence Severity Evaluation Table (only two selected levels shown for illustration).*

8.3 Assessing the Risks for CO_2 Storage

8.3.1 Developing a Risk Register

Compiling a risk register is an important part of risk management and is site-specific. If a risk register already exists (as for the CO_2-EOR project), then the objective is to ensure that it is updated with respect to operations associated with a CO_2 storage project. To establish or update a risk register, the following tasks should be performed.

1. Risk identification: Perform a structured threat-identification process to ensure that the risks recorded in the register are representative, and include all significant risk scenarios.
2. 'Owner' assignment: Assign an 'owner' to each risk, with responsibility and authority to manage that risk.
3. Risk analysis: Conduct a process for understanding the nature and level of risk.

Moderate	Major	Extreme
High level of impact on some communities, or moderate impact on communities area-wide.	High level of impact on communities area-wide.	High impact province-wide.
10–30	100–300	1000

Property/Infrastructure cont.

$1.0 million to $10 million	$10 million to $100 million	$100 million to >$1 billion

Environmental cont.

Measurable changes without loss of function or introduction of new species. Recovery in 1 to 2 years.	Measurable changes to ecosystem and function. Recovery (within historical natural variability) in 3 to 10 years.	Long-term impact and possibly irreversible impact on one or more ecosystem functions. Recovery, if at all, more than 10 years.
5% to 30% of habitat affected in major way. Re-establishment in 1 to 2 years.	30% to 90% of habitat affected in major way. Re-establishment in 3 to 10 years.	Greater than 90% of habitat affected in major way or removed. Re-establishment, if at all, more than 10 years.
Detectable change to population but no impact on viability. Recovery in 1 to 2 years.	Detectable change in population size, and impact on population viability. Recovery (within historical natural variability) in 3 to 10 years.	Local extinctions imminent, or population no longer viable. Recovery, if at all, more than 10 years.

Table 8-2. *cont.*

4. Risk evaluation: Assess the probability of occurrence and the severity of potential consequences for each individual risk, and evaluate the corresponding risk level. The risk level can be determined qualitatively or quantitatively, depending on the amount and quality of data available. The risks should be ranked as insignificant, contingent acceptable, or unacceptable.
    A. Insignificant risks are broadly regarded as acceptable.
    B. Contingent acceptable risks may be accepted, provided the cost and effort required for further treatment is disproportionate to the risk reduction gained.
    C. Unacceptable risks are risks that cannot be accepted, except under extraordinary circumstances, and represent a 'showstopper' for the project unless the risk level can be adequately reduced through implementation of additional action.

5. Safeguard identification: For all risks not deemed insignificant, identify the options for risk treatment that would reduce the risk to a broadly acceptable level.
6. Safeguard evaluation: Evaluate the potential risk-reducing effect of identified safeguards and, if relevant, of selected combinations of safeguards.

For transparency and repeatability, it is important that the (risk identification) process, and the reasoning applied to evaluate risks and safeguard performance, be comprehensively documented. This should include documenting risks that have been considered, but considered negligible, or for some other reason are not included in the risk register. The following sections provide a summary of the research activities undertaken within the project and which help develop the processes associated with risk identification, risk analysis, and risk evaluation.

8.3.2 **Identification of Risks**

Two approaches were used to support risk identification. One involved the use of a list of Features, Events, and Processes (FEPs) to trigger recognition of risks for the project. The other approach was based on a scenario analysis, in which a base case scenario was defined, and possible deviations from it were identified and analysed.

FEP-based Risk Identification

In 2001, a Weyburn-specific FEP list was generated from a segment of an FEP list prepared for radioactive waste disposal, but containing FEPs that are relevant to the geosphere: that is, excluding all FEPs specific to radioactive waste and to the engineered barriers of such a disposal system. FEPs specific to CO_2 storage in the Weyburn reservoir were added to this list. Table 8-3 lists major groups of FEPs related to evaluation of the Weyburn CO_2 storage complex. This table provided the initial framework for assessment calculations performed in the deterministic–probabilistic segments of the risk assessment.

In the FEP database, each FEP has a text description and an associated discussion of its relevance to long-term performance of the storage system. Key references from the published literature are included in the description of each FEP, to enable retrieval of more detailed information. The database is internet-enabled, and incorporates hyperlinks to other relevant sources of information (such as reports, websites, maps, photographs, and videos), and it is searchable in a variety of ways, with the potential to provide a 'knowledge base' for the geological storage of carbon dioxide. Note that the FEPs in the database should not be considered in isolation, because they are interrelated. Each of the FEPs may affect the 'System' by influencing another FEP in some way, or by causing a more specific impact on or interaction with another FEP.

System FEPs	Non-System FEPs
Rock properties	Earthquakes
Hydrogeological properties	Cross-formational flow
Geochemical	Hydraulic fracturing
CO_2 properties and transport	EOR-induced seismicity
Other gases	Climate change
Geology	Regulation changes
Abandoned Wells	

Table 8-3. *Summary of FEPs relevant to the Weyburn CO_2 storage system.*

8.3.3 Scenario-based Risk Identification
Base Scenario
The base scenario allows examination of the expected evolution of the storage system itself. For the project, the base scenario referred to the migration of CO_2 from the 75-well pattern of the Weyburn reservoir, both within and beyond the primary strata (Midale Beds) into which the CO_2 was injected.

The elements of the base scenario developed for the project are outlined below. The main reservoir area is that part undergoing EOR using injected CO_2.

1. The geosphere zone is the area within 10 km of the main reservoir.
2. The biosphere extends to a depth of 300 m below ground surface, and includes soil, surface water, and the atmosphere, as well as the contained flora and fauna.
3. The cap rock (the Midale Evaporite) may have natural fractures or discontinuities, but all are isolated or sealed, and cap rock integrity is not impaired.
4. The time-frame for risk analyses starts from the inception of EOR using injected CO_2. Simulations conducted to support the risk analyses were run to 5000 years.
5. All wells in the system model area have been plugged and abandoned following standard procedures applicable at the time of abandonment.

The base scenario includes consideration of, but is not limited to, processes such as hydrodynamics, geochemistry, buoyancy, and density-driven flow, dissolution of CO_2 in water and residual oil, and pressure-temperature changes occurring within formations.

Alternative Scenarios
The following alternative scenarios were identified, and are intended to reflect future possible deviations from the base scenario, and the manner in which the system model area may be affected

1. Engineering options for EOR, such as greater reservoir pressures for CO_2 injection, where overpressurization and cap rock fractures are possible problems, and the employment of a water flush at the end of EOR, which may decrease CO_2 storage potential.
2. Well abandonment options, with emphasis on the impact of different abandonment strategies and their influence on long-term risk of CO_2 leakage.
3. Impact of salt dissolution and subsidence that may lead to development of fractures.
4. Fault activation or re-activation creating a new and fast CO_2 transport pathway.
5. Tectonic activity affecting integrity of bounding seals to contain CO_2.
6. Human intrusion into the reservoir, impacting the containment characteristics.

8.4 Analysis of Risks

The complexity of the system model and its components was managed through application of a rigorous and formal systems analysis approach. To assist in identifying the processes that could be relevant to the evolution or performance of the system, a list of features, events, and processes was developed. Their evaluation and interactions led to a description of how the system might evolve over the time-frame of the assessment. This description formed the foundation for the development of a scenario that described how the system may be *expected* to evolve – the base scenario – in the far future, and other scenarios that describe alternative but feasible futures. Performance assessment early in the project focused primarily on the base scenario and employed deterministic and probabilistic analyses. Risk assessments conducted in the Final Phase of the project employed a systematic, quantitative process, using a formal group of experts to provide judgements that were incorporated into a quantitative risk analysis and management framework. These two approaches to risk analysis are briefly reviewed below, with the emphasis on quantifying the potential for migration of CO_2 above the reservoir.

8.4.1 Deterministic Method

For the deterministic performance assessment analyses, a modelling approach was adopted. This included the recognition of important features, events, and processes, and attention to the evolution of a realistic, defensible, system model. The modelling philosophy was conservative, with 'worst case' assumptions made in constructing conceptual models, and 'worst case' parameter values used in the calculations. The modelling was also robust, and sensitivity analyses were used to identify controlling parameters, that is, those parameters that have the greatest effect on CO_2 migration. If additional investigations were deemed necessary from the results of the sensitivity analyses, parameter uncertainty was then quantified and a probabilistic assessment carried out.

Compositional reservoir simulations, supported by early performance assessment studies, were conducted for a time period of 5000 years, starting from the end of EOR. They provided an initial understanding of CO_2 migration, and established the process and parameters that may be important to modelling the long-term fate of CO_2. These early studies highlighted the importance of processes such as CO_2 diffusion in the oil phase, the phase saturation distribution at the end of EOR, formation water velocities within the reservoir zone, and the strong interplay between the coupled processes of pressure-driven flow, density-driven flow, and diffusion.

8.4.2 Probabilistic Semi-analytical Method

To assess long-term risk in complex problems, a method called probabilistic risk assessment (PRA) is the industry standard. Process-driven problems over long time-frames, such as geological storage of CO_2, are very well suited to analysis using PRA. In the case of the Weyburn study, a unique first-generation program called CQUESTRA [CQ-1] was developed and applied to components of the project to determine the power of the PRA approach.

CQUESTRA is a generalized performance assessment model employing simplifying assumptions and a compartmental approach, and it provides analytical solutions. Rather than using CQUESTRA on a full probabilistic risk analysis study of the 75-well pattern area, a single-pattern reservoir simulation was undertaken to demonstrate the capability and potential of the probabilistic risk assessment method, and its ability to identify key processes and parameters.

8.4.3 Probabilistic Expert-panel Method

As noted above, the method for the probabilistic expert-panel-based risk assessments is based on the RISQUE method (Bowden and Rigg, 2004). This is a systematic, quantitative process that uses a formal group of experts (here, referred to as the expert panel) to provide judgements that are incorporated into a quantitative risk analysis and management framework – one that is defensible with respect to current world best practice. The steps followed by the RISQUE method are not new; neither are the quantitative techniques, which have been in use for some years. The method is unique, however, in its application of techniques to quantify risks, and its generation of quantitative expressions of risk, risk profiles, and cost-benefit relationships. Some of the areas covered include: various aspects of engineering, a wide range of technical areas, safety, community outreach, business reputation, legal culpability, and environmental impacts. The method proceeds as a staged process, summarized as follows.

Stage 1. Establish the context.
Stage 2. Identify the risk.
Stage 3. Analyse the risk.
Stage 4. Develop risk management strategies.
Stage 5. Implement a risk management strategy.

This process is consistent with the process described in the Australian risk management standard (AS/NZS ISO 31000, 2009). The basic approach is to use quantitative techniques to characterize risk in terms of both the likelihood of identified risk events occurring (such as CO_2 escape and inadequate injectivity), and of their consequences (such as environmental damage and loss of life). In quantitative terms, the RISQUE method defines a risk quotient as follows.

Risk Quotient = Likelihood × Consequence

The RISQUE method also utilizes the concept of a Risk Index, which is a measure of the degree to which a relevant risk quotient is above or below the nominally acceptable risk target. A Risk Index of <1.0 means that the risk quotient is lower than the risk target or evaluation criteria, and the risk is considered acceptable. Conversely, a Risk Index >1.0 means that the risk is unacceptable.

Information provided by the expert panel was entered into a spreadsheet to enable the risks to be analysed. A Monte Carlo simulation using Crystal Ball software was used to handle the variables that were input as probability distributions. The model was run for 2000 trial simulations, and with the same sequence of random numbers. This means that the results can be reproduced, and then changes are permitted to be made to the model.

Three levels of confidence were selected as the model outputs, namely, optimistic, planning, and pessimistic. The 80% confidence level was used to represent a suitably conservative estimate of cost which may be used for planning purposes (the planning estimate). The 50% confidence level was used to represent an optimistic estimate of cost. A very high confidence level (95%) was used to represent the pessimistic estimate. This method was an appropriate approach, because it allowed delivery of a transparent risk assessment process that can be understood by the wider community and allow stakeholders to assess whether the CO_2 injection process is safe, measurable, and verifiable, and whether a selected alternative delivers cost-effective greenhouse benefits.

8.5 Evaluation of Risks
8.5.1 Containment Risk Evaluation

Three workshops were conducted over the period 2007–2010 to evaluate the likelihood of scenarios that might result in migration of CO_2 to the biosphere, and the magnitude of the potential consequences in terms of the amount of

CO_2 lost to the biosphere. Given the uncertainty associated with the likelihood of such an event, and the consequence if it occurs, participants were asked to provide two estimates that could be entered into a log normal distribution. For the consequences, participants were asked for their best estimate of the expected median rate of release (that is, one that would not be exceeded in 50% of the cases) and a high estimate rate (one that would not be exceeded in 95% of cases). In addition, the following information was sought.

1. The first year in which release to the biosphere could occur.
2. The likely duration of the release to the biosphere. Duration of release was categorized into two types of event: short-term (high rate), and long-term (low rate).

Table 8-4 was provided to assist participants in identifying appropriate likelihood of events. If there was uncertainty associated with a likelihood estimate, the estimate was quoted plus or minus one or two orders of magnitude.

A brief description of the risk events identified during the workshop, along with the resulting risk evaluation, is presented in Table 8-5. The events have been grouped by the initiating event.

8.5.2 Containment Risk Profile

In order to compare the relative significance of different containment risks, the amount of CO_2 lost was multiplied by the likelihood of the occurrence for each

Qualitative Description	Order of Magnitude Annual Probability	Basis
Certain	1 (or 0.999, 99.9%)	Certain, or as near to as makes no difference
Almost certain	0.2–0.9	One or more incidents of a similar nature have occurred here
Highly probable	0.1	A previous incident of a similar nature has occurred here
Possible	0.01	Could have occurred already without intervention
Unlikely	0.001	Recorded recently elsewhere
Very unlikely	1×10^{-4}	It has happened elsewhere
Highly improbable	1×10^{-5}	Published information exists, but in a slightly different context
Almost impossible	1×10^{-6}	No published information on a similar case

Table 8-4. *Guide to assist in the quantification of likelihood.*

Risk event	Description	Potential annual mass reaching biosphere (t/a)
Naturally occurring events		
Reactivation of existing fractures and faults by natural seismicity.	Potential reactivation of Souris Fault due to natural seismicity. The fault is in contact with the Midale reservoir.	30 t/a (50% CL) 60 t/a (95% CL)
Creation of new fractures or faults by natural seismicity.	An earthquake capable of creating a new fault and penetrating the carbonate strata, thereby providing a permeable pathway to the surface, would need to be of a magnitude significantly higher than that required to reactivate an existing fault.	
Loss of well integrity due to natural seismicity.	Damage to cement sheath as a result of natural seismicity. There are estimated to be 500 (50% CL) to 600 (95% CL) old wells in the Weyburn project area.	
Creation of new fractures or faults by natural seismicity.	An earthquake capable of creating a new fault and penetrating the carbonate strata, thus providing a permeable pathway to the surface, would need to be of a magnitude significantly higher than that required to reactivate an existing fault.	
Lateral migration of CO_2.	Based on lateral flow occurring outside of the legally defined Weyburn project area. Assumed that CO_2 remains in geosphere but that it leaves the storage complex.	3 t/a (50% CL) 6 t/a (95% CL)
Salt dissolution affects integrity of reservoir.	Evaporite units above reservoir may undergo subsurface dissolution, leading to collapse structures in overlying strata.	
Events resulting from EOR		
Leakage through a network of minor faults which cut across one aquitard.	EOR operation-induced pressure and temperature stress effects allow leakage along existing minor fault networks. 10 (50% CL) to 15 (95% CL) minor fault networks may occur in the Weyburn project area.	3 t/a (50% CL) 6 t/a (95% CL)
Leakage through faults that extend from the reservoir to the biosphere.	EOR operation-induced pressure and temperature stress effects may lead to long-term leakage along existing faults. It is considered plausible that there may be 1 (50% CL) to 3 (95% CL), as yet undetected, 'through faults' in the Weyburn project area.	30 t/a (50% CL) 60 t/a (95% CL)
Chemical variations enhance existing fractures.	Chemical reactions between CO_2 and the reservoir and overlying cap rock may result in precipitation of new mineral phases, or dissolution of host rock minerals along transport pathways. Cap rock partial dissolution could lead to vertical migration of CO_2.	1×10^{-6} (50% CL) 1×10^{-5} (95% CL)

Pressure variations reactivate fractures.	Injection of CO_2 causes pressure stress effects that may reactivate existing fractures.	
Pressure variations lead to new fractures.	Injection of CO_2 causes pressure stress effects that may form new fractures.	
Temperature variations reactivate fractures.	Injection of CO_2 causes temperature stress effects that may reactivate existing fractures.	
Pressure variations lead to new fractures.	The injection of CO_2 into the reservoir causes temperature stress effects that may lead to new fractures.	
Events related to migration along wellbore		
Microfractures and micro-annuli in the cement in old wells.	There are estimated to be 500 (50% CL) to 600 (95% CL) old wells in the Weyburn project area.	3.65×10^{-5} t/a (50% CL) 3.65×10^{-4} t/a (95% CL)
Microfractures and micro-annuli in the cement in new wells.		
Corrosion of well casing.	Estimated that corrosion could affect 100 (50% CL) to 150 (95% CL) wells.	3.65×10^{-3} t/a (50% CL) 3.65×10^{-2} t/a (95% CL)
Undetected faulty wells.	Estimated to be 5 (50% CL), 6 (95% CL) undetected faulty wells.	3.65×10^{-3} t/a (50% CL) 3.65×10^{-2} t/a (95% CL)
Cement channelling in old wells.	Estimated to occur in 15 (50% CL), 20 (95% CL) old wells.	3.65×10^{-4} t/a (50% CL) 3.65×10^{-3} t/a (95% CL)
Old wells outside Weyburn lease area.	Estimated to be 500 (50% CL), 600 (95% CL) old wells near Weyburn project area.	3.65×10^{-5} t/a (50% CL) 3.65×10^{-4} (95% CL)
CL = Confidence level.		

Table 8-5. *Guide to assist in the quantification of likelihood.*

event or pathway. This number was intended to assist in identifying whether there is any particular set of events that influences the overall project risk more than others, and where controls may need to be developed.

Figure 8-5 shows the risk quotients for each of the identified risk events. The right hand column represents the total containment risk quotient for the project. The graph shows the risk quotients on a log scale and therefore each horizontal gridline on the graph represents an order of magnitude difference from the lines above and below.

The different shades of the green bars on the graph represent the risk quotients at different levels of confidence. The 50% confidence level (CL 50%) estimate represents an optimistic assessment of the risk quotient, or one that will be exceeded in 50% of cases. The CL 95% represents a conservative estimate of risk (one that will only be exceeded in 5% of cases), and the CL 80%, a level that is typically used by organizations in planning. The magnitude of the difference between the CL 95% and CL 50% estimates represents the degree of uncertainty in the consequence and/or likelihood estimates.

The column on the right side of the risk profile of Figure 8-5 shows that the total containment risk (calculated as the sum of all individual risk events to the left) is around two orders of magnitude lower than the target acceptable level of containment risk (indicated by the solid green horizontal line). All individual risk events pose acceptable risk with respect to the defined target, with wells generally posing the highest risk to containment. The highest risk event ('well-casing corrosion') still presents a relatively low risk, being about 10 to 30 times less than the target risk for single events (indicated by the dashed green horizontal line).

8.5.3 Effectiveness Risk Evaluation

The following events were evaluated as risks to the effective storage of 30 Mt of CO_2 within the confines of the Weyburn unit.
1. Reduced injectivity into the reservoir.
2. Lack of reservoir capacity.
3. Inadequate source of CO_2.
4. Change of project economics.
5. Lateral migration out of the Weyburn unit.
6. Premature closure.

In addition, a risk to the effectiveness of the Weyburn EOR operation in gaining carbon credits was identified as the inability to verify the quantity of CO_2 stored within the Weyburn unit.

This risk of reduced injectivity relates to the inability to input the target volume of CO_2 into the reservoir as a result of reduced injectivity in the wells. This event was considered highly improbable (1×10^{-5} ± one order of magnitude) in the context of the Weyburn project, due to the number of wells and injection characteristics observed to date. It was agreed that if injectivity was a problem, then 10% (CL 50%) and 15% (CL 95%) of the target volume may not be injected into the reservoir.

The lack of reservoir capacity was not considered applicable to the project since the target storage complex is laterally unconfined in the east-west direction, and the 30 Mt target (for this risk assessment) falls well below the total capacity of the storage complex.

The possibility of not having an adequate source of CO_2 to fill the storage complex to its target capacity is also considered highly unlikely (1×10^{-4} ± one order of magnitude). In determining the likelihood of this risk, the following points should be noted.
1. Currently there is a contract with Dakota Gasification Company to supply the CO_2.

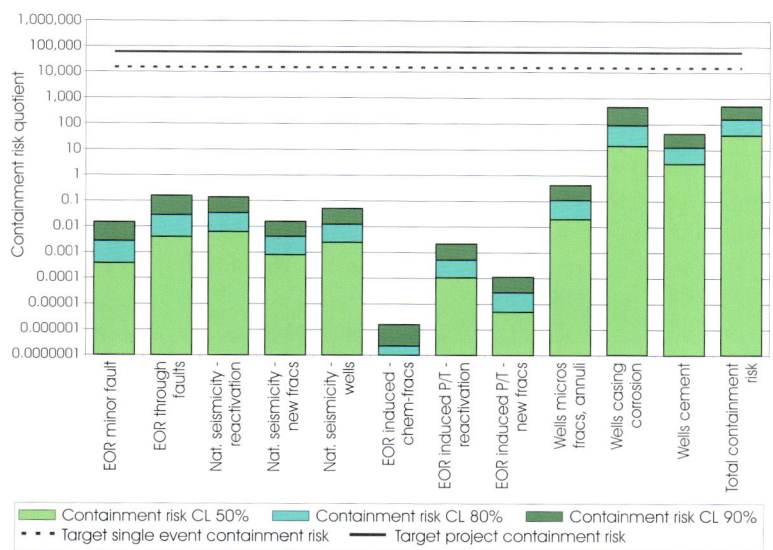

Figure 8-5. *Comparison of event containment risk.*

2. Should the contract with Dakota Gasification Company be cancelled, then it is likely that another operator and source of CO_2 could be located to fill the reservoir to its target capacity.
3. Almost 50% of the target storage volume has already been injected into the site.

In the event that the source did fail, the percentage of the target volume that may not be stored was assessed to be in the range of 20% (CL 50%) to 30% (CL 95%).

The likelihood of a change in project economics leading to the target storage volume not being met was considered to be relatively high (0.03 ± one order of magnitude). Such a change in project economics could be brought about for a variety of reasons including the following.

1. Changes in cost and revenue structures, or the injection rate of CO_2 as oil extraction reduces, or a change in project economics during the transition from EOR to storage.
2. A change to regulations or policy (for example, well abandonment requirements, taxes, cost of carbon).
3. A change to the value of storing CO_2 (for example, CO_2 may have a higher value elsewhere, such as in another EOR project).
4. Regulators not allowing EOR to be recognized as a CO_2 storage project.

It was estimated that 50% (CL 50%) to 70% (CL 95%) of the target volume may not be stored, in the event of a change in project economics. Although 50% of the target storage volume has already been reached, it was recognized that there is even a potential for other project-economics-associated uses of CO_2 to result in extraction of the CO_2 stored to date.

Simulations indicate that the likelihood of CO_2 migrating laterally out of the Weyburn unit in the project period is 1.0. It is estimated that 2% (CL 50%) and 5% (CL 5%) might migrate laterally out of the Weyburn unit within a timeframe of 1000 years.

Note that the fate of the CO_2 that leaves the designated Weyburn storage area is immaterial to the assessment of effectiveness risk. Any loss of containment associated with the CO_2 that moves laterally would occur through the mechanisms identified in the containment risk assessment.

Premature closure of the site may be driven by a number of factors including (1) changes in community perception of CO_2-EOR operations or CO_2 storage projects, (2) changes in regulations allowing access to the site (for example, petroleum lease or operating licenses), (3) requirement for the site to be used to access alternate resources, such as potash, or (4) regulators not allowing EOR to be recognized as a storage project.

The likelihood of premature closure leading to the inability to meet the target storage volume was considered to be 1.0×10^{-2} ± one order of magnitude. If premature closure did occur, the percentage of the target volume not stored was considered to be 10% (CL 50%) and 20% (CL 95%), because 50% of the target volume has already been stored.

The risk of not being able to confirm storage of CO_2 in the Weyburn unit does not relate directly to the effectiveness of storing the target volume of CO_2, but rather to the implications of not being able to confirm that the CO_2 is stored. If the location and quantity of CO_2 stored cannot be confirmed by monitoring, there are implications for the effectiveness of the Weyburn project in generating carbon credits for the storage of CO_2.

Current monitoring methods can detect 10 000 t of CO_2 at depth. If the overburden can be monitored, and used as proof that the CO_2 is contained, then 90% of the total volume is detectable. Further, there is a 100% chance the project will not be able to confirm 5% (CL 50%) to 10% (CL 95%) of the stored CO_2 through monitoring.

8.5.4 Effectiveness Risk Profile

In order to compare different effectiveness risks, the amount of CO_2 that would not be stored if a risk event were to occur was determined for the events discussed above.

Figure 8-6 shows the corresponding effectiveness risk profile. The profile indicates that the total effectiveness risk (shown as the right-hand column on the profile) for the project would be marginally acceptable, being just less than the target at CL 95%. The risk profile shows that three events – 'Lateral migration out of the Weyburn unit', 'Change to project economics', and 'Confirmation of CO_2 stored' – are the highest risk events, contributing almost equally to the total risk. The risk posed by each of these top risk events is about equal to, or just above, the target risk for individual events at CL 95%. The remaining events pose negligible effectiveness risks, being more than three or more orders of magnitude lower than the target.

Given that there is a 100% chance that about 1 Mt of CO_2 will migrate to areas outside the control of the Weyburn operation, the target storage mass should be reduced by 1 Mt to 29 Mt. Similarly, there is a 100% chance that the project will not be able to confirm the presence of 5 to 10% of the CO_2. An effective monitoring plan should be implemented, and allowances made in project feasibility studies for not being able to detect 5 to 10% of the storage volume.

The project has limited ability to manage the project economic drivers. A plan for keeping abreast of events and strategic positioning will have the most influence on the likelihood of these drivers impacting the project.

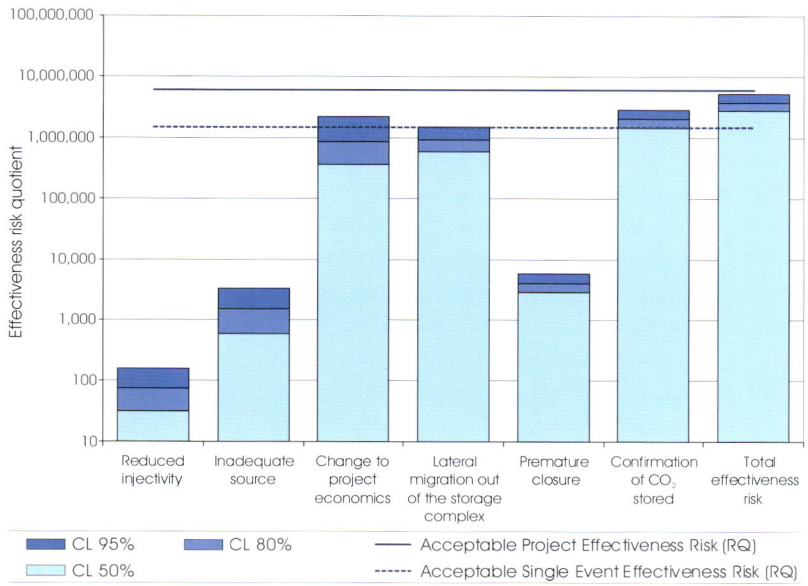

Figure 8-6. *Effectiveness risk profile.*

RISK ASSESSMENT   313

8.5.5 **Biosphere Risk Evaluation**

Biosphere risk assessment addresses the requirement that the project will not pose a threat to the community, or to community assets in the biosphere. This requires a sound understanding of surface and near-surface processes, as well as a strong understanding of social and community values, needs, and expectations, and any potential impacts on lifestyles, recreation, public amenity, heritage, First Nation assets, infrastructure, industry, and local government.

Several workshops were conducted with a selection of biosphere experts in various disciplines relevant to the biosphere risk assessment. Securing technical specialists for the biosphere workshop proved difficult, and some areas were not well-represented in several workshops (for example, hydrochemistry, ecology, agriculture, and soil science). As a consequence, the assessment relied heavily on a few individuals with cross-discipline knowledge. Therefore, it was concluded that some aspects of the science needed to be checked with specialists in the missing areas in order to improve the quality of the information that is used in the assessment and, therefore, the credibility of the biosphere risk assessment results. In particular, better information was required on the changes to groundwater and surface water chemistry as a result of the addition of CO_2.

The geosphere risk assessment results indicate that, generally, it is very unlikely that substantial masses of CO_2 will move beyond the geosphere into the biosphere. However, there are several possible pathways that could potentially allow movement of CO_2 into and throughout the biosphere. Elevated levels of risk would occur where those pathways come into contact with parts of the biosphere (including rivers, groundwater, soil, and atmosphere) and organisms and activities associated with them. The key potential pathways for movement of CO_2 from the geosphere to the biosphere that have been identified at the Weyburn project are faults, fractures, and wells.

Currently, the faults are considered to be quite 'tight' (that is, they have very low permeability) and have very low capacity to act as CO_2 conduits. However, increased pressure due to CO_2 injection during EOR could cause some faults to 'open up' (become more permeable) and thus become temporary conduits for CO_2 flow. Although fracture sets occur within the area, they are considered to be quite 'tight'. Induced temperatures and pressures due to EOR activities could potentially reactivate fractures, and thus generate new fracture systems. EOR activities could also potentially increase fracture set permeability through geochemical changes. Wells are potentially important pathways for CO_2 to reach the biosphere because, in most cases, they could create a direct connection between the geosphere and the biosphere.

Potential cause-and-effect relationships were identified for CO_2 movement within the biosphere. Generally, these relationships considered the effect that

CO_2, moving along identified pathways (fractures and wells, for example), could have on components of the biosphere, such as groundwater, rivers, soils, and air. Assets that are in contact with the changed biosphere (for example, the environment, people, and infrastructure) then may be affected.

To aid in understanding these cause-and-effect relationships, the following studies were commissioned for the biosphere risk assessment: (1) basic surface water modelling, (2) basic groundwater modelling, and (3) basic assessment of the transportation of CO_2 in soils.

It is important to note that the specialists were asked to conduct basic studies, to facilitate order-of-magnitude assessments of the potential risks to the biosphere. The results of the risk assessment should, therefore, be treated as a preliminary guide to the risks associated with the Weyburn project, and not as a definitive assessment of risk.

8.5.6 Biosphere Risk Profile

The consequences associated with potentially significant releases from the geosphere were assessed, using the semi-quantitative risk assessment matrix presented in Table 8-2; and the likelihoods of those consequences occurring were assessed using the likelihood guide shown in Table 8-4.

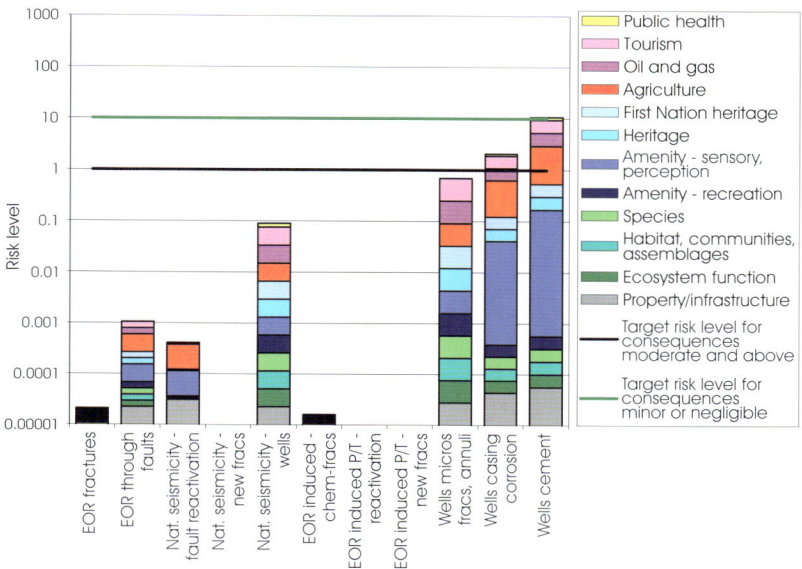

Figure 8-7. *Community assets most at risk for initiating events.*

The likelihoods of CO_2 being released from the geosphere to the biosphere via a given pathway were assessed during the containment-risk workshop. It was assumed that a release of CO_2 had occurred, and so attention was focused on the likelihood of the release affecting an asset in the biosphere. The final outputs of the biosphere risk assessment take into account both the likelihood and the consequences of the event taking place.

Figure 8-7 illustrates the community and environmental assets that are most at risk over 1000 years from the initiating events. The coloured segments within each column show the proportion of the total event risk that is posed to each respective asset.

Figure 8-7 shows that 'amenity and perception' is one of the key risks. The assessment of the biosphere experts was that public perception associated with a change to groundwater chemistry (whether or not the change was associated with the Weyburn project) could be an issue for the project. Figure 8-7 also shows that risks to most assets, including habitat, species, and ecosystem function, are similar in terms of their contribution to the overall risk from a particular initiating event. Risks to agriculture and property/infrastructure are slightly higher than risks to some assets, but still do not significantly dominate the risk profiles associated with particular initiating events. The risks to agriculture and property were driven by the following.

1. The potential for changes of the pH of groundwater, which would alter the effectiveness of agricultural chemicals. The biosphere experts considered these consequences likely to be localized and short-term.
2. The potential for changes in water chemistry which would impact the uses to which a local well owner could put the water.

8.6 Conclusions

Benchmarking studies are invaluable to verifying that important physical processes have been correctly incorporated into the models, and that prediction accuracy has not been compromised between the models.

Significant value can be obtained in the application of semi-analytical models that capture the major element of a CO_2 storage system or complex. Because of their efficient semi-analytical form, these models allow thousands of cases to be solved in minutes, thereby enabling probabilistic risk analyses to be undertaken. In addition, extensive parameter sensitivity analyses allow processes and input parameters to be ranked according to their relative impact on storage effectiveness.

Sensitivity analyses using a semi-analytical model showed that release of CO_2 from the reservoir is strongly influenced by the permeability of the upper Marly unit of the Midale Beds. Restricted flow within the immediate vicinity of a well is one of the key controlling mechanisms for leakage through abandoned wells.

Features, Events and Process (FEP) identification is a very effective method for mapping the key elements in a storage complex, and is a fundamental part of the development of credible scenarios of how a CO_2 storage complex will respond to CO_2 injection.

Risk analysis simulations showed that it was possible to conduct large-scale compositional simulations of the storage complex over a long time-frame: 5000 years, in the case of the Weyburn project. The model built for these simulations was very large and so, given the long solution times for each simulation, did not readily permit sensitivity studies to be conducted.

Semi-analytical solutions use simplified process physics to permit rapid computational speeds. In contrast, compositional simulations are advantageous from the standpoint of complex physical processes associated with CO_2 injection into a hydrocarbon reservoir. These simulations highlighted the importance of (1) processes such as CO_2 diffusion into the oil phase, (2) phase saturations at the end of the EOR and its evolution over the storage time-frame, (3) hydrodynamic formation water velocities within the reservoir zone, and (4) the strong interplay between the coupled processes of pressure-driven flow, density-driven flow, and diffusion.

Compositional simulations of the storage complex allowed (1) an assessment of the fate of the CO_2 within the defined conditions of the storage complex, and (2) an assessment of the validity and accuracy of scaling up which will be necessary to conduct large compositional simulations of the storage process, especially over the long time-frames required for risk assessment.

While simulations using large complex models can provide valuable insight into the behaviour of the storage complex, caution should be exercised when analysing the results. These models are subject to considerable uncertainties due to sparse data sets, and to the associated interpolations necessary to develop them. Predictions of the long-term distribution of injected CO_2 have not been adequately verified on a field scale in a wide range of geological settings, and hence predictions from 1000-year simulations should be viewed with caution.

The use of expert panels can be an effective method for the development of a risk register for a project, and to elicit judgements about risk. The formal structure followed by the panels also ensured that all decisions regarding features, events and/or processes were documented.

Currently, there are no widely accepted conventions or standards that define acceptability in terms of risk. Nevertheless, following the thorough review by an expert panel, the project was successful in establishing risk targets to assess the acceptability of a risk. For instance, a risk target was set that at least 99% of the total injected mass of CO_2 should be retained in the geosphere after 1000 years, which is the same target proposed by the Intergovernmental Panel on Climate Change (IPCC).

Across a wide spectrum of risk events, the highest estimated loss of CO_2 was judged to be via corroded well casing, where losses would be in the range of 8 to 15 kt of CO_2 over a 1000-year period.

Relatively speaking, well leakage pathways pose by far the greatest risk for CO_2 leakage out of the geosphere – approximately 3.5 orders of magnitude higher than the risk posed by natural events, such as earthquakes, or by EOR activities.

The geosphere risk assessment has provided a good understanding of the ability of the reservoir to receive and contain the target mass of CO_2 for a very long time. In terms of risk acceptability for the geosphere, it is reasonable to conclude that 30 Mt of CO_2 will remain safely stored for more than the 1000-year assessment period. The containment risk is very low, being 2 to 3 orders of magnitude less than the established containment risk target. The risk of non-containment is therefore deemed acceptable.

For CO_2-EOR projects, an additional risk relates to the transition to a CO_2 storage operation, and for the Weyburn project this was defined as an effectiveness risk – 'the ability of a project to demonstrate that there is a target volume of 30 Mt of CO_2 (in the case of the Weyburn project) stored within the confines of the storage complex, assuming the containment risk targets are met'. Effectiveness risk allowed external elements such as project economics, CO_2 supply, and the effectiveness of plume monitoring, to be included in the overall assessment.

For the Weyburn project, the effectiveness risk is only just less than the established effectiveness risk target, and is therefore considered to be marginal. It is recommended that a strategy be developed to reduce effectiveness risk. The four main drivers included in a change to storage project economics, the most influential parameter included in the effectiveness risk assessment, were as follows.

1. Transition from EOR to storage.
2. Regulatory intervention.
3. Competition for CO_2 resources.
4. Policy on being able to claim carbon credits for CO_2 storage associated with EOR.

As there is limited ability for a CO_2-EOR operator to manage any of these drivers, with perhaps the exception of the transition process, a plan for keeping abreast of events and strategic positioning will have the most influence on the likelihood of these drivers impacting the project.

It is difficult to conduct a quantitative assessment of biosphere risks when the purpose of the risk assessment is to assess the impact to the intrinsic environment, which for the Weyburn project covers a very large geographical area, or the social value of a range of different environmental or community assets, such as plant species or lakes.

The development of a consequences matrix for the biosphere and community assets of most importance to stakeholders was as effective as conducting a semi-quantitative risk assessment using a panel of biosphere experts.

It is important to understand the key questions and potential concerns of the community, and the environmental and community assets that are valued by the community, and these elements should be included in a biosphere risk assessment program.

On the basis of the information available to the biosphere risk assessment, the Weyburn project was considered to pose an acceptable level of risk from a public safety perspective.

The biosphere component that was assessed as highest risk, but still below an acceptable risk target for CO_2 releases from the geosphere, was the Weyburn Valley aquifer.

The biosphere risk assessment identified one of the highest consequences possibly arising in relation to a future CO_2 storage project (following the end of the CO_2-EOR operations) as being the public perception of issues associated with unrelated changes to groundwater chemistry.

8.7 Recommendations

A wide view must be taken of possible events and associated consequences in the identification of risk for a CO_2 storage project. The need to identify potential unplanned events and determine what could happen as a result of these events, requires a different mind set from that typically required for performance assessment. It is impractical to conduct detailed performance modelling to address every conceivable scenario associated with every conceivable risk. The risk assessment, therefore, should be used to direct research towards addressing the 'what if' scenarios that are considered to potentially give rise to significant consequences, and which pose the greatest uncertainty to the ability of a project to achieve its objectives.

If multiple approaches are adopted within a risk assessment study, benchmarking studies and validation of the analysis techniques are highly recommended.

Stakeholder involvement early in the risk assessment process is very important to identify the community assets and concerns to be included in the risk assessment.

Full field risk assessments for complex, geological environments typical of CO_2 storage projects – whether pursued using large-scale simulation approaches or expert judgement approaches using panels – are intensive, time-consuming efforts that need to be resourced appropriately to ensure a successful outcome.

Risk assessment studies should proceed in advance of, or at the very least in tandem with, the detailed geological characterization studies that are undertaken for a CO_2 geological storage project, and should have multiple cycles of review as new, updated information is obtained within the project.

Wells form the pathway that poses the highest risk to the biosphere. Further, well failures result mostly from casing corrosion and cement problems. Therefore, more detailed modelling of time-dependent well failure processes, such as annular cement degradation and corrosion of the well casing, is required. This is supported by simulations conducted for risk analyses, by the well integrity research reported in Chapter 7, and by simulation studies designed to assess the potential migration of CO_2 above the storage complex. Attention needs to be directed to well cement conditions, the continuity and size of channels, and potential flow rates. For well casings, the focus should be on corroded annuli and a more complete evaluation of well records and a survey of high risk wells. And by no means last is a requirement for further comprehensive evaluation of current and future aquifers containing potable groundwater, with reference to residential properties and developments.

Biosphere risk assessments for CO_2 storage projects are particularly challenging and significant effort is required to identify and to engage the appropriate expertise to enable defensible likelihood and consequence judgements.

Appropriate resource allocation is necessary for specialist studies to support biosphere risk assessments. These specialist studies should include (1) groundwater and surface water modelling, especially including considerations of the interactions between surface and groundwater and soils, and (2) research directed to a better understanding of the ecology of the area around a project to allow more reliable impact assessment.

9

COMMUNITY OUTREACH

Summary

The importance of having an emergency response plan in place before a crisis arises was clearly demonstrated at the Weyburn-Midale project, when the project had to deal with very public allegations that the facility was not secure, and that CO_2 was escaping.

Because a detailed plan of response had been worked out in advance, operators were able to deal with the allegations swiftly and effectively. Details of the plan and its implementation are given in the Crisis Communications section of this chapter.

Community outreach, however, involves much more than crisis management. The Weyburn-Midale project is the world's first extensive study of CO_2 storage in connection with EOR in depleted oilfields. In communication about it, it has been necessary to clearly differentiate between the EOR operations and the research project.

Outreach has involved a continuing program of public education and participation in community events. There is communication with individual landowners, and public information sessions are held at regular intervals to keep the community up to date on developments.

A strategic communications plan was developed to establish key messages and to plan education about CO_2 storage.

Educational aspects of the outreach program included establishing an interpretive centre in Weyburn City Hall, developing a public display booth which travels to relevant trade shows and conferences, launching a website with linkage to other relevant websites, producing printed materials, and developing a mailing list which is used on an on-going basis to distribute news and materials to non-governmental organizations (NGOs) in many countries. Details of these initiatives are given in this chapter.

Authors: N. Sacuta (Petroleum Technology Research Centre, Regina, Saskatchewan), and **F.M. Mourits** (Natural Resources Canada, Ottawa, Ontario).

9.1 Introduction

Communication between the principal stakeholders was well-established by the two oilfield operators prior to the start of CO_2 injection and research. With respect to the research, these stakeholders included the partnering oil companies, landowners, and members of the extended communities around the towns of Weyburn and Midale. In view of the work done by the two oilfield operators predating the start of research in 2000, focus here will be on the processes, structures, and activities put into place by the Petroleum Technology Research Centre, with respect to both external and internal communications.

As stated in the summary, a strategic communications plan was developed to establish key messages and to plan education about CO_2 storage. It was recognized that it would be important to clearly differentiate between the EOR operations and the research project, and also to acknowledge the different players within each activity.

As with most oilfield operations, engagement with local stakeholders has been maintained by both Apache and Cenovus, to deal with such concerns as surface access for wells and pipelines, the interests of holders of mineral rights, and daily oilfield activities. These communications are most often direct and face-to-face. They include the following.

1. Use of landsmen in negotiations for surface access, and for preliminary discussions about mineral rights.
2. Regular contact with landowners through an established liaison officer, who provides information about operations, is freely available to answer inquiries, and manages community engagement.
3. Participation in community events in the Weyburn and Midale areas, including such public functions as hockey tournaments.
4. Annual or semi-annual breakfast- or dinner-hosted information sessions about oilfield operations. These events provide community members with updates, and the opportunity to raise any issues not covered in one-to-one consultations.
5. Research displays hosted by the Petroleum Technology Research Centre that were sometimes a part of these events.

Communities with a long history of oilfield operations have a heightened awareness and a greater knowledge of the technologies involved in oil production than do non-oilfield communities. When CO_2 was first injected into the Weyburn and Midale oilfields, both operators knew that local people were already aware of the methods employed to extend the lives of the fields. Because of this, CO_2 could be placed within that knowledge chain, and was understood as one of many elements that could be used.

With many people in the neighbourhood working in the oil patch, apprehension about the use of CO_2 was generally not expressed. Discussion with landowners indicated few reservations about the injection technologies involved. Residents and community officials also expressed significant pride when the research side of the project began to gain prominence globally, and when visits to the operations increased.

As the research project became better known, public discussions about it began to focus on long-term CO_2 storage, and there was an increased emphasis on education about the subject. Both the oilfield operators and the members of the project attracted the attention of the local community, and of national and international audiences, who wanted to learn more about the project.

Other CO_2-EOR operators may find similar experiences when introducing carbon dioxide as a solvent in their operations. Oil patch audiences tend to be less apprehensive about EOR technologies and their use, and see CO_2 as a useful tool in prolonging field life. However, when transition to storage is planned, education and engagement should be regarded as a necessary part of the outreach. Particular attention must be paid to providing assurance of long-term containment.

Two best-practices manuals dealing with public outreach and communications have recently been published, and may be useful. They are: *Public Outreach and Education for Carbon Storage Projects* (U.S. Department of Energy, National Energy Technology Laboratory, 2009b) and *Guidelines for Community Engagement in Carbon Dioxide Capture, Transport and Storage Projects* (World Resources Institute, 2010). These studies offer extensive communications planning templates, case studies, and outlines to aid in the development of effective public outreach and communications activities for fully integrated projects (capture, transport, and storage of CO_2). Here, we address only the experiences of the Weyburn-Midale project, and offer direction and recommendations arising from the specifics of that project to communicators organizing consortia-driven research programs.

9.2 Communications and the Project

The structure of the project included (1) a Lead Sponsors Executive Committee, which managed the funding decisions for individual research projects, (2) a technical component managed by a Research Theme Lead Group, made up of the principal scientists involved in the research, and (3) a non-technical component managed by a project integrator.

The non-technical component comprised three themes, including a public communications and outreach theme, which was headed by a communications specialist, and supported by a Communications Experts Advisory Panel, responsible for providing advice and direction regarding communications activities. The advisory panel included communications officials from the major sponsors: specifically from the governments of Canada and Saskatchewan, the two oilfield operators, the U.S. Department of Energy, the Petroleum Technology Research Centre, and private sector partners such as Schlumberger Carbon Services. The communications and outreach tasks included developing and implementing a strategic communications plan that would establish key messages and focus on public education on CO_2 storage in general, while leaving local outreach to the oilfield operators, except where there were appropriate synergies between the communication goals of the research project and the oil companies.

The strategic communications plan implemented a number of activities over the life of the project, as listed below.

1. Developed an educational website about CO_2 storage. This helped to inform the local communities and the wider Canadian and international audiences about the science and technologies involved in CO_2 storage. The website (www.ccs101.ca) was launched as a general communications vehicle, aimed at public education on CCS, and continues to expand.
2. Launched a *CCS 101* public display booth, in association with the website. The specific target was trade shows for organizations such as science teachers' associations, environmental trade shows, oil and gas shows, and open house events for other Canadian CCS projects.
3. Published material for use at the booth, and for the website, as well as similar information sites, such as the website of the Plains CO_2 Reduction Partnership (http://www.undeerc.org/PCOR/default.aspx) and the IEAGHG Programme (www.ieaghg.org).
4. Connected with NGOs and groups active in the CCS industry (such as Bellona, Global CCS Institute, Pembina Institute), to link our website and educational materials with their websites. Compiled a list of stakeholders to be contacted with news about the project.
5. Established an interpretive centre at Weyburn City Hall (see Figure 9-1). The objective was to inform the local population about the extensive research being done into CO_2 storage in association with both the Weyburn and Midale EOR operations. The centre includes displays about the history of the two communities and adjacent oilfields and about how oil is extracted using CO_2, and information about the measurement, monitoring, and verification technologies used as part of the research project. In order to enhance liaisons within the community and to promote use of the centre, the City of Weyburn and local schools were offered tours of the display when it was launched.

6. Conducted focus groups in Saskatchewan and Alberta, where CO_2 storage projects are operating or in the process of development. The objective was to gauge the extent of public knowledge, and the effectiveness of the newly created website at addressing knowledge gaps. Results from the focus groups will be used to update and improve the website.

In addition to these specific outreach projects, the Petroleum Technology Research Centre worked with both oilfield operators to develop guidelines for requests for tours of the CO_2 injection facilities. About 15 tours per year are run, often including presentations on the research program.

9.3 Risk Assessment Meeting

As part of implementing the risk-assessment program for the project, researchers coordinated with the oilfield operators and the communications panel to meet with a broad cross-section of people from the Weyburn and Midale communities at a half-day risk assessment workshop in the City of Weyburn in April, 2009. The purpose of the workshop was to provide a broad overview of the research program for local stakeholders, while at the same time asking the

Figure 9-1. *The Weyburn Energy Centre, established at Weyburn City Hall.*

stakeholders what they considered to be key assets in their community. After identifying those assets, participants were asked to break into smaller groups and discuss what perceived threats might be posed by the project to those assets, and what questions they would like addressed in the risk assessment.

Local residents attending included public figures (the mayors of the communities, councillors, and municipal reeves) as well as the general public (ranging from the head of the local Big Brothers and Sisters Association of Weyburn, to oil patch workers). As already indicated, the breadth of knowledge in the community about EOR operations was quite significant, but the meeting did provide background on the storage of CO_2 that was less well-known. This education helped residents come up with 'worst case scenarios' and perceived risks to the important local assets that they identified.

Chapter 8 details many of the assets that were identified during this session, as well as the risk to these assets as determined by the assessment. What was most telling about this meeting, from a communications perspective, was a comment by one of the local politicians who expressed the main concern of many of the participants: not that there were real or significant threats arising from the storage of CO_2 in the reservoirs, but rather that 'the perception could be created that storage is dangerous' and so this would make Weyburn a less desirable destination to visit and live.

It was the public perception of risk that was of main concern to residents in the meeting, not risk in itself. The potential loss in property values and portrayal of the community as dangerous were the main concerns for those in attendance. This was a telling observation, one that would be borne out just two years later, and it is to this incident that we now turn.

9.4 Crisis Communications

In January 2011, a press conference was held in Regina, Saskatchewan, by a farm couple claiming that carbon dioxide from the Weyburn oilfield operations was leaking onto their farm, which was 1.4 km to the southwest of the CO_2-EOR operations of Cenovus Energy. The press conference had been organized with the cooperation of EcoJustice (the former Sierra Legal Defense Fund) and the accusations were based on a scientific report prepared by an independent contractor, who claimed that anomalous CO_2 readings were found in the soil samples taken on the complainants' property in October 2010. The report also claimed that the carbon isotopic signature of the CO_2 matched that of the CO_2 being injected.

As manager of the research project, the Petroleum Technology Research Centre immediately put into action an incident-response plan that used, as its basis, a number of assumptions and resources that were developed ahead of the incident. It was clear that multi-player consortia CO_2-EOR projects must clearly

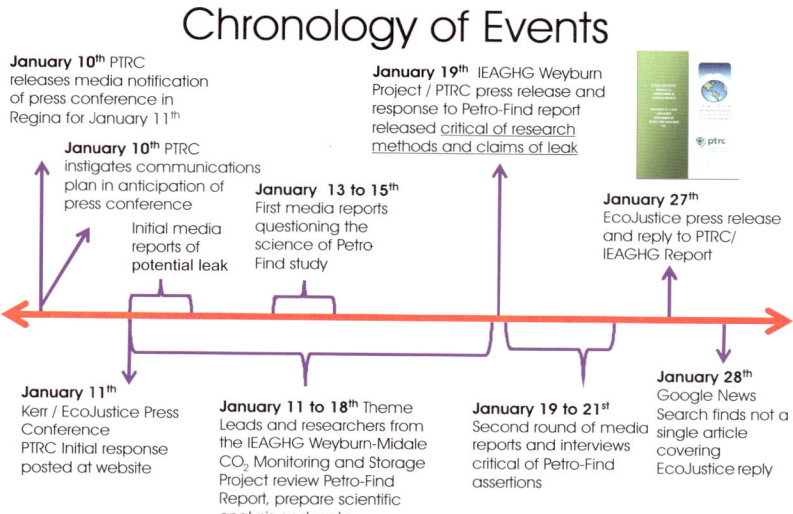

Figure 9-2. *Key events in the Weyburn incident response.*

define roles and responsibilities ahead of any incident. An examination of these resources, and the process put into place, may be of particular value for other projects facing similar situations. Figure 9-2 shows a timeline of the events. The main points in the response to the claims of a leak at Weyburn are as follows.

IDENTIFY KEY PERSONNEL: In particular, this means both a general and a science-specific spokesperson with past experience in dealing with media, and with some level of media training. The centre had very little lead time ahead of the actual press event (24 h) but managed to have both its CEO and the project manager (with a Ph.D. in geology) available for media interviews, and to explain the approach of the project to the claims of a leak, and the process put in place to examine them.

ESTABLISH COMMUNICATIONS AND SCIENTIFIC GROUPS: These groups should be in place to address issues and to quickly offer input as decisions are made. In this case, the Research Theme Leads involved in the program and the Communications Experts Advisory Panel were engaged quickly, and a response plan was put into play, to deal with the allegations. The scientific report of the claimant was immediately reviewed and a plan for a scientific response prepared. The advisory panel met daily via teleconference to discuss media opportunities, to share information, and to develop the media rollout plan.

IDENTIFY KEY STAKEHOLDERS: These are people who need to be contacted about the planned course of action. In the case of the project, the database included all sponsors that had provided funding for the project (10 corporate entities and the governments of Canada, the United States, Saskatchewan, and Alberta). The database also included catalogued media articles as they appeared (both online and in print), and downloadable documents pertinent to the response being prepared. Partners were also provided with contact information for the communications manager at the centre, should they have queries or require updates.

DEFINE THE ROLES OF THE MAIN PARTICIPANTS: The claims about Weyburn were directed towards the oilfield operator, in a complaint filed with the regulator (in this case the Saskatchewan Ministry of Energy and Resources). As a formal grievance process, this made the Petroleum Technology Research Centre a third party, and therefore unable to interfere, despite being manager of the research project. However, as manager of an extensive research program examining the safety of CO_2 storage, the regulator approached both the Centre and the project as the logical consultative group to examine scientific claims of a leak, and the Centre responded accordingly. Defining separate roles in this process was an important step in dealing with the media, since it was important that they understand the difference between EOR operations and an independent research project associated with storing CO_2.

UNDERSTAND THE REPERCUSSIONS OF THE INCIDENT: In this case, an NGO was on record as fundamentally opposed to CCS. This meant that the intended audience for publicizing the claims of a leak at Weyburn was not necessarily the local community. Rather, there was a wider national audience within Canada (in particular in Western Canada, where there were six CCS projects in the advanced planning stages) and globally (where public opposition to CCS, particularly in Europe, was growing). The simple raising of safety concerns was, in itself, a strategy for creating doubt globally. In January 2011, state politicians in the Netherlands, in the process of publicly debating a bill on the storage of CO_2, quoted media reports about Weyburn as examples of safety concerns. In fact, the Netherlands banned on-shore storage shortly thereafter, even before knowing the official scientific response of the Petroleum Technology Research Centre. With the wider repercussions of the Weyburn allegation clear, the Centre developed an expanded distribution list that included most of the major NGOs that support CCS (GCCSI in Australia, Bellona in Europe, Natural Resources Defense Council in the United States, Pembina Institute in Canada) as well as some of those who oppose it (Greenpeace Europe), and kept these groups informed on the progress of the investigation.

IDENTIFY KEY MEDIA: These should be identified ahead of results being released, and should definitely include individuals who understand and write clearly about science. With the numbers of journalists shrinking at major newspapers each year, and in particular the numbers of reporters specifically dealing with science, it is important when trying to get scientifically specific information out, to know which reporters have the time and experience to clearly relay important information to the public. In the initial reports about Weyburn, a CanWest article that went around the globe read 'Carbon Capture and Storage Project Captured, Spewed Dead Animals'. While there is little that can be done to correct such sensationalism, the Centre was able to identify competent journalists in the larger media outlets. As the lead researchers analysed the science of the complainants' report, small bits of information were made available to those journalists to raise pertinent scientific questions about the allegations ahead of the Centre's response. Identifying competent journalists, and being proactive in passing along information to them, is a key component to effective incident response. Also, it is crucial to make sure that the science is clear and easily explained to these chosen journalists – preferably within the context of some narrative.

DEVELOP BASELINE COMMUNICATIONS DATA: An extensive database of baseline soil gas and well water information was extremely important in making a comparative analysis with the claimed anomalous readings. The baselines also helped to create a narrative for the region itself, and to establish that the readings on the land in question were not unusual.

As CO_2 storage projects become more common, the national and international need for communication about their safety and operations will become less necessary, and communication on a local level will again become the most important issue. The Weyburn project, as a bellwether project and currently by far the largest in the world in terms of annual and total accumulated carbon dioxide stored, found that communications carried an unusually heavy degree of international media exposure and response during this incident. While it is important to remember that all CO_2 storage projects in these early days of implementation and study are inextricably bound together, as the science progresses and reports such as this one reveal effective technologies and best practices, communications will hopefully be less politically charged. The project will continue to maintain and expand its legacy materials – printable resources, the website, and the display in the city of Weyburn – to better inform the general public about CO_2 storage.

REFERENCES

Advanced Resources International and Melzer Consulting, 2010. Optimization of CO_2 Storage in CO_2 Enhanced Oil Recovery Projects. Report prepared for UK Department of Energy and Climate Change (DECC), Office of Carbon Capture & Storage. Available on-line at http://www.decc.gov.uk.

Alkalali, A.I. and B.J. Rostron, 2003. Basin-scale analysis of variable-density groundwater flow: Nisku aquifer, Western Canada Basin. *Journal of Geochemical Exploration* 78-79, 313–316.

Al-Marzoug, A., F.A. Neves, J.J. Kim and E.L. Nebrija, 2006. P-wave anisotropy from azimuthal AVO and velocity estimates using 3D seismic data from Saudi Arabia. *Geophysics* 71, E7–E11.

Alnes, H., O. Eiken and T. Stenvold, 2008. Monitoring gas production and CO_2 injection at the Sleipner field using time-lapse gravimetry. Geophysics 73(6), 155–161.

Annunziatellis, A., S.E. Beaubien, S. Bigi, G. Ciotoli, M. Coltella and S. Lombardi, 2008. Gas migration along fault systems and through the vadose zone in the Latera caldera (central Italy): Implications for CO_2 geological storage. *International Journal of Greenhouse Gas Control* 2/3, 353–372.

AS/NZS ISO 31000, 2009. Risk Management – Principles and Guidelines.

Bachu, S., 1995. Flow of variable-density formation water in deep sloping aquifers: review of methods of representation with case studies. *Journal of Hydrology* 164, 19–39.

Bachu, S. and B. Hitchon, 1996. Regional-scale flow of formation waters in the Williston Basin. *American Association of Petroleum Geologists Bulletin* 80, 248–264.

Bachu, S. and T. Watson, 2008. Review of failures for wells used for CO_2 and acid gas injection in Alberta, Canada. Proceedings of 9th International Conference on Greenhouse Gas Technologies (GHGT-9), Washington DC, U.S.A., November 16-20.

Baines, S.J. and R.H. Worden, 2004. The long-term fate of CO_2 in the subsurface: natural analogues for CO_2 storage. *In* S.J. Baines and R.H. Worden (Editors), *Geological Storage of Carbon Dioxide*, 59–85. Geological Society, London, Special Publications, 233.

Barenblatt, G.I., I.P. Zheltov and I.N. Kochina, 1960. Basic concepts in the theory of homogenous liquids in fissured rocks. *Journal of Applied Mathematics and Mechanics* 24, 1286–1303.

Basham, P.W., D.H. Weichert, F.M. Anglin and M.J. Berry, 1982. New Probabilistic Strong Seismic Ground Motion Maps of Canada. Energy, Mines and Resources Canada, Earth Physics Branch. Open File 82-33.

Batzle, M. and Z. Wang, 1992. Seismic properties of pore fluids. *Geophysics* 57, 1396–1408.

Beliveau, D., D.A. Payne and M. Mundry, 1991. Analysis of the waterflood response of a naturally fractured reservoir. Paper SPE-022946-MS, presented at the 66th SPE Annual Technical Conference and Exhibition.

Bell, J.S. and E.A. Babcock, 1986. The stress regime of the Western Canadian Basin and implications for hydrocarbon production. *Bulletin of Canadian Petroleum Geology* 34, 364–378.

Bell, J.S., P.R. Price and P.J. McLellan, 1994. In-situ stress in the Western Canada Sedimentary Basin. *In* G.D. Mossop and I. Shetsen (Compilers), *Geological Atlas of the Western Canada Sedimentary Basin*, 439–446. Canadian Society of Petroleum Geologists and Alberta Research Council.

Berg, R.R., W.D. DeMis and A.R. Mitsdarffer, 1994. Hydrodynamic effects on Mission Canyon (Mississippian) oil accumulations, Billings Nose area, North Dakota. *American Association of Petroleum Geologists Bulletin* 78, 501–518.

Binley, A., S. Henry-Poulter and B. Shaw, 1996. Examination of solute transport in an undisturbed soil column using electrical resistance tomography. *Water Resources Research* 32, 763–769.

Bowden, A.R. and A. Rigg, 2004. Assessing risk in CO_2 storage projects. *The APPEA Journal*, 2004, 677–701.

Brady, J.L., J.L. Hare, J.F. Ferguson, J.E. Seibert, F.J. Klopping, T. Chen and T. Niebauer, 2006. Results of the world's first 4D microgravity surveillance of a waterflood – Prudhoe Bay, Alaska. SPE Paper 101762, 2006 SPE Annual Technical Conference and Exhibition.

Bredehoeft, J.D., C.E. Neuzil and P.C.D. Milly, 1983. Regional Flow in the Dakota Aquifer: A study of the role of confining layers. U.S. Geological Survey Water Supply Paper 2237.

Brown, L., 2002. Integration of rock physics and reservoir simulation for the interpretation of time-lapse seismic data at Weyburn Field, Saskatchewan. Ph.D. thesis, Department of Geophysics, Colorado School of Mines, Golden, Colorado. http://geophysics.mines.edu/rcp/theses/BrownLT_2002_thesis_RCP.pdf.

Bunge, R.J., 2000. Midale reservoir fracture characterization using integrated well and seismic data, Weyburn Field, Saskatchewan. M.Sc. thesis, Department of Geophysics, Colorado School of Mines, Golden, Colorado.

Burrowes, O.G., 2001. Investigating CO_2 storage potential of carbonate rocks during tertiary recovery from a billion barrel oil field, Weyburn Saskatchewan: Part 2 – Reservoir geology (IEA GHG Weyburn CO_2 Monitoring and Storage Project). *In* Summary of Investigations 2001, Volume 1, Saskatchewan Geological Survey, Saskatchewan Energy Mines, Miscellaneous Report 2001-4.1, 64–71.

Busby, J.F., B.A. Kimball and J.S. Downey, 1995. Geochemistry of Water in Aquifers and Confining Units of the Northern Great Plains in Parts of Montana, North Dakota, South Dakota, and Wyoming. U.S. Geological Survey Professional Paper 1402-F.

Calvert, D.G., J.F. Heathman and J.E. Griffith, 1995. Plug cementing: horizontal to vertical conditions. Paper SPE 30514, presented at the SPE Annual Technical Conference and Exhibition, Dallas, Texas, October 22-25.

Cardona, R., E. Ozkan and M. Batzle, 2002. Pressure-transient experiments and the elastic characterization of fractures. 2002 SEG Annual Meeting, Expanded Abstracts.

Celia, M., J. Nordbotten and S. Bachu, 2008. Risk of leakage versus depth of injection in geological storage. Proceedings of 9th International Conference on Greenhouse Gas Technologies (GHGT-9), Washington DC, U.S.A., November 16-20.

Cey, B.D., S.L. Barbour and M.J. Hendry, 2001. Osmotic flow through a Cretaceous clay in southern Saskatchewan. *Canadian Geotechnical Journal* 38, 1025–1033.

Chadwick, R.A., R. Arts, C. Bernstone, F. May, S. Thibeau and P. Zweigel, 2008. Best Practice for the Storage of CO_2 in Saline Aquifers. British Geological Survey Occasional Publication No. 14.

Christiansen, E.A., 1992. Pleistocene stratigraphy of the Saskatoon area, Saskatchewan, Canada: an update. *Canadian Journal of Earth Sciences* 29, 1767–1778.

Ciotoli, G., M. Guerra, S. Lombardi and E. Vittori, 1998. Soil gas survey for tracing seismogenic faults: A case study in the Fucino basin, central Italy. *Journal of Geophysical Research* 103, 781–794.

Daley, T.M., L.R. Myer, J.E. Peterson, E.L. Majer and G.M. Hoversten, 2007. Time-lapse crosswell seismic and VSP monitoring of injected CO_2 in a brine aquifer. *Environmental Geology* 54, 1657–1665.

Daley, T.M., J.B. Ajo-Franklin and C. Doughty, 2012. Ultrasonic velocity and attenuation during CO_2 injection into water-saturated porous sandstone: Measurements using difference seismic tomography. *International Journal of Greenhouse Gas Control. [in press]*.

De Mille, G., J.R. Shouldice and H.W. Nelson, 1964. Collapse structures related to evaporites of the Prairie Formation, Saskatchewan. *Geological Society of America Bulletin* 69, 307–316.

DeMis, W.D., 1995. Effect of cross-basinal hydrodynamic flow on oil accumulations and oil migration history of the Bakken-Madison Petroleum System, Williston Basin, North America. *In* L.D.V. Hunter and R.A. Schalla (Editors), *Seventh International Williston Basin Symposium*, 291–301. Montana Geological Society.

Deutsch, C.V. and A.G. Journel, 1998. GSLIB, Geostatistical Software Library and Users Guide. Oxford University Press.

Dietrich, J.R., J.A. Majorowicz and M.D. Thomas, 1999. Williston Basin Profile, Southeast Saskatchewan and Southwest Manitoba: a window on basement–sedimentary cover interaction. Geological Survey of Canada, Open File 3824.

DNV, 2010. CO2QUALSTORE – Guideline for Selection and Qualification of Sites and Projects for CO_2 Storage. DNV Report No. 2009-1425.

DNV, 2011. CO2WELLS – Guideline for the Risk Management of Existing Wells at CO_2 Geological Storage Sites. DNV Report No. 2011-0448.

DNV, 2012. DNV Recommended Practice DNV-RP-J203 – Geological Storage of Carbon Dioxide. April 2012.
http://exchange.dnv.com/publishing/Codes/ToC_edition.asp#Recommended Practices.

Downey, J.S., 1984. Geohydrology of the Madison and Associated Aquifers in Parts of Montana, North Dakota, South Dakota, and Wyoming. U.S. Geological Survey Professional Paper 1273-G.

Downey, J.S., 1986. Geohydrology of Bedrock Aquifers in the Northern Great Plains in Parts of Montana, North Dakota, South Dakota, and Wyoming. U.S. Geological Survey Professional Paper 1402-E.

Downey, J.S., J.F. Busby and G.A. Dinwiddie, 1987. Regional aquifers and petroleum in the Williston Basin region of the United States. *In* M. Longman (Editor), *Williston Basin: Anatomy of a Cratonic Oil Province*, 299–312. Symposium of the Rocky Mountain Association of Geologists, Denver, Colorado.

Duguid, A., 2008. An estimate of the time to degrade the cement sheath in a well exposed to carbonated brine. Proceedings of the 9th International Conference on Greenhouse Gas Technologies, Washington, DC, U.S.A., November 16-20.

Duxbury, A., D. White, C. Samson, S. Hall, J. Wookey and J.-M. Kendall, 2011. Fracture mapping using seismic AVOA analysis at the Weyburn CO_2 storage site. *Geophysics*, [in press].

Edmonds, A.C. and A.P. Moroney, 1998. Geology, development and optimization of the Weyburn Unit, southeastern Saskatchewan. *In* J.R. Hogg (Editor), *Oil and Gas Pools of the Western Canada Sedimentary Basin*, 1–12. Canadian Society of Petroleum Geologists, Special Publication S-51.

England, W.A., A.S. Mackenzie, D.M. Mann and T.M. Quigley, 1987. The movement and entrapment of petroleum fluids in the subsurface. *Journal of the Geological Society of London* 144, 327–347.

European Commission, 2009. DIRECTIVE 2009/31/EC of the European Parliament and of the Council of 23 April 2009 on the Geological Storage of Carbon Dioxide and Amending Council Directive 85/337/EEC, European Parliament and Council Directives 2000/60/EC, 2001/80/EC, 2004/35/EC, 2006/12/EC, 2008/1/EC and Regulation (EC) No 1013/2006. http://eur-lex.europa.eu/LexUriServ/LexUriServ.do?uri=OJ:L:2009:140:0114:0135: EN:PDF.

European Commission, 2011a. Implementation of Directive 2009/31/EC on the Geological Storage of Carbon Dioxide, Guidance Document 1 – CO_2 Storage Life Cycle Risk Management Framework.
http://ec.europa.eu/clima/policies/lowcarbon/ccs/implementation/docs/gd1_en.pdf.

European Commission, 2011b. Implementation of Directive 2009/31/EC on the Geological Storage of Carbon Dioxide, Guidance Document 2 – Characterization of the Storage Complex, CO_2 Stream Composition, Monitoring and Corrective Measures.
http://ec.europa.eu/clima/policies/lowcarbon/ccs/implementation/docs/gd2_en.pdf.

Gassmann, F., 1951. Uber die elastizitat poroser medien. *Vierteljahrsschrift der Naturforschenden Gesellschaft in Zurich* 96, 1–23.

Global CCS Institute, 2011. The Global Status of CCS: 2011. Canberra, Australia. http://www.globalccsinstitute.com/publications/global-status-ccs-2011/online/26846.

Goodway, B., M. Perez and C. Abaco, 2010. Seismic petrophysics and isotropic-anisotropic AVO methods for unconventional gas exploration. *The Leading Edge* 29, 1500–1508.

Goussev, S., L. Griffith, J. Peirce and A. Cordsen, 2004. Using enhanced HRAM anomalies to correlated faults between 2-D seismic lines. Proceedings and Abstracts, Canadian Society of Petroleum Geologists Annual Convention, Calgary, Alberta.

Gray, D., G. Roberts and K. Head, 2002. Recent advances in determination of fracture strike and crack density from P-wave seismic data. *The Leading Edge* 21, 280–285.

Gunaltun, Y., 1991. Carbon dioxide corrosion in oil wells. Paper SPE 21330, presented at the SPE Middle East Oil Show, Bahrain, November 16-19.

Hall, S.A. and J-M. Kendall, 2003. Fracture characterization at Valhall: Application of P-wave amplitude variation with offset and azimuth (AVOA) analysis to a 3D ocean-bottom data set. *Geophysics* 68, 1150–1160.

Hannon, N., 1987. Subsurface water flow patterns in the Canadian sector of the Williston Basin. *In* M. Longman (Editor), *Williston Basin: Anatomy of a Cratonic Oil Province*, 313–321. Symposium of the Rocky Mountain Association of Geologists, Denver, Colorado.

Hanor, J.S., 1994. Origin of saline fluids in sedimentary basins. *In* J. Parnell (Editor), *Geofluids: Origin, Migration and Evolution of Fluids in Sedimentary Basins*, 151–174. Geological Society Special Publication No. 78.

Hassanzadeh, H., M. Pooladi-Darvish, A.M. Elsharkawy, D. Keith and Y. Leonenko, 2007. Predicting PVT data for CO_2–brine mixtures for black-oil simulation of CO_2 storage. *International Journal of Greenhouse Gas Control* 2, 65–77.

Hermanrud, C., H. Nordgård Bolås, H. Hansen, O. Eiken, J. Lippard and G.M.G. Teige, 2010. CO_2 storage capacity below structural spill point in the Utsira Formation. AAPG Search and Discovery Article, 80091.

Hitchon, B., 1996. Rapid evaluation of the hydrochemistry of a sedimentary basin using only 'standard' formation water analyses: example from the Canadian portion of the Williston Basin. *Applied Geochemistry* 11, 789–795.

Hobson, G.D., 1954. Some Fundamentals of Petroleum Geology. Oxford University Press.

Hoek, E., C. Carranza-Torres and B. Corkum, 2002. Hoek-Brown Criterion – 2002 Edition. Proceedings of 5th North American Rock Mechanics Symposium and 17th Tunneling Association of Canada Conference (NARMS-TAC), Toronto, 1, 267–273.

Iampen, H.T., 2003. The genesis and evolution of pre-Mississippian brines in the Williston Basin, Canada-USA. M.Sc. thesis, Department of Earth and Atmospheric Sciences, University of Alberta, Edmonton.

IEA Greenhouse Gas R&D Programme, 2009. A Review of the International State of the Art in Risk Assessment Guidelines and Proposed Terminology for Use in CO_2 Geological Storage, 2009/7.

IPCC, 2006. 2006 IPCC Guidelines for National Greenhouse Gas Inventories, Vol. 2, Chapter 5. Carbon Dioxide Transport, Injection and Geological Storage by S. Holloway, A. Karimjee, M. Akai, R. Pipatti, and K. Rypdal. http://www.ipcc-nggip.iges.or.jp/public/2006gl/vol2.html/pdf/2_Volume2/V2_5_Ch5_CCS.pdf.

Jenner, E., 2002. Azimuthal AVO: Methodology and data examples. *The Leading Edge* 21, 782–786.

Johnson, J.W., 2008. Structural, compositional, and process controls on the dynamic mass partitioning and spatial distribution of CO_2 trapping mechanisms: Impact on isolation performance. Proceedings of 9th International Conference on Greenhouse Gas Technologies (GHGT-9), Washington DC, U.S.A., November 16-20.

Johnson, J.W., J.J. Nitao and K.G. Knauss, 2004. Reactive transport modeling of CO_2 storage in saline aquifers to elucidate fundamental processes, trapping mechanisms, & sequestration partitioning. *In* S.J. Baines and R.H. Worden (Editors), *Geological Storage of Carbon Dioxide*, 107–128. Geological Society, London, Special Publications, 233.

Johnson, J.W., J.J. Nitao and J.P. Morris, 2005. Reactive transport modeling of cap rock integrity during natural and engineered CO_2 storage. *In* D.C. Thomas and S.M. Benson (Editors), *Carbon Dioxide Capture for Storage in Deep Geologic Formations*, Vol. 2, 787–813.

Kaldi, J.G. and C.M. Gibson-Poole (Editors), 2008. Storage Capacity Estimation, Site Selection, and Characterization for CO_2 Storage Projects. CO2CRC Report No: RPT08-1001.

Kemna, A., A. Binley, A. Ramirez and W. Daily, 2000. Complex resistivity tomography for environmental applications. *Chemical Engineering Journal* 77, 11–18.

Kent, D.M. and J.E. Christopher, 1994. Geological history of the Williston Basin and Sweetgrass Arch. *In* G.D. Mossop and I. Shetsen (Compilers), *Geological Atlas of the Western Canada Sedimentary Basin*, 421–430. Canadian Society of Petroleum Geologists and Alberta Research Council.

Khan, D.K., C.V. Deutsch, C.A. Mendoza and B.J. Rostron, 2007. Approximate sensitivity coefficients for integrating hydraulic head data into geological models. *Journal of Hydrology* 347, 460–473.

Kharaka, Y.K., W.D. Gunter, P.K. Aggarwal, E.H. Perkins and J.D. De Braal, 1988. SOLMINEQ.88: A computer program for geochemical modeling of water-rock interactions. U.S. Geological Survey Water-Resources Investigations Report 88-4227.

Kiessling, D., C. Schmidt-Hattenberger, H. Schuett, F. Schilling, K. Krueger, B. Schoebel, E. Danckwardt, J. Kummerow and the CO2SINK Group, 2010. Geoelectrical methods for monitoring geological CO_2 storage: First results from crosshole and surface-downhole measurements from the CO2SINK test site at Ketzin (Germany). *International Journal of Greenhouse Gas Control* 4, 816–826.

Kutchko, B.G., B.R. Strazisar, G.V. Lowry, D.A. Dzombak and N. Thaulow, 2008. Rate of CO_2 attack on hydrated Class H well cement under geologic sequestration conditions. *Environmental Science and Technology* 42 (16), 4787–4792.

Lake, J. and S. Whittaker, 2006. Occurrences of CO_2 within southwest Saskatchewan: Natural analogues to the Weyburn CO_2 injection site. Summary of Investigations 2006, Vol. 1, Saskatchewan Geological Survey, Miscellaneous Report 2006-4.1. [CD-ROM, Paper A-5].

Lane, D.M., 1987. Subsurface Carbon Dioxide in Saskatchewan: Sources and potential use in enhanced oil recovery. Saskatchewan Energy and Mines, Open File Report 87-2.

Lee, H., 1962. The Technical and Economic Aspects of Helium Production in Saskatchewan, Canada. Saskatchewan Department of Mineral Resources, Report 72.

Lewicki, J.L., D. Bergfeld, C. Cardellini, G. Chiodini, D. Granieri, N. Varley and C. Werner, 2005. Comparative soil CO_2 flux measurements and geostatistical estimation methods on Masaya volcano, Nicaragua. *Bulletin Volcanologique* 68, 76–90.

Li, G., G. Burrowes, E. Majer and T. Davis, 2001. Weyburn field horizontal-to-horizontal crosswell seismic profiling: Part 3 – Interpretation. Society of Exploration Geophysicists 2001 Annual Meeting, San Antonio, Expanded Abstracts.

Li, J. and Z. Duan, 2011. A thermodynamic model for the prediction of phase equilibria and speciation in the H_2O-CO_2-NaCl-$CaCO_3$-$CaSO_4$ system from 0 to 250C, 1 to 1000 bar with NaCl concentrations up to halite saturation. *Geochimica et Cosmochimica Acta* 75, 4351–4376.

Liard, J., J. Huang, J. Silliker, D. Jobin, S. Wang and A. Doherty, 2011. Detecting groundwater storage change using micro-gravity survey in Waterloo Moraine. Proceedings, Geohydro 2011.

Lindeberg, E., B. van der Meer, A. Moen, D. Wessel-Berg and A. Ghaderi, 2000. Fluid and Core Properties and Reservoir Simulation. Final Report – Work Area 2 (Reservoir). Saline Aquifer CO_2 Storage (SACS) Consortium.

Ma, J. and I. Morozov, 2010. AVO modelling of pressure-saturation effects in Weyburn CO_2 sequestration. *The Leading Edge* 29, 178–183.

Maathuis, H. and L.H. Thorleifson, 2000. Potential Impact of Climate Change on Prairie Groundwater Supplies: Review of current knowledge. Saskatchewan Research Council, SRC Publication 11304-2E00.

Maxwell, S.C. and T.I. Urbancic, 2001. The role of passive microseismic monitoring in the instrumented oil field. *The Leading Edge* 20, 636–639.

McLellan, P.J., 1996. Assessing the risk of wellbore instability in horizontal and inclined wells. *Journal of Canadian Petroleum Technology* 35, 21–32.

McLellan, P.J., K. Lawrence and K. Cormier, 1992. A multiple-zone acid treatment of a horizontal well, Midale, Saskatchewan. *Journal of Canadian Petroleum Technology* 31, 71–82.

Meneley, W.A., 1983. Hydrogeology of the Eastend to Ravenscrag Formations in Southern Saskatchewan. W.A. Meneley Consultants Ltd., Saskatoon. Report 0089-001.

Metropolis, N., A. Rosenbluth, M. Rosenbluth, A. Teller and E. Teller, 1953. Equation of state calculations by fast computing machines. *Journal of Chemical Physics* 1, 1087–1092.

Morris, J.P., Y. Hai, W. Foxall and W. McNab, 2011. In Salah CO_2 storage JIP: Hydromechanical simulations of surface uplift due to CO_2 injection at In Salah. Proceedings of the 10th International Conference on Greenhouse Gas Technologies, September 19-23, 2010, Amsterdam, Netherlands. *Energy Procedia* 4, 3269–3275.

Mosegaard, K. and A. Tarantola, 1995. Monte Carlo sampling of solutions to inverse problems. *Journal of Geophysical Research* 100, No. B7, 12431–12447.

Nelson, E.B. and D. Guillot, 2006. Well Cementing. Second Edition, Schlumberger, Sugar Land, Texas, U.S.A.

Nemeth, B., A. Prugger, S. Halabura and T. Danyluk, 2002. Benefits of 3D poststack depth migration: Case study from the potash belt of Saskatchewan. Proceedings of Canadian Society of Exploration Geophysicists meeting, Calgary.

Neuzil, C.E., 1993. Low fluid pressure within the Pierre Shale: a transient response to erosion. *Water Resources Research* 29, 2007–2020.

Neuzil, C.E., J.D. Bredehoeft and R.G. Wolfe, 1984. Leakage and fracture permeability in the Cretaceous shales confining the Dakota aquifer. *In* D.G. Jorgenson and D.C. Signor (Editors), *Proceedings of the First C.V. Thesis Conference on Geohydrology*, 113–120. National Water Well Association, Dublin, Ohio.

Niebauer, T.M., G.S. Sasagawa, J.E. Faller, R. Hilt and F. Klopping, 1995. A new generation of absolute gravimeters. *Metrologia* 32, No. 3, 159–180.

NIST, 2010. On-line Isothermal Property Calculator. http://webbook.nist.gov/chemistry/fluid/.

Norton, D., 1984. Theory of hydrothermal systems. *Annual Review of Earth & Planetary Sciences* 12, 155–177.

Onishi, K., T. Ueyama, T. Matsuoka, D. Nobuoka, H. Saito, H. Azuma and Z. Xue, 2009. Application of crosswell seismic tomography using difference analysis with data normalization to monitor CO_2 flooding in an aquifer. *International Journal of Greenhouse Gas Control* 3, 311–321.

Osadetz, K.G., P.K. Hannigan, L.D. Stasiuk, B.P. Kohn, P. O'Sullivan, S. Feinstein, R.A. Everitt, C.F. Gilboy and R.K. Bezys, 1998. Williston Basin thermotectonics: variations in heat flow and hydrocarbon generation. *In* J. Christopher, C. Gilboy, D. Patterson and S. Bend (Editors), *Eighth International Williston Basin Symposium*, 147–165. Saskatchewan Geological Society, Special Publication No. 13.

Parkhurst, D.L. and C.A.J. Appelo, 1999. User's Guide to PHREEQC (version 2) – A Computer Program for Speciation, Batch-reaction, One-dimensional Transport, and Inverse Geochemical Calculations. U.S. Geological Survey Water-Resources Investigations Report 99-4259.

Pearce, J.M., S. Holloway, H. Wacker, M.K. Nelis, C. Rochelle and K. Batenan, 1996. Natural occurrences as analogues for the geological disposal of carbon dioxide. *Energy Conversion and Management* 37, 1123–1128.

Peterson, R., 1954. Studies of the Bearpaw Shale at a dam site in Saskatchewan. *In* Proceedings of the American Society of Civil Engineers, Soil Mechanics and Foundation Division, Vol. 80, Separate 476.

Ramirez, A., W. Daily and R. Newmark, 1995. Electrical resistance tomography for steam injection monitoring and process control. *Journal of Environmental Engineering and Geophysics* 10, 39–51.

Rawlings, C.G., N.R. Barton, S.C. Bandis, M.A. Addis and M.S. Gutierriez, 1993. Laboratory and numerical discontinuum modeling of wellbore stability. *Journal of Petroleum Technology* 45 (11), 1086–1092.

Rechard, R.P., 1999. Historical relationship between performance assessment for radioactive waste disposal and other types of risk assessment. *Risk Analysis* 19, 763–807.

Risk, D., L. Kellman and H. Beltrami, 2002. Soil CO_2 production and surface flux at four climate observatories in eastern Canada. *Global Biogeochemical Cycles* 16, No. 4, 1122, 69-1 – 69-12.

Risk, D., L. Kellman, H. Beltrami and A. Diochon, 2008. In-situ incubations by root exclusion highlight the climatic sensitivity of soil organic matter pools. *Environment Research Letters* 3 (2008) 044004. doi:10.1088/1748-9326/3/4/044004.

Roberts, L.W. and J.A. Godfrey, 1994. History of Bravo Dome CO_2-discovery and early development. *In* J. Ai-Ilen and A.L. Peterson (Editors), Geologic Activities in the 90s, 9–12. New Mexico Bureau of Mines and Mineral Resources, Bulletin No. 150.

Rochelle, C.A., I. Czernichowski-Lauriol and A.E. Milodowski, 2004. How chemical reactions can affect CO_2 sequestration in geological formations. *In* S.J. Baines and R.H. Worden (Editors), *Geological Storage of Carbon Dioxide*, 87–106. Geological Society, London, Special Publications, 233.

Rostron, B.J. and C. Holmden, 2000. Fingerprinting formation-waters using stable isotopes, Midale Area, Williston Basin, Canada. *Journal of Geochemical Exploration* 69-70, 219–223.

Rüger, A. and I. Tsvankin, 1997. Using AVO for fracture detection: Analytic basis and practical solutions. *The Leading Edge* 16, 1429–1434.

Rutherford, S.R. and R.H. Williams, 1989. Amplitude-versus-offset variations in gas sands. *Geophysics* 54, 680–688.

Rutqvist, J., D. Vasco and L. Myer, 2010. Coupled reservoir-geomechanical analysis of CO_2 injection and ground deformations at In Salah, Algeria. *International Journal of Greenhouse Gas Control* 4, 225–230.

Saito, H., D. Nobuoka, H. Azuma, D. Tanase and Z. Xue, 2008. Time-lapse crosswell seismic tomography for monitoring CO_2 geological sequestration at Nagaoka pilot project site. *Journal of the Mining and Materials Processing Institute of Japan* 124, 78–86.

Saskatchewan Environment, 2006. Saskatchewan's Drinking Water Quality Standards and Objectives (summarized). Saskatchewan Environment, EPB207. http://www.se.gov.sk.ca/environment/protection/water/Drinking_Water_Standards_post.pdf.

Schremp, F. and G. Roberson, 1975. Effect of supercritical carbon dioxide on construction materials. *SPE Journal* 15 (3), 227–233. SPE-4667-PA.

Sherlock, D., A. Toomey, M. Hoversten, E. Gasperikova and K. Dodds, 2006. Gravity monitoring of CO_2 storage in a depleted gas field: a sensitivity study. *Exploration Geophysics* 37, 37–43.

Shuey, R.T., 1985. A simplification of Zoeppritz equations. *Geophysics* 50, 609–614.

Singh, V., A.J. Cavanagh, H. Hansen, B. Nazarian, M. Iding and P. Ringrose, 2010. Reservoir modeling of CO_2 plume behavior calibrated against monitoring data from Sleipner, Norway. SPE 134891.

Slater, L., M.D. Zaidman, A.M. Binley and L.J. West, 1997. Electrical imaging of saline tracer migration for the investigation of unsaturated zone transport mechanisms. *Hydrology and Earth System Sciences* Vol. 1, 291–302.

Smith, D.G. and J.R. Pullen, 1967. Hummingbird structure of southeast Saskatchewan. *Bulletin of Canadian Petroleum Geology* 15, 468–482.

Span, R. and W. Wagner, 1996. A new equation of state for carbon dioxide covering the fluid region from the triple-point temperature to 1100 K at pressures up to 800 MPa. *Journal of Physical Chemistry Reference Data* 25, 1509–1596.

Spetzler, J., Z. Zue, H. Saito and O. Nishizawa, 2007. Case story: time-lapse seismic crosswell monitoring of CO_2 injected in an onshore sandstone aquifer. *Geophysical Journal International* 172, 214–225.

Stauffer, M.R. and D.J. Gendzwill, 1987. Fractures in the northern plains, stream patterns, and the midcontinent stress field. *Canadian Journal of Earth Sciences* 24, 1086–1097.

Strazisar, B., B. Kutchko and N. Huerta, 2008. Chemical reactions of wellbore cement under CO_2 storage conditions: Effects of cement additives. Proceedings of 9th International Conference on Greenhouse Gas Technologies (GHGT-9), Washington DC, U.S.A., November 16-20.

Toop, D.C., 1992. Hydrogeology and Hydrogeochemistry of South-Central Saskatchewan. M.Sc. thesis, Department of Earth and Atmospheric Sciences, University of Alberta, Edmonton.

Toop, D.C. and J. Toth, 1995a. Hydrogeology and the distribution of oil pools, south-central Saskatchewan. *In* L.D.V. Hunter and R.A. Schalla (Editors), *1995 Guidebook: 7th International Williston Basin Symposium*, 303–312. Montana Geological Society, Billings, Montana.

Toop, D.C. and J. Toth, 1995b. Hydrogeological characterization of formation waters using ionic ratios, south-central Saskatchewan. *In* L.D.V. Hunter and R.A. Schalla (Editors), *1995 Guidebook: 7th International Williston Basin Symposium*, 313–319. Montana Geological Society, Billings, Montana.

Toth, J., 1978. Gravity-induced cross-formational flow of formation fluids, Red Earth region, Alberta, Canada: Analysis, patterns, and evolution. *Water Resources Research* 14, 805–843.

Toth, J., and T.F. Corbet, 1986. Post-Paleocene evolution of regional groundwater flow-systems and their relation to petroleum accumulations, Taber area, southern Alberta, Canada. *Bulletin of Canadian Petroleum Geology* 34, 339–363.

Trium, 2011. Site Assessment (SW-30-05-13-W2M), near Weyburn, Saskatchewan. Trium Environmental Inc. and Chemistry Matters, Project Number: T10-167-CEN.

U.S. Department of Energy, National Energy Technology Laboratory, 2009a. Best Practices for: Monitoring, Verification and Accounting of CO_2 Stored in Deep Geologic Formations. http://www.netl.doe.gov/technologies/carbon_seq/refshelf/MVA_Document.pdf.

U.S. Department of Energy, National Energy Technology Laboratory, 2009b. Best Practices For: Public Outreach and Education for Carbon Storage Projects. United States Department of Energy, Rept. DOE/NETL-2009/1391.
http://www.netl.doe.gov/technologies/carbon_seq/refshelf/BPM_PublicOutreach.pdf.

U.S. Department of Energy, National Energy Technology Laboratory, 2010. Best Practices for: Site Screening, Site Selection, and Initial Characterization for CO_2 Storage in Deep Geologic Formations.
http://www.netl.doe.gov/technologies/carbon_seq/refshelf/BPM-SiteScreening.pdf.

U.S. Department of Energy, National Energy Technology Laboratory, 2011. Best Practices for: Risk Analysis and Simulation for Geological Storage of CO_2.

http://www.netl.doe.gov/technologies/carbon_seq/refshelf/BPM_RiskAnalysisSimulation.pdf.

U.S. Environmental Protection Agency, 2010. 40 CFR Parts 124, 144, 145, 146, and 147, Federal Requirements Under the Underground Injection Control (UIC) Program for Carbon Dioxide (CO_2) Geologic Sequestration (GS) Wells, Federal Register / Vol. 75, No. 237 / Friday, December 10, 2010 / Rules and Regulations, 77230–77303. http://www.gpo.gov/fdsys/pkg/FR-2010-12-10/pdf/2010-29954.pdf.

van der Kamp, G., 1985. Yield Estimates for the Estevan Valley Aquifer System Using a Finite Element Model. Saskatchewan Research Council, Publication R-844-4-C-85.

van der Kamp, G., H. Maathuis and W.A. Meneley, 1986. Bulk hydraulic conductivity of a thick till overlying a buried-valley aquifer near Weyburn, Saskatchewan. Proceedings of Third Canadian Hydrogeological Conference, IAH-CNC, 94–99. Saskatoon, April 20-23.

Van Stempvoort, D. and M. Simpson, 1994. Hydrogeology of the Southeast Aquifer Management Plan Area. Volume I: Text. Saskatchewan Research Council, Publication R-1220-9-E-94.

Vasco, D.W., A. Rucci, A. Ferretti, F. Novali, R. Bissell, P. Ringrose, A. Mathieson and I. Wright, 2010. Satellite-based measurements of surface deformation reveal fluid flow associated with the geological storage of carbon dioxide. LBNL-3018E. *Geophysical Research Letters* 37, L03303.

Verdon, J.P., J-M. Kendall, D.J. White and D.A. Angus, 2011. Linking microseismic event observations with geomechanical models to minimise the risks of storing CO_2 in geological formations. *Earth and Planetary Science Letters* 305, 143–152.

Verdon, J.P., D.J. White, J-M. Kendall, D.A. Angus, J. Fisher and T. Urbancic, 2010. Passive seismic monitoring of carbon dioxide storage at Weyburn. *The Leading Edge* 29, 200–206.

Vigrass, L, A. Jessop and B. Brunskill, 2007. Regina Geothermal Project. In *Summary of Investigations* 2007, Volume 1, Saskatchewan Geological Survey, Saskatchewan Industry Resources, Miscellaneous Report 2007-4.1. CD-ROM, Paper A-2.

Watson, T. and S. Bachu, 2008. Identification of wells with high CO_2-leakage in mature oil fields developed for CO_2 enhanced oil recovery. Paper SPE 112924 presented at the SPE Improved Oil Recovery Symposium, Tulsa, OK, April 19-23.

Watson, T. and S. Bachu, 2009. Evaluation of the potential for gas and CO_2 leakage along wellbores. *SPE Drilling & Completion* 24 (1), 115–126. [SPE-106817-PA].

Wegelin, A., 1984. Geology and reservoir properties of the Weyburn Field, southeastern Saskatchewan. *In* J.A. Lorsong and M.A. Wilson (Editors), *Oil and Gas in Saskatchewan*, 71–82. Saskatchewan Geological Society Special Publication No. 7.

Wegelin, A., 1987. Reservoir characteristics of the Weyburn Field, southeastern Saskatchewan. *Journal of Canadian Petroleum Technology* 26, 60–66.

White, D.J., G. Burrowes, T. Davis, Z. Hajnal, K. Hirsche, I. Hutcheon, E. Majer, B. Rostron and S. Whittaker, 2004. Greenhouse gas sequestration in abandoned oil reservoirs: The International Energy Agency Weyburn pilot project. *GSA Today* 14 (7), 4–10.

White, W.S., D.G. Calvert and J.M. Barker, 1992. A laboratory study of cement and resin plugs placed with thru-tubing dump bailers. Paper SPE 24574, presented at the SPE Annual Technical Conference and Exhibition, Washington, USA, October 4-7.

Wilson, M. and M. Monea (Editors), 2004. IEA GHG Weyburn CO_2 Monitoring & Storage Project Summary Report 2000-2004. From the proceedings of the 7th International Conference on Greenhouse Gas Control Technologies, September 5-9, 2004, Vancouver, Canada. Petroleum Technology Research Centre, Regina.

Wittrup, M.B. and T.K. Kyser, 1990. The petrogenesis of brines in Devonian potash deposits of western Canada. *Chemical Geology* 82, 103–128.

Wolterneek, T., 2010. Anhydrite cap rock dissolution behavior: Mechanisms and implications for sealing capacity. M.Sc. Thesis, Department of Earth Sciences, Utrecht University, Netherlands.

World Resources Institute, 2008. CCS Guidelines: Guidelines for Carbon Dioxide Capture, Transport, and Storage. S.M. Forbes, P. Verma, T.E. Curry, S.J. Friedmann and S.M. Wade (Editors), Washington, DC, U.S.A.
http://pdf.wri.org/ccs_guidelines.pdf.

World Resources Institute, 2010. Guidelines for Community Engagement in Carbon Dioxide Capture, Transport, and Storage Projects. S.M. Forbes, F. Almendra and M. Ziegler (Editors), Washington, DC, U.S.A.
http://pdf.wri.org/ccs_and_community_engagement.pdf.

INDEX

Page numbers in **boldface** indicate citations of main themes. Page numbers in *italic* refer to illustrations (photographs and line drawings).

'live oil'
    tests conducted, **152**
'Red Beds'. *see* Watrous Formation

3D seismic data
    salt dissolution, 26
3D seismic monitoring
    without a baseline, 207
3D time-lapse seismic results
    used in history matching, **243**

acid gas wells, 269
aeromagnetic data, 19
aeromagnetic map, *20*
Alida aquifer
    salinity map, *42*
Alida Beds, 31, 53
alkalinity
    change over time, *145*
    of formation water, **144**
anhydrite, 113
    solubility, *114*
anhydrite cements
    in Midale Beds, 31
    Watrous Formation, 31
anhydrite dissolution
    during fracture flow experiments, 231
argon-plus-oxygen
    in soil gas, *132–33*, *135*
AVA cross-plot
    from well-log based models, *198*
AVOA-determined orientations
    and shear-wave splitting, 64

Baker Atlas High-Resolution Vertilog™, 276
Bakken. *see* Mississippian
Basal Cambrian Sandstone. *see* Cambrian

Bearpaw Formation, 31, 34, 38
    defined as 'other' seal, 52
    seal properties, **55, 57**
Bearpaw Mountains, 67
Belly River aquifer
    migration pathways, *96*
best-practice manuals, 323
biosphere
    risk assessment, **295**
    risk assessment approach, *298*
    risk evaluation, **314**
    risk evaluation criteria, 297
    risk profile, **315**
Bravo Dome, 113
brine. *see* formation water
buoyancy, 91

calcite dissolution, **111**
calcium
    change in formation water over time, *149*
    in formation waters, **148**
    Marly coreflood results, *221*
    MCMC inversion results, *249*
    Vuggy coreflood results, *220*
Cambrian
    natural CO_2 accumulations, 67
capillary number equation, 91
capillary pressure, 92
cap rock. *see also* seal
    anisotropy orientation, *65*
    AVOA inversion results, *63*
    modelling, 60
    near-offset AVO response, 61–2
carbonate mass transfer rates
    during wormhole development, 222
carbonate rocks
    key to history matching, 217

343

Carboniferous. *see* Mississippian
carbon isotopes
    ^{14}C in soil gas CO_2, 142
    and solubility trapping, 143
    change in formation water over time, *150*
    CO_2 in soil gas, **141**
    in CO_2, HCO_3 and carbonates, *143*
    in formation water bicarbonate, **148**
cased-hole logging
    for new injection wells, 272
cased wells
    evaluation for integrity, **261**
casing
    examination of cement behind, 271
cement
    examination behind casing, 271
cement evaluation logs, 272
    for zonal isolation, 262
cementing
    best practices, 266
    for well conversion, 269
cementing program, **266**
cement plugs, 260
    and wellbore integrity, 259
    for open-hole abandonment, 270
cement sampling, 277
cement sheath pressure transient testing, **276**
Charles Formation, **15**, 26–7
    Frobisher Member, 15
    Midale Evaporite Zone, 15
    Midale Member, 15
    Poplar Beds, 15
    Ratcliffe Beds, 15
chloride
    change in formation water over time, *237*
City of Weyburn
    water source, 38
    weather charts, *130*
Classification of CCS projects, 4
CO_2
    acoustic impedance change with depth, *161*
    density change with depth, *161*
    distribution 2004-10, *88*
    distribution to 2040, *86*
    flux variations in soil gas, *141*
    in Duperow Formation, *67*
    in soil gas, *132, 135–37, 139–40*
    mass v depth, *176*
    mechanical dispersion map, *87*
    physical properties, **160**
    Vp change with depth, *161*
CO_2-formation water mixtures
    Vp-versus-depth plots, *161*
CO_2 migration, **83, 90**
CO_2 migration modelling, **91**
    hydrodynamic case, **96**
    hydrostatic case, **95**
CO_2 modelling source term, 94
CO_2 saturation
    discrimination from pressure by seismic, **196**
CO_2 saturations
    depth relations, *163*
CO_2 storage
    risk assessment, **300**
CO_2 storage site
    well penetration, 91
coarse-grid simulation, **86**
collapse structures, 26
Colorado aquitard
    map of potential breach points, *102*
Colorado Group, 31, 101
    defined as 'other' seal, 52
    seal properties, **55**
communications
    crisis procedures, **326**
    in community outreach, *322*
    key events to a crisis response, *327*
Compendex, 291
Computer Modelling Group, 88
containment, **111**
coreflood experiments, **218**
    experimental conditions, 219
    simulation model, **223**
Cretaceous
    Bearpaw Formation, 31, 34, **55**
    Colorado Group, 31, **55**

Joli Fou Formation, 31, **55**
Mannville Group, 40
Newcastle aquifer, 40
Second White Speckled Shale, 193
Cretaceous aquifers, **35**
crosswell geophysical methods, **186**
crosswell tomographic methods
    Frio, Texas, 186
    Ketzin, Germany, 186
    Nagaoka, Japan, 186

dawsonite, 112
definitions
    AVOA, 9, 58
    FEP, 302
    history matching, 217
    InSAR, 172
    LEERT, 174
    PRA, 305
    Risk Index, 306
    seals, 52
    storage complex, 13
    water driving force, 40
Devonian
    Duperow Formation, 67
    natural CO_2 accumulations, **67**
    Prairie Evaporite, 26
    salt beds, 26
dolomite dissolution
    during flow-through experiment, *234*
dolomitization
    Midale Beds, 31
downhole monitoring methods, **180**
drilling mud
    for well abandonment, 270
drilling tour reports
    for well integrity evaluation, 260
dry rock frame parameters, **164**
dual porosity model, 236, **240**
    simulation results, *240*
Duperow Formation, 67
    analogue for Midale Beds, 68
    structure map, *67*

earthquakes, 32
Eastend Formation, 35, 38–9
effectiveness risk profile, *313*
electrical tomography methods, **174**
Empress Formation
    hydrogeology, 37
Eocene
    timing of basin flow system, 34
Estevan Valley aquifer system, 36–7
EU CCS Directive, 90
European CCS Directive, 14, 52
evaporites. *see also* salt dissolution
evaporite seals, **113**
event containment risk, *311*
Expro borehole camera, 276

fill-spill pathways, **91**
fine-grid simulation, **86**
FloGrid®, 88
fluid inclusion studies, 53
foreland basin, 16, 18
formation water, **142**
    alkalinity, **144**
    buoyancy correction, 46
    calcium, **148**
    carbon isotopes in bicarbonate, **148**
    change in alkalinity over time, *145*
    change in Ca over time, *149*
    change in Cl and Na over time, *237*
    change in downhole pH over time, *147*
    change in pH over time, *146*
    change in $\delta^{13}C$ over time, *150*
    driving force, **45**
    Estevan Valley aquifer, 37
    groundwater composition, 122
    Hitchcock aquifer, 39
    inter-till aquifers, 39
    map of wells sampled, *144*
    Mesozoic aquifers, **44**
    Mississippian aquifers, **44**
    Paleozoic aquifers, **43**
    pH, **147**

salinity cross-section, *43*
stochastic inversion, **247**
surficial aquifers, 39
Tableland aquifer, 39
Weyburn Valley aquifer, 38
fracture flow experiments, **228**
fracture healing
    Midale Evaporite, 115
fractures
    and salt solution, 64
    AVOA seismic analysis, 58
    Frobisher Evaporite, 30
    healing by anhydrite, 65
    in Weyburn Field, 64
    Midale Evaporite, 29
Frio, Texas, 186
Frobisher aquifer
    salinity map, *42*
Frobisher Beds, 31, 53
Frobisher Evaporite, 22, 30
    rock mechanical properties, **72**
    seal at Weyburn Field, 28
Frobisher Member, **15**

GEM®, **83**, 236
geochemical research, **74**
geological models, **22**, 24
geology, **16**
    study areas, **10**
geomechanical-fluid flow model, **182**
geomechanics, **70**
geophysical monitoring
    general observations, **156**
geophysical monitoring methods
    3D VSP surveys, 189
    crosswell (seismic or electrical), 186
    forward seismic, **177**
    InSAR, 172
    LEERT, 174
    overview, 156
    passive seismic, **180**
    time-lapse 3D seismic, 192
    time-lapse gravity, 175
    vertical seismic profiling, 189

geophysical research, **61**
    anisotropy results, 65
    AVOA of cap rock, *63*
    AVOA of Watrous Formation, *62*
    fracture mapping, **58**
    near-offset AVO response, 61–2
    Ratcliffe Beds, *60*
    seismic cross-section, *19*
    stack images from salt solution, *59*
    Watrous Formation, *60*
geosphere
    risk assessment, **294**
    risk evaluation criteria, 296
glacial tills
    seal properties, **58**
Global CCS Institute
    classification of CCS projects, 4
grain density
    Midale Evaporite, 53
Gravelbourg Formation, 65, 194
groundwater, **121**
    composition, 122
    Piper plot, *123*
    surface casing, **263**
    wells sampled, *121*
Gutenburg-Richter graph, 33
gypsum
    during fracture flow experiments, 231
    solubility, *114*

helium
    in CO_2 natural gas, 67
    in soil gas, *134*
history matching, 83
    carbonate rocks, 217
    field-scale simulations, **235**
    flow chart, *243*
Horner-type plots, 40
hydraulic head distribution maps. *see* maps
    (under individual aquifers)
hydrocarbon pore volume, 89
hydrocarbons
    wells sampled, *151*
hydrochemistry. *see* formation water

hydrogeology, **33**
    chart of shallow aquifers, *36*
    Empress Formation, *37*
    study areas, **10**
hydrostratigraphy
    chart of regional units, *17*

illite
    in Midale Beds, 248
induced elevated pressures, 257
In Salah Field, 173, 176
InSAR, **172**
in-situ stress regime, **70**
invasion percolation modelling, 91

joints
    bedrock map, 22
Joli Fou Formation, 31
    defined as 'other' seal, 52
    seal properties, **55**
Jurassic
    hydrogeology, 40
    Lower Gravelbourg Formation, 194
    Vanguard Group, 31, **55**
    Watrous Formation, 194
Jurassic aquifer
    fluid flow directions, 47
    migration pathways, *96*, *98–100*
    salinity map, *42*

kaolinite
    in Midale Beds, 248
Ketzin, Germany, 186

Laramide orogeny, 34
leakage. *see also* well integrity
Lea Park Formation
    seal properties, **57**
lineament mapping, 19
lineaments
    regional map, 21
lithology
    chart of shallow aquifers, *36*
Little Rocky Mountains, 67

magnesium
    Marly coreflood results, *221*
    MCMC inversion results, *249*
    Vuggy coreflood results, *220*
magnetic anomalies, 19
Mannville aquifer
    fluid flow directions, 47
    fluid velocities, 50
    migration pathways, *96*, *98–100*
    permeability field, *50*
    potentiometric surface map, *46*
    salinity map, *42*
Mannville Group
    major aquifer, 40
maps
    3D mesh simulation, *103*
    aeromagnetic, *20*
    Bakken seismic traveltime, *194*
    bedrock joints, *22*
    CO_2 dispersion, *87*
    CO_2 distribution, *244*
    CO_2 distribution 2004-10, *88*
    CO_2 distribution to 2040, *86*
    CO_2-EOR patterns, *13*
    Duperow structure, *67*
    formation water sampling, *144*
    geology study areas, *12*
    groundwater wells sampled, *121*
    hydrogeology study areas, *12*
    lineaments, *22*
    Mannville permeability field, *50*
    Mannville potentiometric surface, *46*
    Midale Evaporite isopach, *30*
    Midale potentiometric surface, *45*
    migration pathways (1 micron case), *98*
    migration pathways (3 micron case), *99*
    migration pathways (8 micron case), *100*
    migration pathways for regional
        aquifers, *96*
    Missippian subcrop edges, *29*
    Prairie Evaporite isopach, *25*
    P-wave amplitude difference, *244*
    regional lineaments, *21*
    regional stress regimes, *71*

seismic amplitude difference, *193*
seismic interval traveltimes, *195*
shallow aquifers, *37*
soil gas composition, *136–37*
soil gas sampling sites, *128*
solid-liquid-gas behind casing, *278*
time-lapse seismic amplitude differences, *209*
Watrous aquitard breach points, *105*
Watrous Formation isopach, 32
Watrous Formation (potential breaching), *104*
wells sampled for oil and gas, *151*
Weyburn Field (potential breach points), *102*
Weyburn Field, PSInSAR image, *172*
Markov Chain Monte Carlo methodology, **246**
Marly. *see also* Midale Beds
Marly unit
   acoustic impedance *v* pressure, *170*
   bulk and shear moduli, *164*
   bulk modulus *v* pressure, *165*
   coreflood results, *221*
   dry-frame density, *164*
   flow-through results, *234*
   fracture flow experiments, **229, 233**
   map of seismic amplitude difference, *193*
   mineralogy, **219**, 248
   mineral volume fractions, 219
   porosity, 164
   rock mechanical properties, **72**
   saturated density v pressure, *170*
   shear modulus v pressure, *165*
   simulation of wormhole development, *228*
   simulation results of corefloods, **226**
   Vp *v* pressure, *170*
   wormhole development, *223*
mass partitioning, **105**
   closed system, *108*
   open system, *108*
MaxPERF® tool, 277
   perforation positions, *283*

mechanical dispersion, **85**
Mesozoic
   hydrostratigraphy, 43
Mesozoic aquifers
   fluid flow directions, 47
   height of trap columns, 95
meteoric waters
   in Williston Basin, **34**
   penetration into basin, 44
micritization
   Midale Beds, 31
microseismicity, **181**
   at Weyburn Field, *182*
Midale aquifer
   hydrogeology, **51**
   permeability field, *49*
   potentiometric surface map, *45*
   salinity map, *42*
Midale Beds, 16, 22, 27, 31, 53
   anhydrite cements, 31
   dolomitization, 31
   Duperow Fm. analogue, 68
   Marly unit, **28**
   micritization, 31
   sonic and density logs, *60*
   Three Fingers zone, 30
   vertical stress gradient, 70
   Vuggy unit, **27**
Midale Evaporite, 29, 113
   fracture flow experiments, **229–30**, 231
   fracture healing, 115
   fractures, 29, 60
   gas permeability, 52
   grain density, 53
   isopach map, *30*
   primary seal, 52
   primary seal properties, **52**
   rock mechanical properties, **72**
   seal at Weyburn Field, 28
   sulphur isotopes, anhydrite, 29
Midale Evaporite Zone, **15**
Midale Field
   geological mapping, **27**

oil production, 5
oil reserves, 4
Midale Member, **15**
Midale reservoir
    P-S converted wave section, *197*
migration pathways
    1 micron case, *98*
    3 micron case, *99*
    8 micron case, *100*
mineral dissolution
    during fracture flow experiments, 233
mineral trapping, **143**
Mississippian, **27**
    Bakken Formation, 194–5
    Bakken seismic map, *194*
    hydrostratigraphy, **41**
    Tilston Beds, 22
Mississippian stratigraphy, *16*
model
    for coreflood experiments, **223**
models and software
    CQUESTRA, 305
    Crystal Ball, 306
    Darcy-scale continuum model, 223
    EQ3/6, 220
    FloGrid®, 88
    GEM®, 83, 236
    geological, **22**, 24
    geomechanical-fluid flow, 182
    GOCAD™, 23
    Modelbuilder®, 88
    NIST, 161
    NUFT, 224
    Petrel™, 23
    PHREEQC, 248
    risk analysis, **304**
    STRAPP, 166
    Surfer™, 21
Montana
    Bearpaw Mountains, 67

Nagaoka, Japan, 186
natural analogues
    comparison diagrams, 69

Newcastle aquifer
    fluid flow directions, 47
    migration pathways, *96, 98–100*
    salinity map, *42*
Newcastle Formation
    major aquifer, 40
nitrogen
    in CO_2 natural gas, 67
    in Duperow Formation, 67
    in soil gas, *132–33, 135–37, 139*

oil and gas
    wells sampled, *151*
oil staining
    Frobisher Evaporite, 30
Opsens Solutions Inc., 276
optical profilometer, 229
Otway Basin, 175
Oungre Evaporite, 23
overburden
    monitoring, **193**
oxygen
    in soil gas, *135–37, 139*
oxygen activation logs, 273

Paleozoic
    hydrostratigraphy, **41**
Paleozoic succession, 25
Penetrators Canada Inc., 277
permeability
    increase during flow-through
        experiments, 233
    Midale Beds, 27
    Midale Evaporite, 53
    Watrous Formation, 55
    wellbore system, *275*
permeability distribution
    anisotropy, 49
permeability-seismic inversion
    flow diagram, *246*
pH
    and wormhole development,
        225, 227–28
    change in downhole values, *147*

change over time, *146*
Marly coreflood results, *221*
MCMC inversion results, *249*
of formation water, **147**
Vuggy coreflood results, *220*
Pierre Shale. *see* Bearpaw Formation
Poplar aquifer
    salinity map, *42*
Poplar Beds, **15**, 28, 31, 53
porosity
    Midale Beds, 27
    Midale Evaporite, 53
    Watrous Formation, 54
Portland-based oilfield cements, 266
post-closure monitoring, 274
post-EOR-CO_2 storage, **90**
potable water. *see* groundwater
potash mines
    formation water inflow, 34
potentiometric surface maps. *see* maps and individual units
Pozzolan-Portland cement blends, 267
Prairie Evaporite, **26**
    isopach map, *25*
Precambrian basement
    location of earthquakes, 33
pressure data
    preparation, 40
pressure-depth plots
    Weyburn Field, *48*
probabilistic risk assessment, **305**
produced hydrocarbons, **151**

Quaternary
    hydrogeology, 39
Quaternary aquifers
    lithological cross-section, *38*

radioactive tracer logs, 273
radon
    in soil gas, *134*
Ratcliff aquifer
    salinity map, *42*

Ratcliffe Beds, **15**, 28, 31, 53
    sonic and density logs, *60*
Ravenscrag Formation, 35, 38–9
reactive transport model
    constrained by formation water data, **235**
reactive transport modelling, 80
    process resolution, *81*
Regina (population), 13
regional aquifers
    migration pathways, *96*
regional hydrostratigraphy, *17*
regional seismic cross-section, *19*
regional stratigraphy, **17**
regional structure, 18
remote sensing techniques, 172
risk
    quantification guide, 307
risk assessment, **290**
    biosphere, **295**
    community assets most at risk, *315*
    geosphere, **294**
    key performance indicators, *294*
    RISQUE method, 294
    system model, *295*
risk evaluation, **306**
risk evaluation criteria, **296**
risk identification
    using FEPs, **302**
Risk Index, 306
risk management process, *292*
risk profile, **307**
risk register
    development, **300**
RISQUE method, 294, 305
river basins, 13
rock-fluid system
    geophysical characterization, **160**
rock mechanical properties, **71**
    data table, 73
RST log, *282*

salinity maps
    major aquifers, 42

salt dissolution, **25**
    and fractures, 65
    geophysical stack images, *59*
sandstone
    physical properties, 162
Saskatchewan Drinking Water Standards, 122
Saskatchewan Ministry of Energy and Resources, 263
Saskatchewan Provincial Stratigraphic Chart, *17*
saturated rock properties, **168**
Schlumberger Isolation Scanner, 276
    mapping behind casing, *278*
Schlumberger Reservoir Saturation Tool (RST), 276
Schlumberger Sonic Scanner, 276
    identification of cement top, *280*
seal integrity, **113**
seals, **113**
    definitions, **52**
    Mesozoic strata, **31**
    Midale Evaporite, 22
    Weyburn Field, **28**
Second White Speckled Shale, 194–95
seismic AVOA analysis, **58**
seismic cross-section, *19*
    Weyburn Field, *194*
seismic data
    repeatability, 205
seismic results
    simulation and history matching, **242**
siliciclastic rocks
    CO_2 trapping, 143
    in Cambrian, 67
single-pattern models, 84
single porosity model, 238–39
Sleipner, 95, 175
sodium
    change in formation water over time, *237*
soil gas, **124**
    analytical methods, *126*
    carbon isotopes, 128
    CO_2 content and flux, 13*2, 135–37, 139*

sampling methods, *125*
    sampling sites, *128*
soil gas migration testing, 272
solubility trapping, **143**
Souris River Fault, 19
squeezing
    recommendations, 271
storage capacity, **105**
storage complex
    definition, **13**
STRAPP, 166–68
stratigraphy
    chart of regional units, *17*
    chart of shallow aquifers, *36*
structurally closed systems, **109**
structurally open systems, **109**
sub-Mesozoic unconformity, 15–6, 31, 53, *195*
    subcrop map, *29*
sulphur isotopes
    anhydrite, Midale Beds, 29
supercritical CO_2
    general properties, 165
surface casing, *263*, 269
surface-casing vent, 267–68
surface-casing vent flow, 272

Tableland-Hitchcock aquifer, 39
tables
    biosphere risk evaluation, 300
    carbon isotopes in soil gas CO_2, 142
    classification of community assets, 299
    gas composition, 152
    groundwater composition, 122
    guide to risk quantification, 309
    migration model parameters, 92
    parameters for leaky aquitards, 94
    quantification of risk, 307
    rock mechanical properties, 73
    storage capacity model parameters, 106
    till aquifers (hydraulic properties), 59
    Watrous Fm. porosity and permeability, 56

Watrous Formation mineralogy, 54
Watrous Formation porosity, 55
temperature
    and CO_2 in soil gas, *140*
    effect on anhydrite solubility, *114*
temperature logs, 273
Tertiary
    aquifers, **35**
    intrusions, 67
Three Fingers zone. *see* Midale Beds
TIC. *see* total inorganic carbon
till aquifers
    hydraulic properties, 59
Tilston aquifer
    salinity map, *42*
Tilston Beds, 22
time-lapse 3D seismic methods, **192**
time-lapse seismic data
    stochastic inversion, **247**
total inorganic carbon
    Marly coreflood results, *221*
    MCMC inversion results, *249*
    Vuggy coreflood results, *220*
Triassic
    Watrous Formation, 15, 28

unstable dissolution fronts. *see* wormholes
U.S. Environmental Protection Agency, 14, 52, 90

Vanguard Group, 31
    defined as 'other' seal, 52
    seal properties, **55**
variable-density flow
    data preparation, 40
vertical pressure profiles, **47**
Viking Formation. *see* Newcastle Formation
VSP. *see* vertical seismic profile
Vuggy. *see also* Midale Beds
Vuggy unit
    bulk and shear moduli, 164
    coreflood results, *220*
    dry-frame density, 164
    fracture flow experiments, **229–31**, *232*
    mineralogy, **218**, 248
    mineral volume fractions, *219*
    porosity, 164
    rock mechanical properties, **72**
    simulation of wormhole development, *227*
    simulation results of corefloods, **225**
    wormhole development, 223

water driving force, 40
Watrous aquitard
    breach criteria, 104
Watrous Formation, **15**, 18, 28, **31**
    anisotropy orientation, *65*
    anisotropy results, **61**
    AVOA inversion results, *62*
    fractures, 60
    hydraulic fracture tests, 71
    hydrogeology, **50**
    isopach map, *32*
    map of potential breaching, *104*
    mineralogy, **54**
    permeability for modelling, 55
    porosity, 55
    porosity and permeability, 56
    porosity for modelling, 55
    sonic and density logs, *60*
    ultimate seal, 52
    ultimate seal properties, **53**
weather
    precipitation, temperature, *130*
well abandonment
    cased-hole wells, **270**
    open-hole wells, **270**
well casing, **266**
well conversion, **268**
well integrity
    assessment, **257**
    evaluation of D&A wells, **259**
    monitoring, **272**
well integrity assessment
    key questions, 257

well remediation, **268**
well repairs
    key evaluation criteria, 268
well trajectory
    and corrosion, 266
    and leakage, **265**
Western Canada Sedimentary Basin
    regional stress map, 71
Weyburn City
    Interpretative Centre, 324
Weyburn CO_2
    bulk density v pressure, *167*
    bulk modulus v pressure, *167*
Weyburn Energy Centre, 325
Weyburn Field
    3D model of calcite dissolution, *112*
    3D time-lapse seismic monitoring, **192**
    change in Ca in formation water over time, *149*
    change in downhole pH over time, *147*
    change in formation water alkalinity, *145*
    change in formation water pH, *146*
    CO_2 injection and microseismicity, *182*
    coreflood experiments, **218**
    crosswell seismic reflection image, *188*
    crosswell seismic survey geometry, *187*
    crosswell seismic surveys, 186
    definitions of seals, 52
    effectiveness of risk evaluation, **310**
    fold values for identical stacks, *206*
    fracture permeability, **229**
    fracture potential, *184*
    geological mapping, **27**
    horizontal compressive stress, 164
    horizontal crosswell tomogram, *187*
    lithographic framework, **20**
    map of potential breach points, *102*
    microseismic events, *183*
    microseismicity, **181**
    Midale permeability field, *49*
    minimum horizontal stress, 164
    Mississippian stratigraphy, *16*
    near-surface cross-section, *38*

oil production, *5*
oil reserves, 4
overburden pressure, 164
Poisson's ratio, 164
pore pressure changes, *204*
pressure-depth plots, *48*
PSInSAR map, *172*
range of well types, *264*
reflection stack sections, *190*
salinity for modelling, 166
seals, **28**
seismic cross-section, *194*
solubility of anhydrite, *114*
synthetic seismic field traces, *178*
synthetic seismic offsets from logs, *178–80*
synthetic seismogram, *190*
system model for risk assessment, *295*
time-lapse seismic amplitude differences, *209*
Watrous aquitard breach points, *105*
wellbore database, 263
wellbore statistics, *265*
well integrity field test, **274**
Weyburn oil
    and CO_2 saturation, *168*
    bulk modulus *v* CO_2 saturation, *169*
    bulk modulus *v* pressure, *167–68*
    density *v* CO_2 saturation, *169*
    density *v* pressure, *167–68*
Weyburn Valley aquifer, 38
Williston Basin, 12, 32, 293
    geology, 16
    hydrogeology, **34**
    hydrostratigraphy, **41**
    natural CO_2 accumulations, 67
wormholes, **221**
    breakthrough time, 225
    during fracture flow experiments, 232